Workbook with Integrated Review Worksheets

Margaret H. Moore
University of Southern Maine

Timothy D. Britt
Jackson State Community College

Mathematics All Around

Sixth Edition

Thomas L. Pirnot
Kutztown University of Pennsylvania

in collaboration with

Margaret H. Moore
University of Southern Maine

Pearson

Reproduced by Pearson from electronic files supplied by the author.

Publishing as Pearson, 330 Hudson Street, NY, NY 10013

1 17

Pearson

ISBN-13: 978-0-13-446235-6
ISBN-10: 0-13-446235-1

CONTENTS

INTEGRATED REVIEW WORKSHEETS

Chapter 1: Problem Solving

Section 1.1 Problem Solving

Objective 1 – Use Polya's method to solve problems.
Objective 2 – State and apply fundamental problem-solving strategies.
Objective 3 – Apply basic mathematical principles to problem solving.
Objective 4 – Use the Three-Way-Principle to learn mathematical ideas.

Objective 1 – Use Polya's method to solve problems.

Key Terms
Polya's method

Summary
George Polya developed a strategy for solving problems that will help with solving problems in this course – and beyond! Polya's method includes four steps:

1. Understand the Problem
   In this step, clearly articulate your understanding of the problem. Sometimes you have to make some assumptions, so this is the place to make that clear. As an example, if you were solving a problem where you needed to determine the amount of money earned in a month, you would need to specify the number of days in that month.

2. Devise a plan
   There are many strategies you can use to solve problems. In the next section we will discuss several of them. As you progress through the text, you will learn many more. It is important to keep in mind that you have choices when solving problems, and that you and a peer could solve the same problem using different strategies or plans.

3. Carry out your plan
   Now it's time to get to work solving the problem!

4. Check your answer.
   Does your answer make sense? If not, might you have made an error in step 3 or an incorrect assumption in step 1.

Objective 2 – State and apply fundamental problem-solving strategies.

Key Terms

Draw Pictures
Choose Good Names for Unknowns
Be Systematic
Look for Patterns
Try a Simpler Version of the Problem
Guessing is OK
Relate a New Problem to an Older One

Summary

When solving a problem, there are many strategies you can use. In this section, we focus on seven possible strategies. Keep in mind that, although we may select a particular strategy, there are often others that can be employed! The strategies we will focus on are:

Draw Pictures – If possible, draw a picture to represent the conditions that must be satisfied.
Choose Good Names for Unknowns – It is a good practice to name the objects in a problem so that you can remember their meaning easily.
Be Systematic – Approach the problem in an organized, systematic way.
Look for Patterns – Identify any patterns to help answer the question .
Try a Simpler Version of the Problem – With complex problems, it is often helpful to solve a similar, simpler version of the problem. Once you have done this, you can then go back and try to solve the original problem.
Guessing is OK – Sometimes making guesses will help clarify your understanding of the problem.
Relate a New Problem to an Older One – Don't re-invent the wheel! Look to see if the problem you are attempting to solve is similar to one you have solved before.

As you encounter new topics and concepts in the text, you will be reminded to use these strategies. Time spent now working with them will yield benefits later!

Example:

Determine the digit 40 places to the right of the decimal point when 5/27 is expressed as a decimal.

The problem is asking us to find the 40th decimal point, but we don't need to write the 39 that came before it! If we use a calculator, we find that $\frac{5}{27} = 0.185185185...$ so the 40th digit will be either 1, 8, or 5. To determine which it is, let's look for patterns. We notice that there is a 1 is in the 1st, 4th, and 7th place to the right of the decimal. There is an 8 in the 2nd, 5th, and 8th place to the right of the decimal point. Finally, there is a 5 in the 3rd, 6th, and 9th place to the right of the decimal. If we were to extend this pattern for the digit 5, we would expect a 5 in each place that is a multiple of 3, so the 39th place to the right of the decimal would be a 5. The 40th place to the right of the decimal would be the digit which follows 5 which is a 1.

Notes:

Guided Example	***Practice*** (Exercises 9 & 27 pp. 15-16)
A restaurant offers the fish of the day (g)rilled or (f)ried and with a choice of one of three seasonings. Use the Draw Pictures strategy and draw a tree diagram that illustrates the different ways to order the fish of the day. The tree diagram is a very useful graphic that we will use extensively in future chapters. To construct a tree diagram, identify a starting point, and then draw branches from that point listing the possible selections that can be made. Then, at each of the endpoints of the initial branch, we draw another set of branches representing the possible sections at stage two. For our example, the first stage is G or F (grilled or fried) and the second is 1, 2, 3 for the three seasonings. 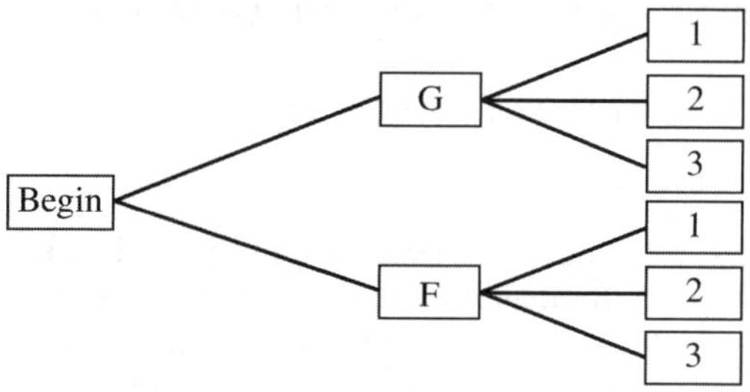 Choices are G1, G2, G3, F1, F2, F3.	1. List the different combinations of heads and tails that can occur when a penny and nickel are flipped. 2. Instead of solving this problem, state a simpler problem and solve it: Using all the letters of the alphabet, how many two-letter codes can we form if we are allowed to use the same letter twice? For example, *ah*, *yy*, *yo*, and *bg* are all allowable pairs. (Note: *bg* is different from *gb*.)

Objective 3 – Apply basic mathematical principles to problem solving.

Key Terms

The Always Principle
The Counterexample Principle
The Order Principle
The Splitting-Hairs Principle
The Analogies Principle

Summary

In addition to problem solving strategies, we need to be aware of basic mathematical principles which provide the foundation for our work. The following principles will be referenced throughout the text:

The Always Principle – In mathematics, if something is true, it is always true, in other words, 100% of the time.
The Counterexample Principle – A single example where a mathematical statement does not work makes the statement false.
The Order Principle – The order in which we do things in mathematics is very important; not doing things in the correct order may lead to an incorrect answer.
The Splitting-Hairs Principle – You should pay careful attention to the meaning of terms you encounter. It may seem as if you are splitting hairs, but in mathematics even though terms are similar, they may not mean the same thing.
The Analogies Principle – You can enhance your understanding of mathematical terms and concepts by associating them with ideas from real life.

Example: In chapter 14 we will define the median of a set of data. What does this term mean to you? Think about the Analogies Principle. The median in a roadway is the strip that divides the two parts of the road. We will learn that the median in a set of data is the point that divides the lower and upper half of the data. For the data set $\{1, 5, 8, 11, 20\}$ the median is the number in the middle, 8. Thinking about the roadway may help you remember your statistics!

Notes:

Guided Examples	***Practice*** (Exercises 33 & 41 p. 16)
Use the splitting hairs principle and explain the differences you see in the following pairs of symbols: 1. $(1,2)$ and $(2,1)$ The order of the numbers is different. In many situations these two symbols would represent different things. 2. $6x$ and $6 \cdot x$ In one case we have a multiplication dot, in the other we do not. In this situation, the two mean the same thing.	1. Provide a counterexample to show that the following is false: $\frac{a}{b}+\frac{c}{d}=\frac{a+c}{b+d}$. 2. Decide whether the two sequences of operations will give the same results: Adding two numbers x and y, then dividing the result by 3; dividing x and y by 3, then adding the results.

Objective 4 – Use the Three-Way Principle to learn mathematical ideas.

Key Terms
Three Way Principle

Summary
When learning a new concept or working on a problem, try to approach the problem verbally (state in your own words), graphically (draw a graph or a diagram), and by example (make examples to illustrate the situation). We call this the Three-Way Principle.

Notes:

Guided Example	***Practice*** (Exercise 71 p. 17)
In a particular game, a card is selected from a fair deck. You win the game if you select a queen or a diamond. In how many ways can you select a winning card? You could solve this by making a diagram of the possible cards in the deck; you could also approach it verbally (there are 13 diamonds, one of which is a queen, and then 3 queens which are not diamonds), and finally you could make up example scenarios to help you visualize the action of drawing a card to determine if it is a winning card.	1. The board of trustees for your school is considering a tuition increase. One faction in the board supports raising the tuition 8% this year and 5% the next. Others want the 5% raise this year and the 8% next year. Some say it doesn't matter because in either case, you will be paying the same in the second year. Does it make any difference as to which plan is implemented? Use the three-way-principle.

Section 1.2 Inductive and Deductive Reasoning

Objective 1 – Use inductive reasoning to make conjectures.
Objective 2 – Give examples of correct and incorrect inductive reasoning.
Objective 3 – Be able to distinguish between inductive and deductive reasoning.

Objective 1 – Use inductive reasoning to make conjectures.

Key Terms
Inductive reasoning
Hypothesis
Conjecture

Summary
We use inductive reasoning when we observe a series of events and then make a general conclusion about the situation at hand. This process is common in scientific inquiry; the conclusion is called a hypothesis or a conjecture. It is important to note that the hypothesis made might or might not be true. In mathematics we try to prove if a conjecture is true based on mathematical facts.

Notes:

Guided Examples	***Practice*** (Exercises 13 & 43 pp. 26 & 28)
Use inductive reasoning to predict the next two symbols in the sequence ⇉↺⇉⇉↺⇉⇉⇉↺⇉⇉⇉ We observe that the number of times the symbol ⇉ appears between each ↺ is increasing by one. Since there were 3 ⇉ in the last complete grouping, inductive reasoning would tell us that there would be 4 in the current grouping, so the next two symbols would be ⇉↺ yielding a sequence of terms ⇉↺⇉⇉↺⇉⇉⇉↺⇉⇉⇉⇉↺	1. Use inductive reasoning to predict the next term in the sequence 3, 6, 12, 24, 48… 2. Use inductive reasoning to solve the following: Committee is to Etimoc as 367768899 is to what number?

Objective 2 – Give examples of correct and incorrect inductive reasoning.

Key Terms
none

Summary
When employing inductive reasoning, we must be careful because sometimes we are led to a false conclusion.

Problem Solving When using inductive reasoning, you are making a conjecture, or educated guess. Keep the *Always Principle* and *Counterexample Principle* in mind; in mathematics a statement is true provided it is true for all cases; if we can find a situation where a conjecture is not true we have identified a counterexample, and thus the statement itself is false.

Notes:

Guided Example	***Practice*** (Exercise 27 p. 26)
You notice that January contains letter a, February contains letter a, March contains letter a. Using inductive reasoning you conclude that each month of the year contains letter a. Find a counterexample to show that this conjecture is not true. June, July, September, October, November, and December would all be counterexamples.	1. Look at the series of numerical equations below. Make a conjecture as to what the next two equations in the pattern are. Check your conjecture with a calculator. a. $1+2=\frac{2\times3}{2}$ b. $1+2+3=\frac{3\times4}{2}$ c. $1+2+3+4=\frac{4\times5}{2}$.

Objective 3 – Be able to distinguish between inductive and deductive reasoning.

Key Terms:
Inductive reasoning
Deductive reasoning

Summary
Recall that inductive reasoning is used when we observe a series of events and then make a general statement, or conjecture. In deductive reasoning, we use accepted facts and general principles to arrive at a specific conclusion. These can be mathematical facts, but we use deductive reasoning in many aspects of our lives.

Example: You are signed up for American Political Theory on Monday and Wednesday at 8:00. You wake up on Wednesday morning, so you conclude that you have American Political Theory class at 8:00. This is an example of deductive reasoning.

Notes:

Guided Example	***Practice*** (Exercises 3 & 9 p. 26)
Identify the type of reasoning in the following example: The last three times you have had the oil changed in your car, it has been done in under one hour. You are taking your car for an oil change and expect that it will take under one hour. This is inductive reasoning. You have experienced three instances where the oil was changed in under one hour, so you conclude that this will happen in the future.	Is each of the following situations an example of inductive or deductive reasoning? 1. As you read the mystery thriller *Gone Girl*, you are keeping track of the author's clues to solve the mystery. 2. The American Conference team has won the Super Bowl the past four times. You expect that the American Conference team will win again this year.

Section 1.3 Estimation

Objective 1 – Use rounding to make estimates. **Objective 2** – Approximate calculations by using compatible numbers. **Objective 3** – Make estimates based on information presented in graphs, maps, and photographs.

Objective 1 – Use rounding to make estimates.

Key Terms
none

Summary
Rounding occurs when we take a number and "bump it up" or "bump it down" to yield a simpler number. To round a number, look to the digit to the right of the digit being rounded. If the digit to the right is 5 or greater, we round up; otherwise, we round down (i.e. leave the digit as is).

Example: Round 3,174 to the nearest ten, nearest hundred, nearest thousand.

Rounding to the nearest ten, we look at the digit to the right, in the ones place. Since 4 is less than 5, we leave the 7 in the tens place unchanged, so the rounded number is 3,170.
Rounding to the nearest hundred, we look at the 7 in the tens place. Because 7 is greater than 5, we round the 1 in the hundreds place to 2, yielding 3,200.
Finally, rounding to the nearest thousand, we look at the 1 in the hundreds place. Since this is less than 5, we leave the 3 in the thousands place unchanged and the rounded value is 3,000.

Notes:

Guided Example	***Practice*** (Exercises 5 & 13 p. 35)
Estimate the expression $4.2 \times 5.8 \div 11.9$ by rounding. Rounding each number to the nearest whole number we obtain $4 \times 6 \div 12$ which is 2.	1. Estimate by rounding to the nearest whole number 4.75×16.3 2. Mike is training for the cross country team. He will run 3.7 miles per day, five days a week. Estimate how many miles he will run during the next six weeks.

Objective 2 – Approximate calculations by using compatible numbers.

Key Terms
Compatible numbers

Summary
To approximate the result of a calculation, we can replace the given numbers with ones that are easier to work with. This procedure is called using compatible numbers. The closer the numbers are to the given ones, the better the approximation will be.

Notes:

Guided Examples	***Practice*** (Exercises 11 & 19 p. 35)
You have gone out to eat with some friends and the waiter has put everything on one bill. Estimate the portion of the bill and 20% tip you owe by rounding your food costs to the nearest dollar and then using compatible numbers to find the 20% tip. You ordered the following: calamari appetizer \$6.79, hamburger entrée \$9.49, iced tea \$2.29 apple crisp \$6.25. Estimating the costs and summing, we find the food costs to be approximately $7+9+2+6=24$. Using compatible numbers, $20\%\times 25=5$ so you owe approximately \$5 for the tip.	Estimate the following: 1. $8.7\%\times 1,024$ 2. Emily is buying plants for spring planting. She bought three packs of petunias for \$2.95 per pack, four packs of vincas for \$1.39 per pack, and a package of potting soil for \$2.79. What is her total bill?

Objective 3 – Make estimates based on information presented in graphs, maps, and photographs.

Key Terms

none

Summary

When looking at a map, graph, or a photograph we can use our tools of estimation to approximate unknown quantities. In the case of a map, you can estimate the distance from one place to another by measuring and comparing with the map's scale. If you are looking at a pie graph or bar graph, estimation is often used to determine the amount of objects in a bar or sector of the graph. Use of rounding and/or compatible numbers can make this task much easier, while at the same time giving a reasonably good estimate of the quantity of interest.

Notes:

Guided Example

Using the map below, estimate the distance a boat would travel on a trip from Port St. Lucie Florida to Charleston, South Carolina, with a stop in Daytona Beach.

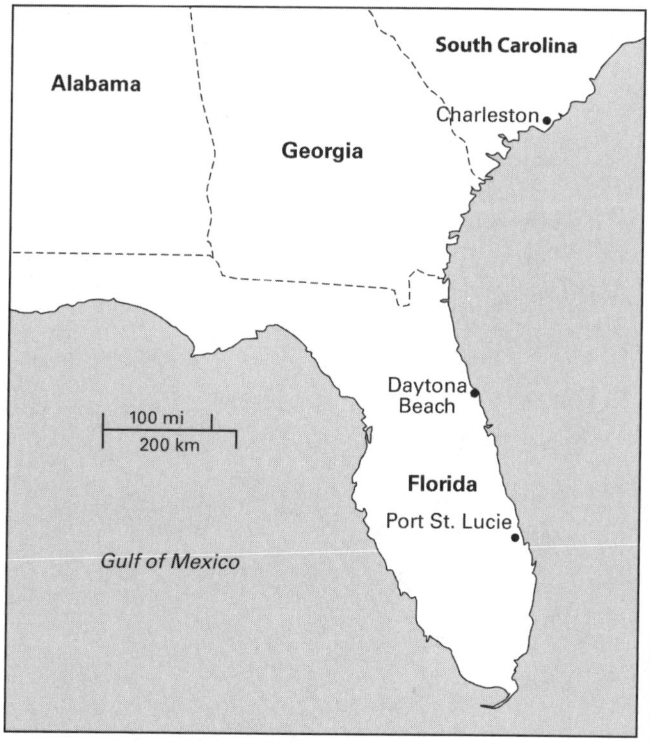

We estimate the distance from Port St. Lucie to Daytona Beach to be $\frac{3}{4}$ inch and from Daytona Beach to Charleston to be approximately $1\frac{1}{2}$ inches. The total trip would be approximately $2\frac{1}{4}$ inches. Looking at the key, $\frac{9}{16}$ inch represents 100 miles, so $2\frac{1}{4}$ or $4\cdot\frac{9}{16}$ inches would represent 4×100 or 400 miles.

Practice (Exercises 41 & 43 pp. 36 & 37)

The following pie chart shows revenues of the federal government in billions of dollars. Use this graph to estimate answers to the questions below. The total revenue is $2,165 billion.

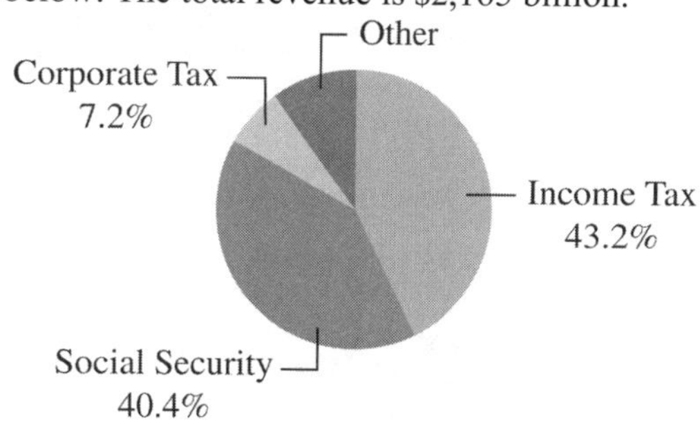

1. How much revenue is derived from Social Security taxes?

2. How much revenue is derived from corporate taxes?

Chapter 1 Answers

Section 1.1

Objective 2	1) HH, HT, TH, TT 2) Possible answer: With three letters there are 9 ways; with five letters there are 25 ways.
Objective 3	1) False; almost any numbers give a counterexample 2) the same.
Objective 4	1)Yes, if the 8% raise is done first, you will pay more over the two years than if the 5% raise is done first.

Section 1.2

Objective 1	1) 96 2) 986763
Objective 2	1) $1+2+3+4+5=\frac{5\times 6}{2}$, $1+2+3+4+5+6=\frac{6\times 7}{2}$
Objective 3	1) deductive 2) inductive

Section 1.3

Objective 1	1) $5\times 16=80$ 2) $4\times 5\times 6=120$
Objective 2	1) $9\%\times 1,000=0.09\times 1,000=90$ 2) $3\times \$3+4\times \$1.50+\$3=\$18$
Objective 3	1) $875 billion dollars 2) $156 billion

Chapter 2: Set Theory

Section 2.1 The Language of Sets

Objective 1 – Specify sets using both listing and set-builder notation.
Objective 2 – Identify when sets are well defined.
Objective 3 – Use the element symbol properly.
Objective 4 – Find the cardinal number of sets.

Objective 1 – Specify sets using both listing and set-builder notation.

Key Terms
Set
Elements
Listing
Natural numbers
Counting numbers
Whole numbers
Integers
Rational numbers
Real numbers
Set builder notation

Summary
Mathematicians refer to a collection of objects a set. The objects in the set are referred to as elements, or members, of the set. We often write a set by listing the elements within braces. The set of primary colors could be written {red, blue, yellow}. We name sets using a capital letter, so we might say C = {red, blue, yellow}.

There are several number sets that should be familiar to you. The set of natural numbers (also called the counting numbers) is often labeled with N. $N = \{1, 2, 3, \ldots\}$. We use the ellipsis notation ... when the set continues in the established pattern, but there is no "largest" natural number. We also use the ellipsis to indicate the continuation of a pattern until a certain point; as an example consider the set $A = \{1, 2, 3, \ldots, 100\}$ which would contain the natural numbers less than or equal to 100. The set of whole numbers $W = \{0, 1, 2, 3, \ldots\}$ is the set of natural numbers with one addition, the number 0. If we take the whole numbers and add the additive inverses (negatives) of the natural numbers, we get the set of Integers. This set is frequently represented by I, and we write $I = \{\ldots, -3, -2, -1, 0, 1, 2, 3, \ldots\}$. The next number set to consider is the set of rational numbers. This set consists of all decimals which terminate or repeat. More formally, a rational number is one that can be written as the ratio of two integers.

Problem Solving The rational numbers are defined as a ratio; remember the *Analogies Principle* tells us to use everyday terms to help remember formal terminology!

We have a more formal way to write the definition of the rational numbers using set builder notation. When using this notation, we identify the common characteristic that all elements of the set share, and use that to describe the set. As an example, we could write

$Q = \{x : x \text{ can be written in the form } \frac{a}{b} \text{ where } a \text{ and } b \text{ are integers and } b \text{ is not zero}\}$. We would read this as

"Q is the set of all x such that x can be written in the form $\frac{a}{b}$ where a and b are integers and b is not equal to 0."

You might ask why we don't use R for rational numbers. The answer is simple; we save R for the set of real numbers! The real numbers include the rational numbers and those decimals that neither terminate nor repeat. An example of a number which is real but not rational is $\sqrt{2}$.

Notes:

Guided Examples	***Practice*** (Exercises 1 & 13 page 48)
Use an alternate way to write each of the following sets: 1. $A = \{10, 20, 30, \ldots, 80, 90\}$ The elements listed in the set are all multiples of 10.The smallest is 10 and the largest 90 so we can write $A = \{x : x$ is a multiple of 10 between 10 and 90, inclusive$\}$ 2. $B = \{x : x$ is the name of one of the thirteen colonies$\}$ B is written using set builder notation. We could write the set using a listing of the elements $B = \{$New Hampshire, Massachusetts, Rhode Island, Connecticut, New York, Pennsylvania, New Jersey, Delaware, Maryland, Virginia, North Carolina, South Carolina, Georgia$\}$.	1. Use set notation to list all the elements of the set described by the natural numbers from 10 to 15 inclusive. 2. Use an alternative method to express the set $\{3, 6, 9, 12\}$

Objective 2 – Identify when sets are well defined.

Key Terms
Well defined
Null set
Empty set
Universal set

Summary
A set is well defined if we are able to determine whether or not an object is an element of that set. The criteria for membership must be stated clearly so that we know if an object meets the criteria or not. Some sets are well defined, but have no elements that meet the criteria for membership. Such a set is called the empty or null set. The empty set can be written { } or using the symbol $\varnothing$. $\{\varnothing\}$ is not a correct way to represent the empty set. When discussing sets it is important to have a frame of reference. The universal set, denoted U, is the set of all elements under consideration in a particular problem or discussion.

Examples:

The rational numbers are well defined because we know that if a number x cannot be written in the form $\frac{a}{b}$ where a and b are integers and $b \neq 0$ x does not belong to the set of rational numbers.

The set $A = \{x : x$ is a natural number that is not a whole number$\}$ is empty because all natural numbers are also whole numbers. We could also write $A = \{\ \}$.

An example of a set that is not well defined is $G = \{x: x$ is a United States President who did a great job$\}$. This set is not well defined because determining if an element belongs to the set depends on the way we interpret *did a great job.*

Notes:

Guided Example	***Practice*** (Exercises 25 & 29 page 48)
Determine if the following set is well defined: If it is well defined, determine if it is empty. $T = \{x : x$ was a top 12 contestant on the 2016 season of The Voice$\}$ This set is well defined. Those contestants who made the top 12 are elements of T. This set is not empty as there are contestants who made the top 12. When discussing set T, the universal set would be all who tried out for the 2016 season. Your author did not make the top 12, but also was not a contestant in that season (maybe next year!).	Determine whether each set is well defined. 1. $\{y : y$ has an interesting job$\}$ 2. $\{1, -3, 5, -7, 9, -11, \ldots\}$

Objective 3 – Use the element symbol properly.

Key Terms
None

Summary
We use the symbol $\in$ to represent the statement *is an element of.* If an object is not an element of a set, we use $\notin$ to show this relationship. The notation red $\in A$ expresses the fact that red is an element of the set A. If red is not an element of set A, we would write red $\notin A$.

Problem Solving It is important to note that to indicate red is an element of the set A we write red $\in A$ and not $\{\text{red}\} \in A$. Keep the *splitting-hairs principle* in mind; notation that looks different may mean something other than you intended!

Example:
World famous cellist Yo-Yo Ma was born in France. Use set builder notation to define a set F of persons born in France and use the element symbol to express the fact that Yo-Yo Ma is an element of that set.

$F = \{x : x \text{ is a person who was born in France}\}$. To express that Yo-Yo Ma is an element of set F, we write Yo-Yo Ma $\in F$.

Notes:

Guided Examples	***Practice*** (Exercises 33 & 39 page 49)
Using appropriate notation to express each of the following: 1. 2 is an element of the set of prime numbers. We will define the set of prime numbers using set builder notation. $P = \{x : x \text{ is a prime number}\}$. We would write $2 \in P$. 2. 1 is not an element of the set of prime numbers. Since 1 is not a prime number, we write $1 \notin P$.	Replace each # with either $\in$ or $\notin$ to express a true statement. 1. Barack Obama # $\{x : x \text{ is a past president of the United States}\}$ 2. $0 \# \varnothing$

Objective 4 – Find the cardinal number of sets.

Key Terms
Cardinal number
Finite
Infinite

Summary
In many applications of set theory it is important to know how many elements are in the set. Sometimes the number of elements is all we need; we don't need to list them. We define the cardinal number of a set to be the number of distinct elements in the set. The notation used for the cardinal number of a set A is n(A). A set is finite if the cardinal number is a whole number. If the cardinal number is not a whole number, we call the set infinite.

Example: Determine the cardinal number of the set A, where $A = \{ x : x$ is a letter in the word mathematics$\}$.

A common mistake is to simply count the letters in the word and incorrectly state n (A)=11. When finding the cardinal number, we look at the number of distinct elements. The letters m, a, and t repeat in the word, but each letter only contributes 1 to the number of distinct letters. The correct statement is that n (A) = 8. This is a finite set, because the number of elements is a whole number.

Notes:

Guided Examples	***Practice*** (Exercises 43 & 47 page 49)
State whether sets B and C below are finite or infinite. If the set is finite, state the cardinal number. 1. $B = \{2, 6, 10, \ldots, 26, 30\}$ B is finite, the ellipsis is used to identify that there are elements between 10 and 26 (they are 14, 18, and 22). The cardinal number is 8 2. $C = \{2, 6, 10, \ldots\}$ C is infinite. The ellipsis indicates that the pattern continues on, and there is no stopping point.	Find n(A) for each of the sets. 1. $A = \{1, 3, 5, 7, \ldots, 11\}$ 2. $A = \{x : x$ is a letter in the word Mississippi$\}$.

Section 2.2 Comparing Sets

Objective 1 – Determine when sets are equal.
Objective 2 – Distinguish between the ideas of "equal" and "equivalent" sets.
Objective 3 – Explain the difference between subsets and proper subsets.
Objective 4 – Use Venn diagrams to illustrate set relationships.
Objective 5 – Determine the number of subsets of a given set.

Objective 1 – Determine when sets are equal.

Key Terms

Equal

Summary

Two sets A and B are equal if they have exactly the same elements. Each element of A must be in B and every element of B must be in A. We represent set equality with the equal sign; if A is equal to B we write $A = B$. If A and B are not equal, we write $A \neq B$. It is important to note that the elements do not need to be written in the same order, and that one set may have repeated elements while an equal set does not have repeats.

Notes:

Guided Example	***Practice*** (Exercises 3 & 7 pages 57)
Determine whether sets B and C are equal. Support your answer. $B = \{x : x$ is a letter in the word tweet$\}$ $C = \{\text{e, t, w}\}$ $B = C$ because they contain the same elements.	Decide whether each pair of sets is equal. Justify your answer. 1. $\{x : x$ is a counting number between 5 and 19 inclusive$\}$ and $\{y : y$ is a rational number between 5 and 19 inclusive$\}$ 2. $\varnothing$ and $\{x : x$ is a living American born before 1800$\}$

Objective 2 – Distinguish between the ideas of "equal" and "equivalent" sets.

Key Terms
Equivalent
One-to-one correspondence

Summary
Two sets that have the same cardinal number are said to be equivalent. If sets A and B are equivalent, we can assign each element to A to an element in B, and no elements are left over. We say that they are in one-to-one correspondence. Two sets that are equal are necessarily equivalent, but the reverse is not true.

Problem Solving We must be very careful when discussing equal and equivalence in set theory. The two words are sometimes used interchangeably in common language, but with regards to mathematics they have different meanings. The *splitting hairs principle* reminds us that in mathematics, just because two terms are similar they usually do not mean the same thing.

Example: Show that the sets {red, white, blue} and {5, 10, 15} are equivalent by putting them in one-to-one correspondence.

To show one-to-one correspondence, we line up the elements in each set. Two examples are shown below:

red	white	blue
↕	↕	↕
5	10	15

red	white	blue
↕	↕	↕
10	5	15

Notes:

Guided Example	***Practice*** (Exercises 17 & 19 page 57)
Determine if the sets A and B below are equal, equivalent, both, or neither. $A = \{x : x$ is a card in a fair deck that has rank 3$\}$ $B = \{x : x$ is a card in a fair deck that is an Ace$\}$ A and B are equivalent because $n(A) = n(B) = 4$. There are four cards in the deck with number 3 and four Aces. They are not equal, because the cards with rank 3 are different than the Aces (just ask a poker player).	Decide whether each pair of sets is equivalent. Justify your answer. 1. $\{x : x$ is a letter in the word song$\}$ and $\{x : x$ is a letter in the word songs$\}$ 2. $\varnothing$ and $\{\varnothing\}$

Objective 3 – Explain the difference between subsets and proper subsets.

Key Terms:
Subset
Proper subset

Summary
In section 2.1 we introduced several number sets. If we look at the whole numbers, $W = \{0, 1, 2, 3, \ldots\}$ and the natural numbers , $N = \{1, 2, 3, \ldots\}$ we observe that every natural number is also a whole number. We say that N is a subset of W. In general, A is a subset of B if every element in A is also an element of B. The notation we use is $A \subseteq B$. If A is not a subset of B, we write $A \not\subseteq B$. In order to show that A is not a subset of B, we need to find at least one element of A that is not an element of B.

A is a proper subset of B if A is a subset of B but A is not equal to B. Another way of thinking about this is in terms of cardinal numbers. A is a proper subset of B if $A \subseteq B$ but $n(A) < n(B)$. In this case we write $A \subset B$.

Problem Solving The *analogies principle* can be used to help us understand and remember the concept of subset and proper subset. The difference between $\subseteq$ and $\subset$ is similar to that between $\leq$ and $<$.

Notes:

Guided Example	***Practice*** (Exercises 9 & 11 on page 57)
Determine if A is a subset of B; if it is determine if it is a proper subset. A = {Georgia, Pennsylvania, Virginia, Vermont} $B = \{x$: x is a state that lies east of the Mississippi River$\}$. We conclude that $A \subseteq B$ because every state listed in A lies east of the Mississippi. Since there are more states east of the Mississippi than those listed in A, $n(A) < n(B)$ and so we can also write $A \subset B$ since A is a proper subset of B.	Decide whether each statement is true or false. Justify your answer. 1. {rose, daisy, sunflower, orchid} $\subset$ {rose, orchid, sunflower, daisy, carnation}. 2. $\{x : x$ is a letter in the word happy$\} \subseteq \{x : x$ is a letter in the word happiness$\}$

Objective 4 – Use Venn diagrams to illustrate set relationships.

Key Terms:
Venn Diagram

Summary
English logician and philosopher John Venn came up with the idea of representing relationships between sets in a relatively simple diagram. We represent the Universal set with a rectangle, and then draw circles within the rectangle to show the relationship between individual sets. If A is a subset of B, then we draw the circle for A within the circle for B.

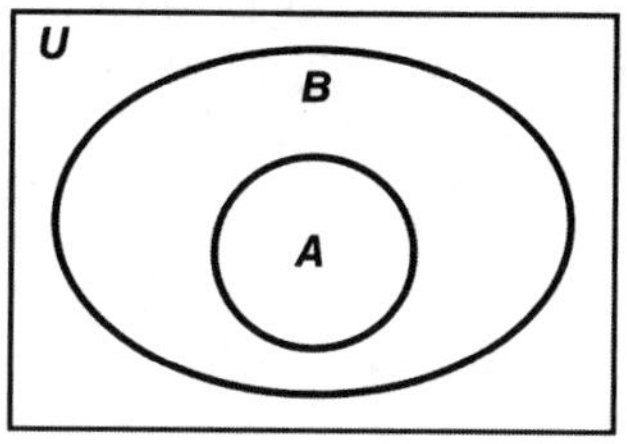

If A is not a subset of B, there must be some elements in A that are not in B. The following Venn diagrams illustrate this situation.

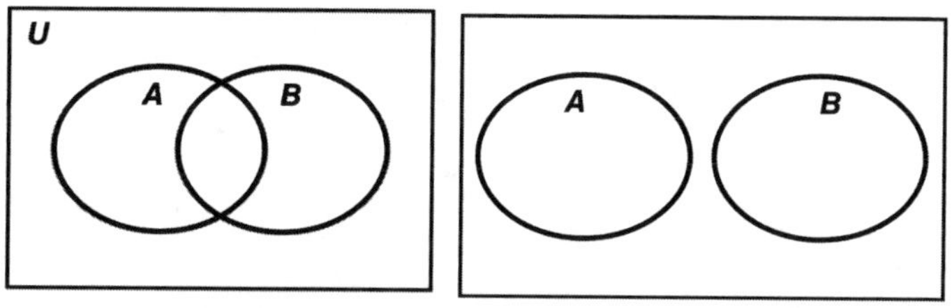

Problem Solving Drawing a Venn diagram is an application of the *draw pictures* strategy. Viewing the sets in the diagram clarifies the relationships.

Notes:

Objective 5 – Determine the number of subsets of a given set.

Key Terms:
none

Summary
In some cases, we don't need to know the subsets and their elements, but rather need to know how many there are. A set with cardinal number *k* has 2^k subsets and $2^k - 1$ proper subsets. It is also helpful to have this number in mind when listing the subsets to ensure that you create a complete listing and haven't forgotten any.

Problem Solving If we need to list all of the subsets of a given set, using the *Be Systematic Strategy* is a good choice. We know the total number of subsets is 2^k, but if we don't approach the problem in an organized way it might be difficult to list all of the subsets.

Notes:

Guided Example	***Practice*** (Exercises 29 & 41 on page 58)
A small hospital has three nurses on the medical/surgical floor, (A)lan, (B)renda, and (C)amden. List all of the possible subsets of the set {A, B, C} and determine how many ways the nurses on the floor could respond to a patient's call. One possible subset is { }. There are three subsets with one element, {A}, {B}. and {C}. To find the subsets of size two, we will first consider those including A, which are {A, B} and {A, C}. One other set of size two would be {B, C}. There is only one subset of size three, namely {A, B, C} Putting these all together, the subsets of {A, B, C} are { } , {A}, {B}, {C}, {A, B}, {A, C}, {B, C} and {A, B, C}. As the formula suggested, there are $2^3 = 8$ subsets. When a patient calls for a nurse, they could send all three nurses, two nurses, one nurse, or decide not to send anyone. These choices are precisely the subset of the set of nurses, so there are $2^3 = 8$ ways to respond to a patient's call.	1. If a set A has five elements, how many subsets does A have? How many of these subsets are proper? 2. Burger King advertises that you can "Have it your way." If their burger can be purchased with from none to eight toppings, such as pickle, onion, tomato, etc., in how many different ways can you order your burger?

Section 2.3 Set Operations

Objective 1 – Perform the set operations of union, intersection, complement, and difference.
Objective 2 – Be able to perform set operations in the proper order.
Objective 3 – Be able to apply DeMorgan's laws in set theory.
Objective 4 – Use the Inclusion-Exclusion Principle to calculate the cardinal number of the union of two sets.
Objective 5 – Use Venn diagrams to prove or disprove set theory statements.

Objective 1 – Perform the set operations of union, intersection, complement, and difference

Key Terms

Union
Intersection
Complement
Difference

Summary

Just as we have operations we perform on numbers, we also have operations we perform on sets. The union of two sets is the merger of the two sets; think about putting the elements from each set together. The formal definition of A union B is $A \cup B = \{x : x$ is an element of A or x is an element of $B\}$. Be careful; the "or" in set theory is an inclusive or, it includes elements in A or B or both.

The intersection of two sets A and B contains those elements which are common to set A and set B. More formally, $A \cap B = \{x : x$ is an element of A and x is an element of $B\}$.

Problem Solving Working with set operations, it is a good idea to keep the *Analogies Principle* in mind. For union, think about joining the two sets together. When you think about intersection, think about the intersection of two streets; it is the point where they share the same pavement!

The complement of set A, denoted A', consists of all elements of the universe that are not in set A. Using set builder notation, we write $A' = \{x : x \in U$ but $x \notin A\}$.

A final operation we consider is the difference of sets B and A. Using set builder notation $B - A = \{x : x$ is a member of B and x is not a member of $A\}$. When finding the difference of sets, think of listing the elements in B, and removing those elements that also belong to A.

Problem Solving Working with set difference, it is very important that you remember the *Order Principle.* As with subtraction of numbers, the order in which you list the sets is very important. $A - B$ is not necessarily the same as $B - A$.

Notes:

Guided Examples	***Practice*** (Exercises 3, 15, & 17 on page 68)
Define sets B and C on universe $U = \{1, 2, 3, \ldots, 10\}$, $B = \{1, 2, 3, 4, 5, 6\}$ and $C = \{2, 4, 6, 7, 8\}$ Determine the following: 1. B' Recall that B' consists of all elements of the universe that are not in B. Since $B = \{1, 2, 3, 4, 5, 6\}$, $B' = \{7, 8, 9, 10\}$ 2. $B' \cap C$ Here we need to consider all elements which are common to B' and C. Looking at the definitions of sets B' and C, we find $B' \cap C = \{7, 8\}$ 3. $B' \cup C$ $B' \cup C$ contains all elements that are in B' or C or both. Merging the two sets we find that $B' \cup C = \{2, 4, 6, 7, 8, 9, 10\}$. 4. $B - C$ To find $B - C$ we start with elements in B and remove any that belong to C. We find $B - C = \{1, 3, 5\}$	For Exercise 1, let $U = \{1, 2, 3, \ldots, 10\}$, $B = \{1, 2, 3, 4, 5, 6\}$, and $C = \{2, 4, 6, 7, 8\}$ Perform the indicated operation. 1. $B \cup C$ For Exercises 2 & 3, consider the universal set U = {*apple, flat-screen TV, hat, satellite radio, fish, sofa, hybrid automobile, potato chip, bread, banana, hammer, pizza*} Let $M = \{x : x \text{ is human-made}\}$, $E = \{y : y \text{ is edible}\}$, $G = \{t : t \text{ grows on a plant}\}$. Find each set. 2. $E - M$ 3. $M' \cap G'$.

Objective 2 – Be able to perform set operations in the proper order.

Key Terms
None

Summary
When you were learning arithmetic, you likely were reminded of the mnemonic "please excuse my dear Aunt Sally". This was intended to remind you that we perform computations in a particular order (parentheses, exponents, multiplication, division, addition, subtraction). In a similar fashion, when working with set operations, we need to be careful that we perform operations in the correct order. The most important thing is to make sure to do things within parentheses first!

Problem Solving. Keep the *Order Principle* in mind; in mathematics we can obtain the wrong answer if we do not perform operations in the correct order.

Notes:

Guided Examples

Define sets *A*, *B*, and *C* on universe U = {0, 1, 2, 3, 4, 5, 6, 7, 8, 9, 10}. A = {1, 6}, B = {2, 4, 6, 8, 10}, C = {1,8}. Perform the indicated operations:

1. $A\cup(B\cap C)$

   We start by looking inside the parentheses to determine the elements in $B\cap C$. The only element common to *B* and *C* is 8, so we write $B\cap C$ = {8}. Now we union $B\cap C$ with *A*, yielding $A\cup(B\cap C)$ = (1, 6, 8}

2. $(A\cap B')\cap C$

   Before we perform the operation inside the parentheses, we need to determine *B'*, which is {0, 1, 3, 5, 7, 9}. Now look to see what elements *B'* has in common with set *A*. It appears that $A\cap B'$ = {1}. Finally, look to see if there are elements common to $A\cap B'$ and C. Our final answer is that $(A\cap B')\cap C$ = {1}.

Practice (Exercises 9 & 11 on page 68)

For these Exercises, let U = {1, 2, 3, …,10}, A = {1, 3, 5, 7, 9}, B = {1, 2, 3, 4, 5, 6}, and C = {2, 4, 6, 7, 8}. Perform the indicated operations:

1. $A\cap(B\cup C)$

2. $(A-B)\cap(A-C)$

Objective 3 – Be able to apply DeMorgan's laws in set theory.

Key Terms
DeMorgan's Laws

Summary
Augustus DeMorgan developed many important results in set theory and logic. Notable among them are the DeMorgan's Laws, which address how to find the complement of a union and also the complement of an intersection. The DeMorgan's Law for the complement of a union states that the complement of a union is the intersection of the complements, or $(A \cup B)' = A' \cap B'$. In the case of the intersection, the DeMorgan's law states $(A \cap B)' = A' \cup B'$. Note that we distribute the complement sign over the sets, and reverse union to intersection, or intersection to union.

Notes:

Objective 4 .Use the Inclusion-Exclusion Principle to calculate the cardinal number of the union of two sets.

Key Terms
Inclusion exclusion principle

Summary
The inclusion exclusion principle is used to determine the cardinal number for the union of two sets. The principle states that $n(A \cup B) = n(A) + n(B) - n(A \cap B)$. Recall that $n(A \cup B)$ is the cardinal number of the union of A and B, that is, the number of elements in A or B or both. With that in mind, you may ask why we subtract $n(A \cap B)$ on the right hand side of the equation. We do this because when we sum $n(A)$ and $n(B)$, we add the elements in $n(A \cap B)$ twice!

Problem Solving To understand the inclusion-exclusion principle, it is helpful to use the *Draw Pictures strategy*. Consider the Venn diagram below where we shade the region corresponding to $A \cup B$ and note that when we add the cardinal number of set A with the cardinal number of set B, we would count region r_3 twice.

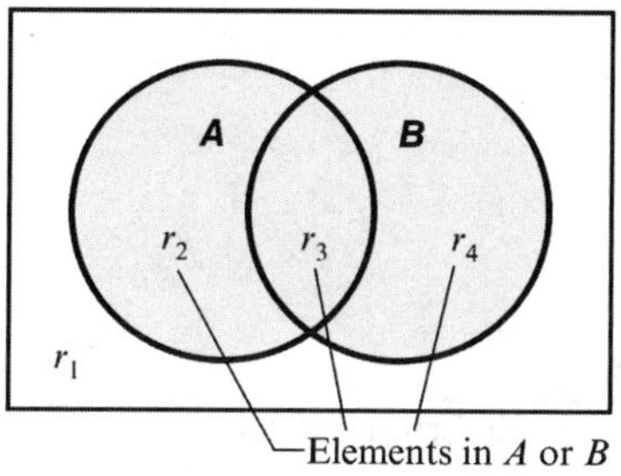

Notes:

Objective 5 – Use Venn diagrams to prove or disprove set theory statements.

Key Terms
none

Summary
It is often desirable to express a set in a different, but equal way. In order to determine if two sets are equal, we can draw Venn diagrams and see if the shaded areas are the same.

Problem Solving When proving or disproving set theory statements, it is helpful to use the *Draw Pictures strategy*. This is particularly true when our statement involves more than two sets, as we see in the guided example below.

Notes:

Guided Example

Use Venn diagrams to determine if the given pair of sets are equal.

$(A \cup C) - B$ and $(A - B) \cup (A - C)$

First we will draw a Venn diagram for $(A \cup C) - B$

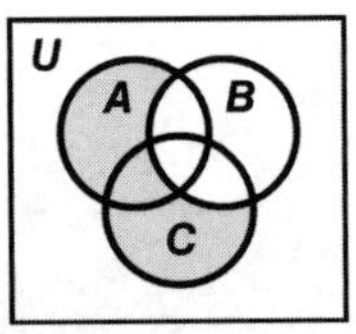

Now consider $(A - B) \cup (A - C)$

In the Venn diagram below, $(A - B)$ is indicated with horizontal lines, and $(A - C)$ is shown with vertical lines. When we find the union of these two regions, we see that it is not the same as the region shaded in the Venn diagram above. We conclude the sets $(A \cup C) - B$ and $(A - B) \cup (A - C)$ are not equal.

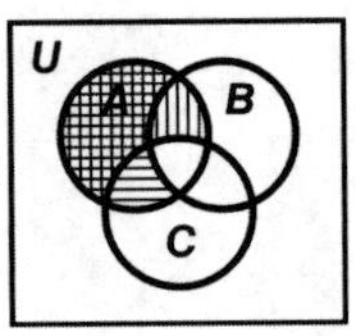

Practice (Exercise 39 p. 68)

1. Use Venn diagrams to determine if the given pairs of sets are equal:

   $(A \cup B')'$ and $A' \cap B$

Section 2.4 Survey Problems

Objective 1 – Label sets in Venn diagrams with various names.
Objective 2 – Use Venn diagrams to solve survey problems.
Objective 3 – Be able to identify contradictory information in survey problems.

Objective 1 – Label sets in Venn diagrams with various names

Key Terms
None

Summary
Venn diagrams are particularly useful to visualize the relationships between various sets. When using a Venn diagram, it is important that we can identify the individual regions. It is helpful to name the regions so that we can refer to them when discussing our solutions. See the diagram in the worked example below to see an example with three sets, W, X, and Y. Note that each separate region should have a distinct name.

Notes:

Guided Example	***Practice*** (Exercises 5 & 7 p. 76)
Describe the set $X \cup Y$ by referring to the numbered regions in the diagram below. 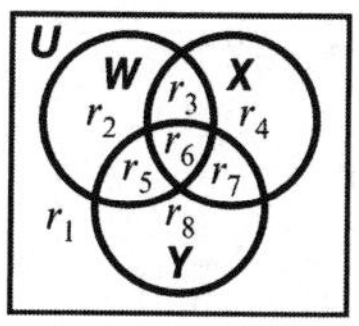 Recall that $X \cup Y$ includes all regions that are in *X* or *Y* or both. Regions 3, 4, 6, and 7 are in *X*; regions 5, 6, 7, and 8 are in set *Y*. Putting these together, we find that $X \cup Y$ includes regions 3, 4, 5, 6, 7, and 8.	Describe each set by referring to the numbered regions in the diagram at left: 1. W 2. $X - W$

Objective 2 – Use Venn diagrams to solve survey problems.

Key Terms
None

Summary
Problem Solving It is important to keep your work organized when solving survey problems. Use of the *Choose Good Names strategy* will help avoid confusion.

Example: The political science department at a university surveyed number of state politicians regarding their perception of college students and gathered the following information:
10 saw students as hardworking and economically advantaged
25 saw students as hardworking
11 believed students were economically advantaged but not hardworking
16 believed students were neither hardworking nor economically advantaged.
How many surveyed believed students were economically advantaged?

To solve this problem, we draw a Venn diagram with two circles. Using the Choose Good Names strategy, we label one circle H for hardworking, and the other E for economically advantaged.

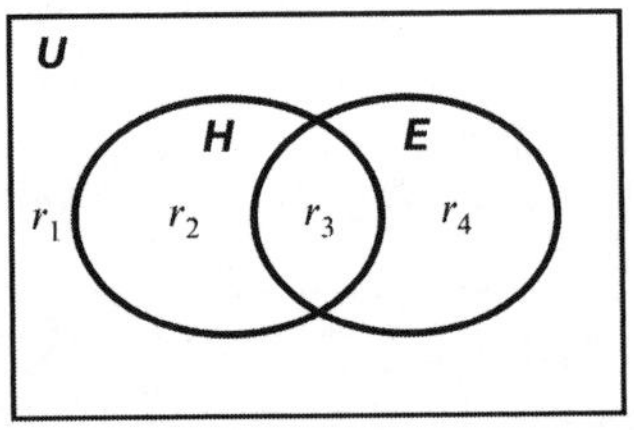

We will start with the information that 10 saw students as hardworking and economically advantaged. This is the cardinal number for $(H \cap E)$ and we put this in region 3. Next, we use the fact that n(*H*) = 25. Since *H* is made up of regions 2 and 3 and region 3 has cardinal number 10, region 2 must have cardinal number $25-10=15$. We are told that 11 believed students were economically advantaged but not hardworking; this is the cardinal number of region 4 because they are in the part of *E* not shared with *H*. Finally, since 16 are not in either circle, the cardinal number of region 1 would be 16. We were asked to determine the number who believed students were economically advantaged, which is *n*(E). Looking at the diagram, $\text{n}(E)=10+11=21$

Notes:

Guided Example	***Practice*** (Exercise 35 p.78)
Five hundred students at your institution were surveyed regarding their experiences during a recent summer. 107 took courses (*C*). 350 stayed in state (*S*). 135 had internships (*I*). 75 stayed in state and took courses. 100 had summer internships in state. 72 took courses and had an internship. 53 took courses, had internships, and stayed in state.	1. A group of young adults who were asked which issues would be important during the next decade gave the following answers: 13 believed that nuclear war, terrorism, and environmental concerns would be important. 43 believed that nuclear war would be important. 17 believed that nuclear war and terrorism would be important. 23 believed that nuclear war or terrorism but not the environment would be important. 28 believed that nuclear war and the

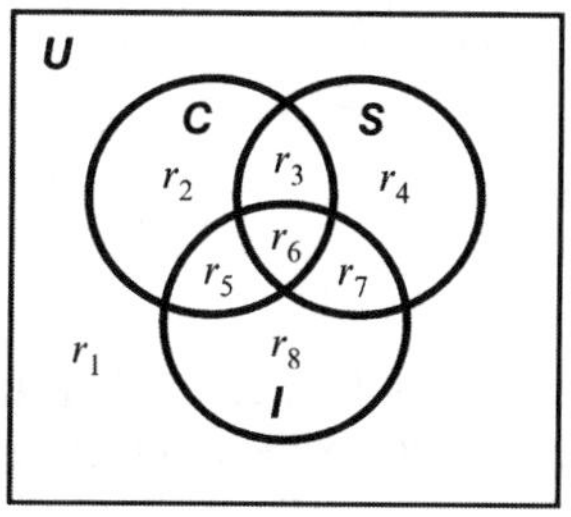

1. Determine the cardinality for each of the 8 regions identified in the Venn diagram.

   Starting with those who took courses, stayed in state, and had internships, we put 53 in region 6.
   Next we use the fact that 75 stayed in state and took courses. These students would comprise regions 3 and 6. Since we already have 53 in region 6, there would be $75-53=22$ in region 3.
   Since 100 had summer internships in state, the cardinal numbers for regions 6 and 7 must sum to 100. We have 53 in region 6, so there must be $100-53=47$ students in region 7.
   With a total of 350 staying in state, we determine the cardinality of region 4 by subtracting the numbers in regions 3, 6, and 7 from 350 yielding $350-122=228$ for the cardinal number of region 4.
   In a similar fashion, we find that region 5 must have $72-53=19$ students. Region 2 would have cardinal number $107-94=13$ and region 8 would contain $135-119=16$ students. To determine the cardinal number of region 1, we sum the numbers for regions 2 – 8 and subtract from 500, yielding 102.

2. How many students took summer courses or had an internship?
   To answer this, we can sum up the cardinal numbers for regions 2, 3, 5, 6, 7, and 8 yielding 170. We could also employ the inclusion exclusion principle from the previous section;

$$\begin{aligned} n(C\cup I) &= n(C)+n(I)-n(C\cap I) \\ &= 107+135-72 \\ &= 170 \end{aligned}$$

environment would be important.

18 believed that terrorism but not nuclear war would be important.
7 believed that only the environment would be a serious issue.
6 believed that none of these issues would be important.

a. How many people were surveyed?
b. How many people believed the environment would be an important issue?
c. How many people believed only terrorism would be an important issue?

Objective 3 – Be able to identify contradictory information in survey problems.

Key Terms:
None

Summary
It is possible to have a situation where the conditions listed in a problem lead to a contradiction. This can be discovered in the construction of a Venn diagram, or through the use of the inclusion-exclusion principle. In this case, we state the conditions are inconsistent.

Problem Solving If you think you have identified a contradiction it is a good idea to use the *Three-Way Principle* to confirm this decision. You can draw a Venn diagram, use inclusion-exclusion principle, or think of examples where the contradiction occurs.

Notes:

Section 2.5 Infinite Sets

Objective 1 – Be able to recognize the difference between finite and infinite sets.
Objective 2 – Show that the set of natural numbers and the set of integers are countable.
Objective 3 – Be able to show that the set of rational numbers is countable.
Objective 4 – Give an example to show that not all infinite cardinal numbers are the same size.

Objective 1 – Be able to recognize the difference between finite and infinite sets.

Key Terms
One-to-one correspondence
Infinite
Countable

Summary
Two sets are in one-to-one correspondence if we can pair the elements of two sets so that every element in one sets is paired with exactly one element in the other set.

Example: Show that the sets {red, white, blue} and {5, 10, 15} are equivalent by putting them in one-to-one correspondence.

To show one-to-one correspondence, we line up the elements in each set as below:

red	white	blue
↕	↕	↕
5	10	15

The sets in the example above are finite because each can be placed in one-to-one correspondence with the set {1, 2, 3}. In general, a set is finite if it can be put in one-to-one correspondence with the set {1, 2, 3 …, n} where n is a natural number.

A set is infinite if we can place it in a one-to-one correspondence with a proper subset of itself. Remember that a proper subset of A is one that has fewer elements than A. If we can put this proper subset in one-to-one correspondence with A, this suggests that each set has an infinite number of elements.

Notes:

Guided Example

Show that the set {3, 5, 7, 9, …} is infinite by placing it in one-to-one correspondence with one of its proper subsets.

{5, 7, 9 ,11…} is a proper subset of the given set. The one-to-one correspondence is illustrated in the diagram below:

3	5	7	9	…
↕	↕	↕	↕	
5	7	9	11	…

In general terms, we match $2n+1$ with $2n+3$.

Practice (Exercises 11 & 15 p. 83)

1. Use the expression below to establish a one-to-one correspondence between the natural umbers and the subset that is described.

   $3n-2$

2. Describe a one-to-one correspondence between the set {7, 10, 13, 16, 19 …} and one of its proper subsets.

Objective 2 – Show that the set of natural numbers and the set of integers are countable.

Key Terms
Countable

Summary
In this section we learn that some sets, while infinite, are called countable. Any set which can be put in one-to-one correspondence with the natural numbers is countable and has cardinal number $\aleph_0$ (pronounced "Al-luf null" or "Al-luf naught"). All finite sets are countable.

The set of natural numbers is countable, as they can be put into one-to-one correspondence with themselves.

To show that the integers are countable, consider the following matching between the natural numbers and integers. Note that for this correspondence, we assume n is a non-negative integer (for example if $n = 2$, $2n = 4$, $-n = -2$ and $2n+1 = 5$).

0	1	−1	2	−2	…	n	$-n$	…
↕	↕	↕	↕	↕		↕	↕	
1	2	3	4	5	…	$2n$	$2n+1$	…

To show that a set is countable, we establish the pairing in the one-to-one correspondence as we did above, and also show the general correspondence.

Notes:

Guided Example

Show that the set {5, 9, 13, 17, …} is countable. Indicate the general correspondence between this set and the set of natural numbers.

The one-to-one correspondence is illustrated in the diagram below:

1	2	3	4	…
↕	↕	↕	↕	
5	9	13	17	…

In general terms, we match n with $4n+1$.

Practice (Exercise 3 p. 83)

1. Show that the set below has cardinal number $\aleph_0$. Be sure to indicate the general correspondence.

   {8, 11, 14, 17, 20,…}

Objective 3 – Be able to show that the set of rational numbers is countable.

Key Terms:
None

Summary
We will show that the positive rational numbers are countable by establishing a one-to-one correspondence with the natural numbers. A similar argument could be made for the negative rational numbers which would, with the inclusion of 0, show that the set of rational numbers is countable. To show that the positive rational numbers are countable, it is helpful to look at the listing of the numbers as arranged in the figure below.

1/1 2/1 3/1 4/1 5/1 . . .
1/2 2/2 3/2 4/2 5/2 . . .
1/3 2/3 3/3 4/3 5/3 . . .
1/4 2/4 3/4 4/4 5/4 . . .
1/5 2/5 3/5 4/5 5/5 . . .
⋮

Notice that the first row of the table contains all positive rational numbers with denominator 1. The second row has those with denominator 2, the third with denominator 3, and so on. Moving through the table as indicated in the pathway, we establish the following one-to-one correspondence between the positive rational numbers and the natural numbers, thus showing that they are countable.

1/1	2/1	1/2	1/3	2/2	3/1	4/1	3/2	2/3	1/4	1/5…
↕	↕	↕	↕		↕	↕	↕	↕	↕	↕
1	2	3	4	skip	5	6	7	8	9	10…

Note that we skipped 2/2 because it was already in the list. We would also skip other names for 1 and positive rational numbers not in simplest form.

Notes:

Guided Example

Show that the set below has cardinal number $\aleph_0$. Be sure to indicate the general correspondence.

$$\left\{1, \tfrac{1}{3}, \tfrac{1}{5}, \tfrac{1}{7}, \tfrac{1}{9}, \ldots\right\}$$

The one-to-one correspondence is illustrated in the diagram below:

1	2	3	4	…
↕	↕	↕	↕	
1	1/3	1/5	1/7	…

In general terms, we match n with $\tfrac{1}{2n-1}$.

Practice (Exercise 7 p.83)

1. Show that the set below has cardinal number $\aleph_0$. Be sure to indicate the general correspondence.

$$\left\{1, \tfrac{1}{2}, \tfrac{1}{3}, \tfrac{1}{4}, \tfrac{1}{5}, \ldots\right\}$$

Objective 4 – Give an example to show that not all infinite cardinal numbers are the same size.

Key Terms:
None

Summary

We have shown that the set of integers and the set of rational numbers are countable. One might conclude that all infinite sets are countable, but that is not the case. There are infinite sets that are not countable, and have cardinal numbers greater than $\aleph_0$. The real numbers are an example, and we use c for the cardinal number of the reals. This representation was given by Georg Cantor. You might think that we use c for Cantor, but actually we use c for the word continuum.

Notes:

Chapter 2 Answers

Section 2.1

Objective 1 1) {10, 11, 12, 13, 14, 15} 2) $\{x : x$ is a multiple of 3 between 3 and 12 inclusive$\}$

Objective 2 1) not well defined 2) well defined

Objective 3 1) $\in$ 2) $\notin$

Objective 4 1) 6 2) 4

Section 2.2

Objective 1 1) not equal 2) equal

Objective 2 1) equivalent 2) not equivalent

Objective 3 1) true 2) false

Objective 5 1) 32, 31 2) $2^8 = 256$

Section 2.3

Objective 1 1) {1, 2, 3, 4, 5, 6, 7, 8} 2) {apple, fish, banana} 3) {fish}

Objective 2 1) {1, 3, 5, 7} 2) {9}

Objective 5 1) equal

Section 2.4

Objective 1 1) 2, 3, 5, 6 2) 4, 7

Objective 2 1) a. 74 b. 45 c. 8

Section 2.5

Objective 1

1)

1	2	3	4	5	…	n	…
↕	↕	↕	↕	↕		↕	
1	4	7	10	13	…	$3n-2$	…

2) Match {7, 10, 13, 16, 19, …} with {10, 13, 16, 19, 22,…}; in general match $3n+4$ with $3n+7$

Objective 2

1)

1	2	3	4	5	…	n	…
↕	↕	↕	↕	↕		↕	
8	11	14	17	20	…	$3n+5$	…

Objective 3

1)

1	2	3	4	5	…	n	…
↕	↕	↕	↕	↕		↕	
1	1/2	1/3	1/4	1/5	…	$1/n$	…

Chapter 3: Logic

Section 3.1 Statements, Connectives, and Quantifiers

Objective 1 –Identify statements in logic.
Objective 2 –Represent statements symbolically using five connectives.
Objective 3 –Recognize the difference between the universal and existential quantifiers.
Objective 4 –Write the negations of quantified statements.

Objective 1 – Identify statements in logic.

Key Terms

Statement

Summary

British mathematician George Boole invented symbolic logic in the 19th century. We will use the structure established by Boole to identify arguments that are logically valid and those that are not. In order to do this, we need to first look at the types of sentences that will be used, called statements. A statement in logic is a declarative sentence that is either true or false. We do not allow questions or statements which are matters of opinion. We represent statements by lowercase letters.

Notes:

Guided Example	***Practice*** (Exercises 3 & 7 p.94)
Identify the following as either logical statements or not logical statements. 1. Today is Tuesday. This is a logical statement. It is true if it happens to be made on a Tuesday and false if made any other day of the week. 2. Tuesdays are the best day of the week. This is not a logical statement, it is matter of opinion. Your author thinks that Saturdays are the best! 3. Is today Tuesday? This is not a logical statement; it is a question.	Determine if the sentences below are logical statements. 1. Come skiing with me this weekend in Tahoe. 2. You will be a millionaire by the time you are 30.

Objective 2 – Represent statements symbolically using five connectives.

Key Terms
Simple statement
Compound statement
Connectives
Negation
Conjunction
Disjunction
Conditional
Biconditional

Summary
The statement "golden retrievers are a breed of dog" is a simple statement. It contains a single idea. We will represent this statement with *g*. "Golden retrievers are not a breed of dog" is a negation of the statement *g* and is written ~g. Negation is one of several types of connectives we can use to combine ideas. If we define another statement *c*: this breed is expensive, we could combine the two ideas with the word "and" and obtain the compound statement "golden retrievers are a breed of dog and this breed is expensive." This is called a conjunction and is represented by $g \wedge c$. If we wish to have the statements connected with the word "or", we would write it as $g \vee c$. This is called a disjunction. We must be very careful when we combine connectives as they can have very different meanings.

Problem Solving; When using multiple connectives, we must keep the *Order Principle* in mind. With the statements defined above $\sim g \wedge c$ is different than $\sim (g \wedge c)$.The first would read "golden retrievers are not a large breed of dog and this breed is expensive" would lead you to think that the breed is not large but it is expensive. The second statement, however, would read "it is not true that golden retrievers are a large breed and they are expensive" and would have a different meaning.

A connective that we will use extensively when evaluating arguments is the conditional. The statement "if it is raining then I will bring my umbrella" is an example of a conditional statement. If we had defined *r*: it is raining and *u*: I will bring my umbrella then we would represent the statement with $r \rightarrow u$. The direction of the arrow in the conditional statement is very important. If we wrote $u \rightarrow r$ the statement would have very different meaning, specifically "if I bring my umbrella then it is raining."

The final connective we address is the biconditional. It is used extensively in mathematics, but less so in popular language. The biconditional is written with a double arrow, and it indicates that the conditional arrow goes both ways. The statement $a \leftrightarrow b$ is read "*a* if and only if *b*" and means $a \rightarrow b$ and $b \rightarrow a$.

Notes:

Guided Example

Consider the simple statements w: the weather clears and f: our flight will take off on time.

1. Express the statement "if the weather clears, our flight will take off on time" using w and f.

   This is a conditional statement; the weather clearing is just after "if" so we write $w \rightarrow f$.

2. What statement is represented by $w \vee \sim f$

   This is a disjunction so we use the word "or" in the compound statement "the weather clears or our flight will not take off on time."

Practice (Exercises 15, 19, & 21 p.94)

For exercises 1 & 2, identify each statement as simple or compound. If a statement is compound, identify the connectives used.

1. Alvaro likes opera.

2. Red sky in morning, sailor take warning.

3. Write the following statement in symbolic form: Global warming will increase or Congress will not pass new energy legislation.

   Use the statements g: global warming will increase and c: Congress will pass new energy legislation.

Objective 3 – Recognize the difference between the universal and existential quantifiers.

Key Terms

universal quantifier
Existential quantifier

Summary

In addition to connectives, we often use quantifiers with logical statements. As the word suggests, quantifiers tell us how many objectives have a particular property. Universal quantifiers state that all objects of a certain type satisfy a given property. Examples of universal quantifiers are *all* and *every*. Existential qualifiers tell us that one or more object satisfies a given property. You will see them written as *some*, *there exists*, or *there is at least one*.

Example:

"Every passenger flying out of O'Hare airport must pass through security" is an example of a statement with a universal quantifier. "At least one passenger boarding the plane will be selected for additional screening" is an example of an existential quantifier. Note that statement would have the same meaning if "at least one passenger" were replaced with "some passengers" or "there exists a passenger".

Notes:

Objective 4 – Write the negations of quantified statements.

Key Terms
Not all are
At least one is not
Not some are
None are
Euler diagrams

Summary
Negating quantifiers can be tricky! Imagine that your parent said, "in order to use the car tonight, you must have all of your homework done." What does that mean? You must do everything! Now, imagine that your parent said "I've changed my mind. What I said earlier isn't true." What would you understand that to mean? It would tell you that some of your homework can be left undone, and you still get to use the car!

To negate the universal qualifier "all", we use the phrases "not all are", "some are not" and "at least one is not." These have the same meaning. To help understand this, we use a diagram called a Euler diagram.

Problem Solving; When negating quantified statements, the *Draw Pictures* strategy can be very useful. The details of the English language can make negating the statement difficult. The Euler diagram is a tool to help organize our thinking.

A Euler diagram uses circles to show relationships between objects in qualified statements.We show below how an Euler diagram can be used with the the example below.

Example: Negate the statement "all players on the team wear Nike sneakers" and show an Euler diagram.

In the first Euler diagram, the players on the team circle is inside the wears Nike circle. When we negate the original statement, some part of the players on the team circle must be outside of the wears Nike circle. The negation of "all players" uses one of the following: "not all do", "some do not" or "at least one does not."

Now we will consider negating the existential quantifier "some."

Example: Negate the existential quantifier "some sports cars are red."

Before we negate this statement, let's remember that "some sports cars are red" is the same as "there is at least one sports car that is red." Negating this we would say "there is not at least one sports car that is red", or more clearly, "no sports cars are red." The Euler diagram below illustrates this negation.

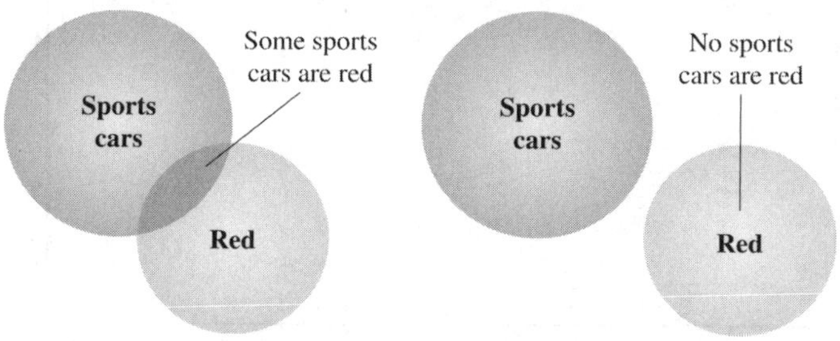

Notes:

Guided Example	***Practice*** (Exercises, 29, 31 p. 95)
Negate each of the following quantified statement. 1. None of the students passed the exam. Be careful! The most common mistake here is to negate *none* with *all*. But in order to negate the statement, all we need is one student who passed! The negation would be "at least one student passed the exam." 2. Some passengers will be selected for additional security screening. Recall that "some passengers" means the same as "at least one passenger." To negate this statement, there would be no passengers selected, so we could say "none of the passengers will be selected for additional security screening."	Negate each quantified statement and then rewrite it in English in an alternative way. 1. All snakes are poisonous. 2. Some personal items are not covered by your renter's insurance.

Section 3.2 Truth Tables

Objective 1 –State the truth tables for the five fundamental connectives.
Objective 2 –Compute truth tables for compound statements.
Objective 3 –Determine when statements are logically equivalent.
Objective 4 –State and apply DeMorgan's laws.

Objective 1 – State the truth tables for the five fundamental connectives.

Key Terms
Truth table

Summary
We have defined a logical statement to be a declarative sentence which is either true or false. But which is it? It depends! This is a particularly good question for compound statements. We will start out this discussion by looking at the following connectives: negation, conjunction, and disjunction. In order to organize our work, we will put the information in a table, called a truth table.

The first truth table we will at is the one for a negation.

p	$\sim p$
T	**F**
F	**T**

all possible truth values for statement p (rows) → logical truth values for $\sim p$ (column $\sim p$)

The original statement, p, can be true or false. The negation of p has the opposite truth value.

When we consider the conjunction and disjunction, there will be two statements to consider, so we need to look at four possibilities as indicated in the tables below. Notice that the conjunction $p \wedge q$ is true in the case where p and q are both true, but is false in the three other cases. This should make intuitive sense; the translation of $p \wedge q$ is "p and q" so we would expect that if the statement is true, both p and q would be true. The truth table for the conjunction is given below:

p	q	$p \wedge q$
T	T	T
T	F	F
F	T	F
F	F	F

all possible ways that p and q can be true or false (rows)

The disjunction $p \vee q$ is translated "p or q" and is true in all cases except where p and q are both false. It is very important to remember that the use of *or* in logic differs from everyday language. You can, in fact, have your cake and eat it too! We use an inclusive *or* which means p or q or both. The truth table for $p \vee q$ below illustrates that the only row in which $p \vee q$ is false is the last one, where each component statement is false.

p	q	$p \vee q$
T	T	T (inclusive *or*)
T	F	T
F	T	T
F	F	F

all possible ways that p and q can be true or false (rows)

Notes:

Objective 2 – Compute truth tables for compound statements.

Key Terms
Tautology

Summary
We have established truth values for three key connectives; the negation, conjunction, and disjunction. Now we combine these in compound statements and construct truth tables to determine if and when the compound statement is true. The tables can be complex, and it is important to keep your work organized. As a starting point, write the simple statements that are involved and all possible combinations of truth values you need to consider. If your compound statement involves *k* simple statements, you need to have 2^k lines in the truth table.

Problem Solving; When constructing a truth table for a compound statement, remember the *Order Principle*. To correctly determine the truth values for a compound statement, you need to address the parts in the correct order. Statements within parentheses should be addressed first!

If you construct a truth table for a compound statement and find that it is always true, the statement is called a tautology.

Example:
Construct a truth table for $\sim(\sim p \vee q)$

Since the statement has two simple components, there will be $2^2 = 4$ lines in the truth table. We work inside the parentheses first, and our first step is to negate *p*. You might ask why we don't look at the disjunction of *p* and *q* first. The reason is that the negation symbol is next to *p*. The negation symbol outside the parentheses will be addressed after the disjunction. Our second step, in the fourth column, is to look at the disjunction of $\sim p$ and *q*. Recall that the disjunction is true in all cases except when both statements are false (which occurs in row 2). Our final step is negate the expression in parentheses, reversing the truth values yields our final answer.

p	q	$\sim p$	$(\sim p \vee q)$	$\sim(\sim p \vee q)$
T	T	F	T	F
T	F	F	F	T
F	T	T	T	F
F	F	T	T	F

Notes:

Guided Example

Construct a truth table for $\sim(p\vee\sim q)\wedge\sim r$

Since this statements involves 3 simple statements, there are $2^3=8$ rows in the truth table. We address the logical operations in a specific order:

1. negate q,
2. determine truth values for the disjunction of p and $\sim q$,
3. negate result in #2
4. negate r
5. determine truth values for the conjunction of $\sim(p\vee\sim q)$ and $\sim r$

See the table for the details on the steps outlined above.

p	q	r	$\sim q$	$p\vee\sim q$	$\sim(p\vee\sim q)$	$\sim r$	$\sim(p\vee\sim q)\wedge\sim r$
T	T	T	F	T	F	F	F
T	T	F	F	T	F	T	F
T	F	T	T	T	F	F	F
T	F	F	T	T	F	T	F
F	T	T	F	F	T	F	F
F	T	F	F	F	T	T	T
F	F	T	T	T	F	F	F
F	F	F	T	T	F	T	F

Practice (Exercises 15 & 19 p. 106)

Construct a truth value for each statement.

1. $p\wedge\sim q$

2. $\sim(p\wedge q)\vee\sim(p\vee q)$

Objective 3 – Determine when statements are logically equivalent.

Key Terms
Logically equivalent

Summary
Two statements are logically equivalent if they include the same simple statements and, when their truth tables are determined, the final columns are identical. This tells us that the statements have the same truth values. They may sound quite different, but logically they are deemed to be equivalent.

Notes:

Guided Example

Determine if the following statements are logically equivalent:
$\sim(\sim p \vee q)$ and $p \wedge \sim q$

In order to answer this question, we need to determine truth values for each of the statements. The truth table for $\sim(\sim p \vee q)$ was done in the previous worksheet:

p	q	$\sim p$	$(\sim p \vee q)$	$\sim(\sim p \vee q)$
T	T	F	T	F
T	F	F	F	T
F	T	T	T	F
F	F	T	T	F

We need to compare the truth values (the last column) with those in the table for $p \wedge \sim q$, below.

p	q	$\sim q$	$(p \wedge \sim q)$
T	T	F	F
T	F	T	T
F	T	F	F
F	F	T	F

Since the truth values in the tables are the same, the statements are logically equivalent.

Practice (Exercises 35 & 37 p.106)

Determine whether the pairs of statements are logically equivalent.

1. $\sim(p \wedge \sim q)$ and $(\sim p) \vee q$

2. $\sim(p \vee \sim q) \wedge \sim(p \vee q)$ and $p \vee (p \wedge q)$

Objective 4 – State and apply DeMorgan's laws.

Key Terms
DeMorgan's Laws

Summary
In previous sections, we discussed the order of operations, and how to evaluate a statement such as $\sim(p \wedge q)$. One might wonder if there is a way to "distribute" the negation over the conjunction - and the answer is yes! Nineteenth century mathematician Augustus DeMorgan established the following laws for logic:

$\sim(p \wedge q)$ is logically equivalent to $(\sim p) \vee (\sim q)$
$\sim(p \vee q)$ is logically equivalent to $(\sim p) \wedge (\sim q)$

Looking back at the guided example for the Objective 3, we found that $\sim(\sim p \vee q)$ and $p \wedge \sim q$ were logically equivalent. Note that we could have used the DeMorgan's law for negation over union to determine this result. Looking at the expression $\sim(\sim p \vee q)$ and using the DeMorgan's law, we would conclude that it is logically equivalent to $\sim(\sim p) \wedge (\sim q)$ which is the same as $p \wedge \sim q$.

Example:
At the coffee shop, the barista asks "do you want cream and sugar in your coffee?" What does an answer of "no" tell the barista?

It indicates that there is at least one of the two that you don't want. You don't want cream or you don't want sugar.

Notes:

Guided Example	***Practice*** (Exercises 29 & 33 p. 106)
Negate the following statement and express using the DeMorgan's laws. I have decided to finish the article and submit it for publication. Negating the statement: it is not true that I have decided to finish the article and submit it for publication. Using the DeMorgan's law to re-express this, we would have: I have decided not to finish the article or not to submit it for publication.	Use the DeMorgan's laws to rewrite the negation of each statement. 1. Bill is tall and thin. 2. Ken qualifies for a rebate or a reduced interest rate.

Section 3.3 The Conditional and Biconditional

Objective 1 – Construct truth tables for conditional statements
Objective 2 – Identify logically equivalent forms of a conditional.
Objective 3 – Use alternative wording to write conditionals.
Objective 4 – Construct truth tables for biconditional statements.

Objective 1 – Construct truth tables for conditional statements

Key Terms
Hypothesis
Conclusion

Summary
In a conditional statement, two simple statements p and q are connected with "if…then…." The placement of the statements is important; the one after *if* is called the hypothesis and the one after *then* is the conclusion. A conditional statement represents a bargain of sorts; if the hypothesis is true the conclusion will follow, but if the hypothesis is not true, anything can happen. We see in the truth table for $p \to q$ below that the only time the conditional statement is deemed to be false is when the hypothesis is true and the conclusion is false (see the last row of the table).

p	q	$p \to q$
T	T	T
T	F	F
F	T	T
F	F	T

Problem Solving Keep the *Order Principle* in mind; the truth value for the conditional depends on the truth values of the hypothesis and conclusion in that order!

Notes:

Guided Example

Construct a truth table for $\sim(p \wedge q) \rightarrow p$

Since we have two simple statements, we need a truth table with 4 rows. We should address the steps in the following order:

1. Determine truth values for $(p \wedge q)$.
2. Negate your result in #1 to get truth values for $\sim(p \wedge q)$.
3. Determine the truth values for the conditional, where $\sim(p \wedge q)$ is the hypothesis and p is the conclusion.

p	q	$(p \wedge q)$	$\sim(p \wedge q)$	$\sim(p \wedge q) \rightarrow p$
T	T	T	F	T
T	F	F	T	T
F	T	F	T	F
F	F	F	T	F

Practice (Exercises 9 & 11 p. 115)

Construct a truth table for each statement.

1. $p \rightarrow \sim q$
2. $\sim(p \rightarrow q)$

Objective 2 – Identify logically equivalent forms of a conditional.

Key Terms

Converse
Inverse
Contrapositive

Summary

If we take a conditional statement $p \to q$ and reverse the hypothesis and conclusion, we obtain a statement which is not logically equivalent to the original one. This new statement is the *converse* of the original statement, and has form $q \to p$. Similarly, if we take a conditional statement $p \to q$ and negate both the hypothesis and conclusion, we obtain a statement with different truth values. This statement is called the *inverse* and has form $\sim p \to \ \sim q$. If we do both, switch the hypothesis and conclusion and negate each, we end up with a statement that is logically equivalent to the original. This statement is called the *contrapositive* and has form $\sim q \to \ \sim p$.

Problem Solving Keep the *Splitting Hairs Principle* in mind. What may seem like a small difference in the conditional, such as changing the order of the statements, yields a statement with different truth values.

Notes:

Guided Example	***Practice*** (Exercises 37 & 39 p. 116)
Determine the converse, inverse, and contrapositive of the statement below and identify which are logically equivalent to it. If you knock on the door, the dog will bark. To guide our work, define statements *p:* you knock on the door and *q*: the dog will bark. The original statement is in the form $p \to q$. The converse of the statement is in the form $q \to p$; if the dog barks, you knocked on the door. This statement is not logically equivalent to the original. The dog may be barking at the cat! The inverse is in the form $\sim p \to \ \sim q$; if you don't knock on the door then the dog will not bark. This statement is not logically equivalent to the original. Just because you don't knock on the door, there is no guarantee that the dog won't bark! The contrapositive is in the form $\sim q \to \ \sim p$ If the dog did not bark, then you did not knock on the door. This statement is logically equivalent to the original.	Determine if the pairs of statements are equivalent. 1. If you activate your cell phone before October 1, then you receive 100 free minutes. If you do not receive 100 free minutes, then you did not activate your cell phone before October 1. 2. If it is raining, then use your headlights. If you use your headlights, then it is raining.

Objective 3 – Use alternative wording to write conditionals.

Key Terms
None

Summary
The conditional form plays a large role in logical arguments. You will find, however, that there are many statements in English that translate to the form $p \rightarrow q$. It is important that you recognize them as conditional statements and that you are able to rewrite them.

A statement written in the form "*q* if *p*" is equivalent to "if *p* then *q*" and thus has the same truth values. The form "*p* only if *q*" is also equivalent to $p \rightarrow q$. Two other ways the conditional is expressed use the terms "sufficient" and "necessary". The statement "*p* is sufficient for *q*" is logically equivalent to $p \rightarrow q$, as is "*q* is necessary for *p*."

Example:
Rewrite the statement "coming to class is sufficient for earning your participation points."

This is in the form statement "*p* is sufficient for *q*" with *p:* coming to class and *q:* earning participation points. Thus we could also write "if you come to class then you earn the participation points."

Notes:

Guided Example	***Practice*** (Exercises 41 & 45 p. 116)
Rewrite the following statement in the form "if … then …." Turning in your project is a necessary condition for earning an A in your math course. The original statement is in the form "*q* is necessary for *p*" with *q*: turning in your project and *p*: earning an A in your math course. We can rewrite this in $p \rightarrow q$ form: "if you earned an A in your math course, then you turned in your project."	Rewrite each statement using the words if… then…. 1. I'll take a break if I finish my workout. 2. You will receive a free cell phone only if you sign up before March 1.

Objective 4 – Construct truth tables for biconditional statements.

Key Terms
None

Summary
When we addressed connectives at the beginning of the chapter, we pointed out that the connective "if and only if" is used extensively in mathematics. We might write "$2x+5=11$ if and only if $x=3$." This tells us that if $2x+5=11$ then $x=3$ and if $x=3$ then $2x+5=11$. The biconditional is true when both p and q have the same truth value, as we see in the table below.

p	q	$p \leftrightarrow q$
T	T	T
T	F	F
F	T	F
F	F	T

Notes:

Guided Example

Construct a truth table for
$(p \vee q) \leftrightarrow (p \rightarrow \sim q)$

Since there are 2 simple statements involved in this biconditional, we need a truth table with 4 rows. The steps we need to follow are:

1. Determine the truth value for $(p \vee q)$
2. Find the truth values for $\sim q$
3. Determine when the conditional $(p \rightarrow \sim q)$ is true.
4. Look to see when $(p \vee q)$ and $(p \rightarrow \sim q)$ have the same truth values; that is when the biconditional $(p \vee q) \leftrightarrow (p \rightarrow \sim q)$ is true.

p	q	$(p \vee q)$	$\sim q$	$(p \rightarrow \sim q)$	$(p \vee q) \leftrightarrow (p \rightarrow \sim q)$
T	T	T	F	F	F
T	F	T	T	T	T
F	T	T	F	T	T
F	F	F	T	T	F

Practice (Exercises 19 & 17 p. 116)

Construct a truth table for each biconditional statement.

1. $(\sim p \rightarrow q) \leftrightarrow (\sim q \rightarrow p)$

2. $(p \vee q) \leftrightarrow (p \vee r)$

Section 3.4 Verifying Arguments

Objective 1 – Use truth tables to show that an argument is valid.
Objective 2 – Use truth tables to verify arguments.
Objective 3 – Recognize some common valid argument forms.
Objective 4 – Recognize some common fallacies.

Objectives 1 & 2– Use truth tables to show that an argument is valid and thus verify the argument.

Key Terms
Argument
Premise
Conclusion
Valid

Summary
An argument consists of a series of statements called premises followed by a single conclusion statement. An argument is deemed to be valid if the conclusion is true whenever all of the premise statements are true. The steps to verify an argument are as follows:

1. Make a truth table with separate columns for each premise and the conclusion.
2. Examine only the lines in the table in which all of the premises are true.
3. If the conclusion is also true for the lines you examined in step 2, the argument is valid.
4. If the conclusion is false for even one of the lines you examined in step 2, then the argument is invalid.

Problem Solving Keep the *Order Principle* in mind when evaluating logical arguments. We must first determine the cases (if any) when all of the premise statement are true, and then determine if the conclusion is true in these cases.

Notes:

Guided Example	***Practice*** (Exercises 19 & 21 p. 124)
Determine if the following argument form is valid or invalid by constructing a truth table:	Determine whether the following forms represent a valid argument.

Guided Example

$q \vee \sim p$

$\underline{p}$

$\therefore p \rightarrow q$

In order to determine if this is a valid argument form, we need to construct a truth table with 4 rows where there is a column for each of the premise statements, and one for the conclusion. The truth table is below:

			Premise 1	**Premise 2**	**Conclusion**
p	q	$\sim p$	$(q \vee \sim p)$	p	$p \rightarrow q$
T	T	F	T	T	T
T	F	F	F	T	F
F	T	T	T	F	T
F	F	T	T	F	T

The argument is valid because in row 1, where both of the premise statements are true, the conclusion is also true. We do not need to consider any of the other rows because at least one of the premise statements is false.

Practice

1. $\sim r$

   $\underline{r \rightarrow q}$

   $\therefore \sim q \wedge r$

2. q

   $\underline{p \vee \sim q}$

   $\therefore \sim (p \wedge q) \rightarrow \ \sim p$

Objective 3 – Recognize some common valid argument forms.

Key Terms
Law of detachment
Law of contraposition
Law of syllogism
Disjunctive syllogism

Summary
There are several forms of arguments that are known to be valid; if we can determine that an argument is in one of these forms, we do not need to construct a truth table, but rather, can name the form of the valid argument. The valid forms are outlined below:

Valid Arguments

Law of Detachment	Law of Contraposition	Law of Syllogism	Disjunctive Syllogism
$p \to q$	$p \to q$	$p \to q$	$p \vee q$
p	$\sim q$	$q \to r$	$\sim p$
$\therefore q$	$\therefore \sim p$	$\therefore p \to r$	$\therefore q$

Notes:

Guided Example

Identify the form of the argument below and determine if it is in one of the established forms for valid arguments.

If you are at the store before 5am on Black Friday then you will get in the store when it opens.
If you are in the store when it opens, then you can get one of the Black Friday specials.
Therefore, if you are at the store before 5 am on Black Friday, then you will be able to get one of the Black Friday specials

Defining the statements p: you are at the store before 5 am on black Friday and q: you get in the store when it opens and r: you get the specials, the form of the argument is

$p \to q$
$q \to r$
$\therefore p \to r$

This is a known valid form; Law of Syllogism. We have verified the argument; it is valid.

Practice (Exercises 5 & 11 pp. 123-124)

Identify the form of each argument and state whether the argument is valid.

1. Anna's new luxury car has zero-gravity seat or hydrophobic windows.
Anna's car does not have zero-gravity seats.
Therefore, this car has hydrophobic windows.

2. If Ben does not buy a season ticket, then he will not be able to buy a playoff ticket.
He is able to buy a playoff ticket.
Therefore, Ben did buy a season ticket.

Objective 4 – Recognize some common fallacies.

Key Terms

Fallacy of the inverse
Fallacy of the converse

Summary

Just as there are established valid forms for arguments, there are common forms of arguments which are not valid. The most common involve the conditional. If you state the conditional and then then conclude the converse or inverse, neither of which is equivalent to the original statement, the argument is not valid. These common fallacies are summarized in the table below:

Invalid Arguments

Fallacy of the Converse	Fallacy of the Inverse
$p \rightarrow q$	$p \rightarrow q$
q	$\sim p$
$\therefore p$	$\therefore \sim q$

Notes:

Guided Example	***Practice*** (Exercises 3 & 13 p.124)
Identify the form of the argument and determine if the argument is valid: If we get 2 feet of snow tonight, then school will be cancelled. School is cancelled. Therefore, we got 2 feet of snow. This argument is in the form "fallacy of the converse." While school may be cancelled, it doesn't necessarily mean that we had snow. Perhaps the heating system went out and that is why we don't have school!	Identify the form of each argument and state whether the argument is valid. 1. If a movie is exciting, then it will gross a lot of money. The movie grossed a lot of money. Therefore, it is exciting. 2. If Phillipe joins the basketball team, then he will not be able to work part-time. Phillipe did not join the basketball team Therefore, he will be able to work part time.

Section 3.5 Using Euler Diagrams to Verify Syllogisms

Objective 1 – Use an Euler diagram to identify a valid syllogism.
Objective 2 – Use an Euler diagram to identify an invalid syllogism.

Objective 1 – Use an Euler diagram to identify a valid syllogism.

Key Terms
Valid
Invalid
Syllogism
Euler diagram
Summary

Summary
In this section we will study arguments called syllogisms; arguments that often contain quantifiers such as all, some, and none. We will continue to have premise statements and conclusions as we did in section 3.4, and our goal will be to determine if the arguments are valid or invalid. Rather than using truth tables to evaluate these arguments, we will construct Euler diagrams. Euler is pronounced "oiler". The Euler diagram will be reminiscent of a Venn diagram, as we will use circles to represent the relationship between components of interest.

Starting with the universal quantifier "all", if a statement indicates "all As are Bs" then we would draw nested circles, with B as the outer circle and A in the inner circle, as drawn below.

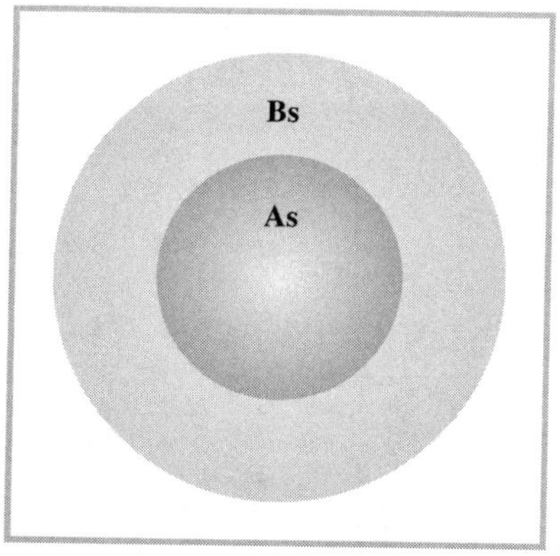

Problem Solving When we construct Euler diagrams to evaluate the validity of an argument, we are using the *Draw Pictures* strategy.

If you have an argument "no As are Bs", we would have the two circles separate from one another, as shown in this diagram:

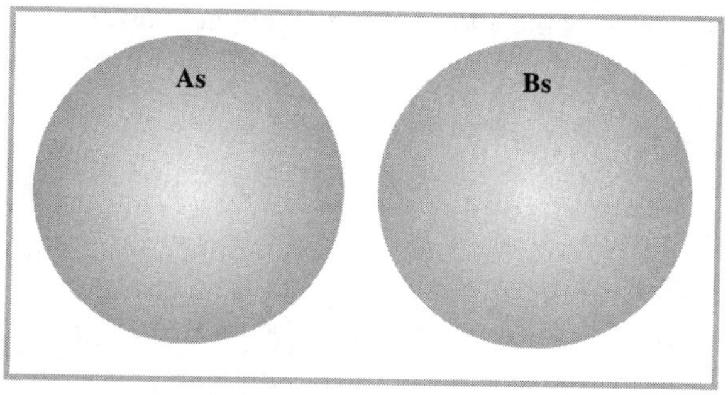

A final case to consider is "Some As are Bs", where the circles overlap in some way. It is possible that the circles are inside one another, but we need to remember that when we have this type of statement, some As may not be Bs.

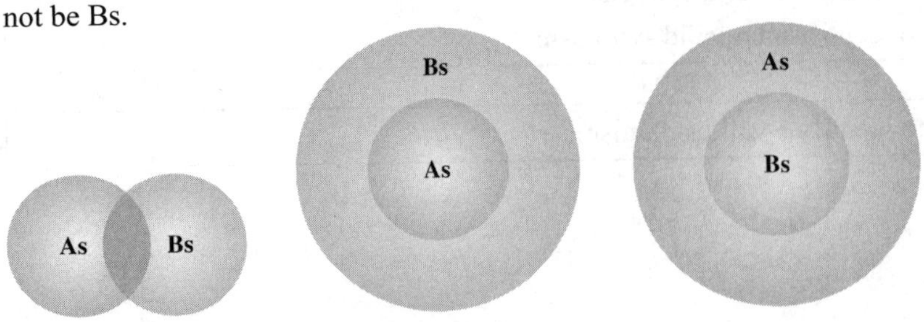

Now we will use these diagrams to determine the validity of a syllogism.

Example:
Use an Euler diagram to determine if the syllogisms below are valid.

All full time students must pay the activities fee.
<u>Colin did not have to pay the activities fee.</u>
∴ Colin is not a full time student.

When we draw the Euler diagram for this argument, the circle for full time students goes inside the activities fee circle. Since Colin does not pay that fee, he is outside of the activities fee circle, and thus not in the full time student circle. There are no other options for where to locate Colin, so this syllogism is valid.

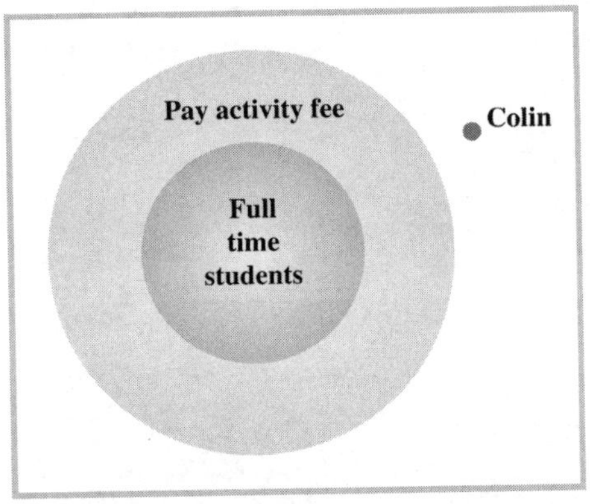

No taxi cabs are allowed on the airport runway.
<u>Shelby's car is a taxi</u>
∴ Shelby's car is not allowed on the airport runway.

In this Euler diagram, the circles for taxi cabs and allowed to drive on the airport runway are disjoint, as we saw above.

Shelby's car must be inside the taxi circle, so it is necessarily outside of the allowed to drive on the runway circle. We determine that this syllogism is valid.

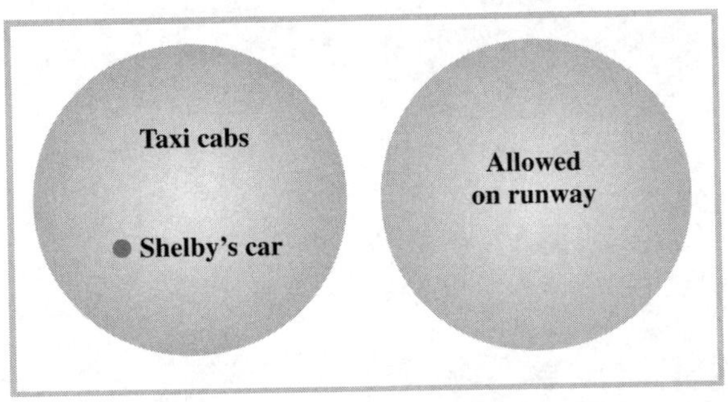

In this worksheet, we have looked only at Euler diagrams for syllogisms that lead to a valid argument. In the next objective we will address cases where this is not true.

Notes:

Guided Example

Use an Euler diagram to show that the syllogism below is valid.

Some restaurants have pool tables.
<u>All restaurants must pass Health inspections</u>
∴ Some places that have pool tables must pass health inspections.

Syllogisms involving some are a bit tricky. In order to determine that the arguments is valid, we must make sure that the conclusion holds for each of the three cases identified above when discussing the existential quantifier *some*.
Consider each of the possible Euler diagrams for this problem, below. They are labeled R for restaurants, H for places that must pass health inspections, and P for places that have pool tables. Note that in each of them, the circle for restaurants is inside the circle for must pass health inspections, and thus in each case there are some places that have pool tables that are subject to health inspections.

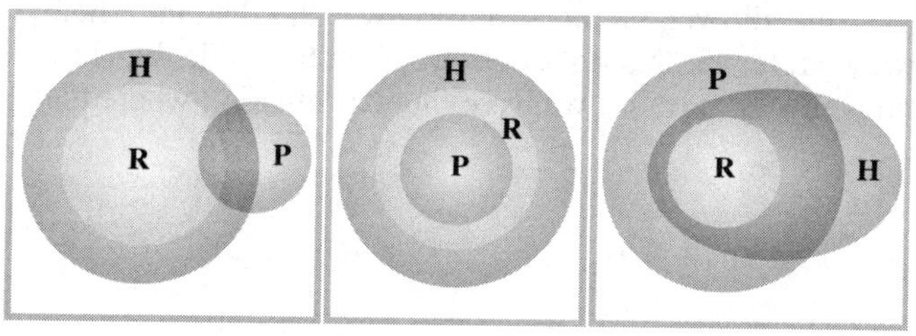

Note that in the examples above, we used a dot for Shelby and for Colin, but in this guided example we did not use a dot because all of the objects of interest were groups.

Practice (Exercises 9 & 15 p. 132)

Use an Euler diagram to determine if the syllogisms below are valid.

1. All politicians are honest
   <u>Walt is not honest.</u>
   ∴ Walt is not a politician.

2. Some mammals are large.
   <u>All large animals are dangerous</u>
   ∴ Some mammals are dangerous.

Objective 2 – Use an Euler diagram to identify an invalid syllogism.

Key Terms
None

Summary
We will now see how an Euler diagram can be used to identify an invalid syllogism. We will use the same basic types of diagrams identified in objective 1 for the quantifiers *all, none*, and *some*. Once we set up the premise statements in the diagram, we will then see if the conclusion is necessarily true.

Problem Solving Keep the *Always Principle and Counter-example* principles in mind. It is not enough to show that the conclusion sometimes holds; if there is even one case where it does not, you have found a counter-example and the argument is invalid.

Example:
Use an Euler diagram to determine if the syllogisms below are valid.

All full time students must pay the activities fee.
Colin is not a full time student.
∴ Colin did not have to pay the activities fee.

This argument, while similar to one we have looked at, differs in an important way. We know the circle for full time students is inside the circle for pays activity fee. Colin is not inside the full time student circle, but that doesn't mean he is outside the activities fee circle. He could be in the spot identified in the Euler diagram below, which provides a counter-example to the conclusion that Colin does not pay the activities fee. The syllogism is invalid.

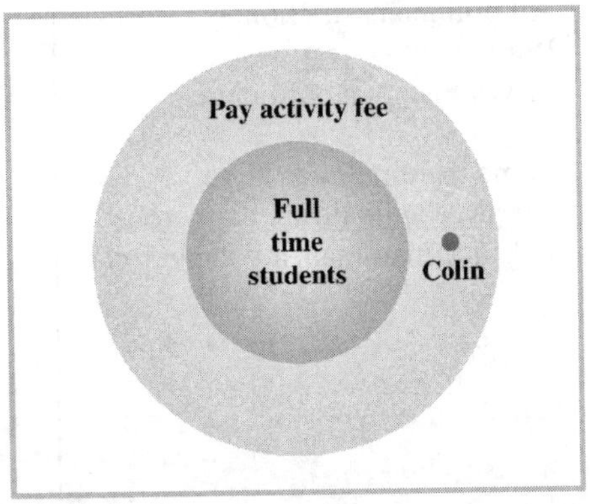

No taxi cabs are allowed on the airport runway.
Shelby's car is not allowed on the airport runway
∴ Shelby's car is a taxi.

In the Euler diagram below we see the circles for taxis and allowed on the airport runway are disjoint. Shelby's car must be outside the allowed to drive on the runway circle, but that does not force it to be inside the taxi circle. It could be as indicated below. This syllogism is invalid because we have found a situation where the premises hold but the conclusion is not true.

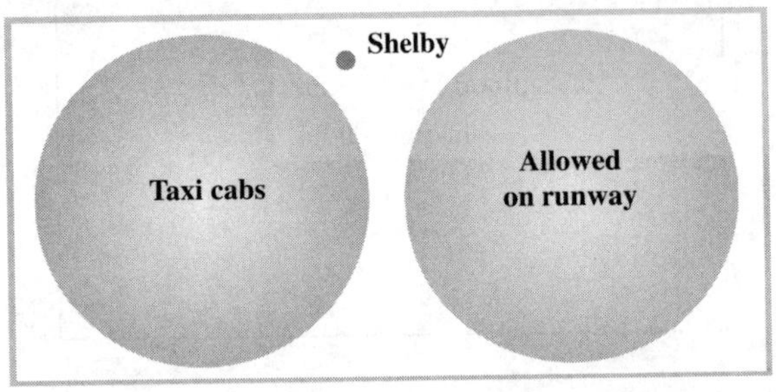

Notes:

Guided Example

Use an Euler diagram to show that the syllogism below is not valid.

Some restaurants have pool tables.
Some restaurants must pass Health <u>inspections</u>
∴ Some places that have pool tables must pass health inspections.

In order to determine if the arguments is valid, we must make consider all possibilities associated with the existential quantifier some. If we find even one case where the conclusion does not hold, we decide the syllogism is invalid.

Consider the Euler diagram below, which satisfies all of the premise statements, but where the conclusion does not hold. This Euler diagram provides a coutner-example and we determine that the argument is invalid.

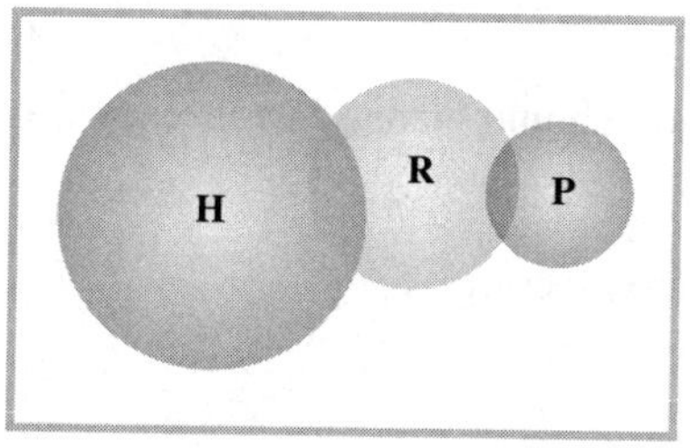

Practice (Exercises 3 & 11 pp. 131 – 132)

Use an Euler diagram to determine if the syllogisms below are valid.

1. All imports are subject to a surcharge.
   <u>This jacket is subject to a surcharge.</u>
   ∴ This jacket is an import.

2. Some large corporations are concerned about the environment.
   <u>General Motors is not a large corporation.</u>
   ∴ General Motors is not concerned about the environment.

Section 3.6 Fuzzy Logic

Objective 1 – Explain the definition of statements in fuzzy logic. **Objective 2** –Be able to work with connectives in fuzzy logic. **Objective 3** –Use fuzzy logic to make decisions.

Objective 1 – Explain the definition of statements in fuzzy logic.

Key Terms
Truth value

Summary
The Aristotelian logic we have studied this far in the chapter has not allowed for any "degrees of truth." We defined a logical statement to be one that was either true or false. Statements such as "Tuesday is the best day of the week" were not allowed in our analyses.

With fuzzy logic, we allow such sentences as statements. In fuzzy logic, a statement is a declarative sentence that has an associated truth value between 0 and 1 (inclusive). Each individual can assign his or her own truth value to the statement.

Notes:

Guided Example	***Practice*** (Exercises 5 & 7 p.138)
We will consider the statement "Tom Brady is the best quarterback of all time" and assign a truth value. This author is a big New England Patriots fan, so would assign a truth value of 0.95. You, of course, may disagree and are welcome to assign any truth value between 0 and 1 (including 0 and 1).	Assign a truth value between 0 and 1 to each fuzzy statement. 1. The United States should increase its foreign aid. 2. Money brings happiness.

Objective 2 – Be able to work with connectives in fuzzy logic.

Key Terms
None

Summary
In order to apply fuzzy logic to real world problems, we must address connectives. We will focus on the negation, disjunction, conjunction, and conditional.

If p is a statement in fuzzy logic, we truth value for the negation of p would have truth value 1 – truth value of p. This is fairly intuitive; if you assign truth value 0.3 to some statement, it would make sense that you assign 0.7 to its negation.

The truth value for the disjunction of p and q, written $p \vee q$, is the maximum of the truth values for p and q. The truth value for the conjunction of p and q, written $p \wedge q$, is the minimum of the truth values for p and q.

Problem Solving; When using multiple connectives, we must keep the *Order Principle* in mind. As always, look at connectives inside of parentheses first!

In order to address the truth values for the conditional, we state without proof that $p \rightarrow q$ is logically equivalent to $(\sim p) \vee q$. You are encouraged to construct truth tables to show that this is true for "non-fuzzy" logic. This allows us to determine the truth values for the conditional using those for the negation and disjunction.

Notes:

Guided Examples	***Practice*** (Exercises 17 & 23 p.138)
Statements p, q, and r have truth values 0.25, 0.84, and 0.60, in that order. Determine the truth values of 1. $p \rightarrow q$ To answer this question, we will re-express the conditional as $(\sim p) \vee q$. The truth value for $\sim p$ is 0.75, thus the truth value for $(\sim p) \vee q$ is the maximum of 0.75 and 0.84 which is 0.84 2. $(p \rightarrow q) \wedge p$ Using the results of our work in #1, we know the truth value for $p \rightarrow q$ is 0.84, so the truth value for $(p \rightarrow q) \wedge p$ is the minimum of 0.84 and 0.25, which is 0.25.	Assume that p has a truth value of 0.27, q has a truth value of 0.64, and r has a truth value of 0.71. Find the truth value of each statement. 1. $p \wedge \sim q$ 2. $q \rightarrow (p \vee r)$

Objective 3 – Use fuzzy logic to make decisions.

Key Terms
Option
Characteristic

Summary
Fuzzy logic can be used to assist in making decisions. To apply this process, we first identify the options that are under consideration. We also identify the characteristics that are important for us to consider when deciding between the options.

The next step is to quantify how important each characteristic is to us, and for this we identify a truth value identified as *i*. A separate set of truth values is established for how well you think each option satisfies the characteristics that are important to you. We denote this with *c*. We will consider the truth values associated with $i \to c$ for each of the characteristics. The idea is to look at the statement "if *fill in the characteristic* is important to you, then *fill in the option* satisfies this condition." In order to do this, we use the equivalent form of the conditional, $(\sim i) \vee c$. To finish making your decision, you calculate the conjunction of all the expressions $(\sim i) \vee c$ for each option. The conjunction will have the minimum truth value. Thus, for each of the characteristics, we find the minimum value for $i \to c$. We then select the option that has the highest final rating.

Notes:

Guided Example

You are trying to decide which of two summer lacrosse programs to attend, LAXRS or STiX. In making your decision, you have identified four characteristics that are important to you: close to home, college recruiting opportunities, skill level, and fun. You have determined the degree to which each camp will satisfy the characteristics you have identified. The information is summarized in the tables below. Use the decision capabilities of fuzzy logic to make the decision on the camp to attend.

First we set up the table with the characteristics that are important, and suggest truth values for each.

Characteristic	Importance
Close to home	0.50
College Recruiting Opportunity	0.85
Skill Level	0.75
Fun	0.80

Next we identify the degree to which each camp will satisfy the characteristics.

	LAXRS	STiX
Close to home	0.90	0.25
College Recruiting Opportunity	0.80	0.75
Skill Level	0.55	0.80
Fun	0.40	0.70

In the following table, we look at each option and compute $(\sim i) \vee c$ for each of the characteristics. To make the decision we look at the minimum value of $(\sim i) \vee c$ for each option, and then selection the option that does the best. It looks like you are headed to STiX camp!

	i	$\sim i$	LAXRS c	LAXRS $(\sim i) \vee c$	STiX c	STiX $(\sim i) \vee c$
Close to home	0.50	0.50	0.90	0.90	0.25	0.50
College Recruiting Opportunity	0.85	0.15	0.80	0.80	0.75	0.75
Skill Level	0.75	0.25	0.55	0.55	0.80	0.80
Fun	0.80	0.20	0.40	0.40	0.70	0.70
Conjunction of all $(\sim i) \vee c$				0.40		0.50

Practice (Exercise 25 p.138)

1. You are trying to decide which of two job offers to accept. The first job is a retail sales trainee for a large consumer products company and the second is a marketing data analyst for a communications company. In the first table, we list the characteristics of a job that are important to you; in the second table we list the degree to which each job satisfies these characteristics. Which job should you take?

Characteristic	Importance
Salary	0.70
Interesting	0.80
Work with people	0.60
Flexible hours	0.75

	Sales Trainee	Data Analyst
Salary	0.60	0.65
Interesting	0.50	0.80
Work with people	0.90	0.30
Flexible hours	0.80	0.60

Chapter 3 Answers

Section 3.1

Objective 1	1) not a statement 2) statement
Objective 2	1) simple 2) compound; if…then 3) $g \vee \sim c$
Objective 4	1) There exists at least one snake that is not poisonous. 2) All personal items are covered by your renter's insurance.

Section 3.2

Objective 2	1) FTFF 2) FTTT
Objective 3	1) yes 2) no
Objective 4	1) Bill is not tall or Bill is not thin. 2) Ken does not qualify for a rebate and he does not qualify for a reduced interest rate.

Section 3.3

Objective 1	1) FTTT 2) FTFF
Objective 2	1) equivalent 2) not equivalent
Objective 3	1) If I finish my workout, then I'll take a break. 2) If you receive a free cell phone, then you signed up before March 1.
Objective 4	1) TTTT 2) TTTTTFFT

Section 3.4

Objectives 1&2	1) invalid 2) valid
Objective 3	1) disjunctive syllogism; valid. 2) law of contraposition; valid.
Objective 4	1) fallacy of the converse; invalid 2) fallacy of the inverse invalid

Section 3.5

Objective 1	1) valid 2) valid
Objective 2	1) invalid 2) invalid

Section 3.6

Objective 1	1) answers will vary 2) answers will vary
Objective 2	1) 0.27 2) 0.71
Objective 3	1) sales trainee (0.50)

Chapter 4: Graph Theory (Networks)

Section 4.1 Graphs, Puzzles, and Map Coloring

Objective 1 – Use graph terminology correctly.
Objective 2 – Apply Euler's theorem to graph tracing.
Objective 3 – Model real-life situations using graphs.
Objective 4 – Use Fleury's theorem to find Euler circuits.
Objective 5 – Utilize graph coloring to simplify a problem.

Objective 1 – Use graph terminology correctly.

Key Terms

Graph
Vertices
Edges
Trace
Connected
Bridge
Networks
Degree
Odd
Even

Summary

A graph consists of a finite set of points and figures that join pairs of points. We call the points vertices and the figures connecting vertices are called edges. Generally speaking, we label vertices with capital letters. Graphs may appear different but relay the same information. The graph below has two representations, but the same information is provided. Note that in each case, there is an edge between A and D.

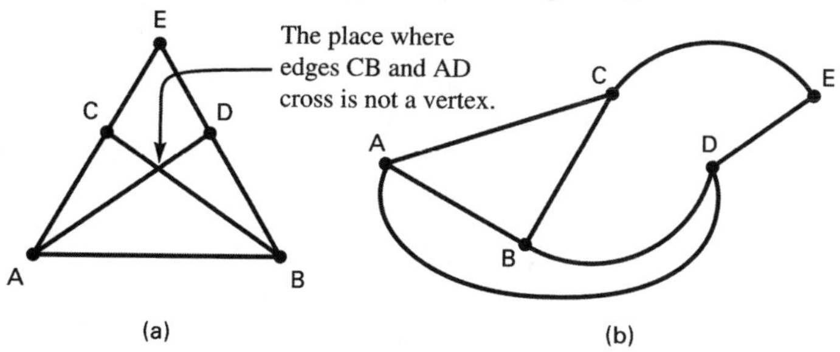

A graph is connected if it is possible to travel from any vertex to any other vertex by moving along the edges. Connected graphs are also called networks. If there is an edge which, if removed, caused a connected graph to no longer be connected, the edge is called a bridge. A graph can be traced if we can begin at a vertex and travel over the entire graph without lifting our pencil and without traveling over an edge more than once.

The degree of a vertex is the number of edges that meet at that vertex. We classify vertices as even or odd based on their degree. We will later link the degree of the vertex to properties of the graph.

Notes:

Guided Example	***Practice*** (Exercises 3 & 21 pp.156-7)
Consider the graph below. Is it connected? Can it be traced? Identify an edge that is a bridge. Determine which vertices are even and which are odd. 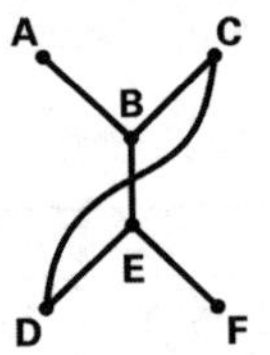 This graph is connected because we can travel from a vertex to any other vertex by moving along successive edges. It does not appear that we can trace it; if we start at vertex C, we cannot traverse each edge without traveling over the edge between A and B or E and F more than one time. If we start at A or F, in order to travel over each edge, we would have to travel an edge more than once. The edge from A to B is a bridge; if it is removed the graph is no longer connected. Vertices A, B, E, and F are odd since each is the endpoint of an odd number of edges. Vertices C and D are even.	1. Determine if the graph below is connected. Which vertices are odd? Which vertices are even? 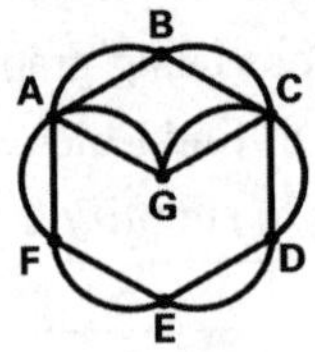 2. Given an example of the graph described, or explain why you think it may be impossible. A connected graph with one even vertex and four odd vertices.

Objective 2 – Apply Euler's theorem to graph tracing.

Key Terms

Euler's Theorem
Path
Length
Euler path
Euler circuit
Eulerian graph

Summary

Recall that a graph can be traced if we start at a vertex and travel over all of the edges of the graph, traversing each edge exactly once. Euler determined the conditions under which a graph can be traced. Specifically, Euler's theorem states that a graph can be traced if it is connected and has zero or two odd vertices. If a graph has two odd vertices, the tracing must begin at one of these and end at the other. If all of the vertices are even, the tracing must begin and end at the same vertex.

There are several other key terms in graph theory that are important to have in our vocabulary. A path in a graph is a series of consecutive edges in which no edge is repeated. The length of a path is the number of edges that are traveled upon. If a path contains all of the edges in the graph, we call it an Euler path. If an Euler path starts and ends at the same vertex, it is called an Euler circuit. If you have a graph with all even vertices, it will contain an Euler circuit and it is called an Eulerian graph.

Notes:

Guided Example

Can the graph below be traced? If not, explain why.

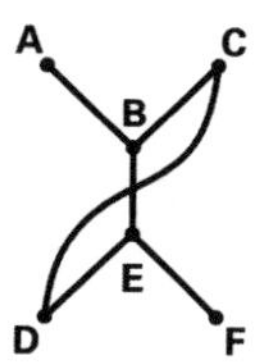

This graph cannot be traced. Vertices A, B, E, and F are odd and vertices C and D are even. Euler's theorem tells us that the graph can be traced if it has zero or two odd vertices. This graph has 4 odd vertices and thus cannot be traced. This confirms our findings in the previous worksheet.

Practice (Exercises 9 & 11 p.156)

Use Euler's theorem to decide whether the graphs below can be traced. If the graph cannot be traced, tell which condition of the theorem fails.

1.

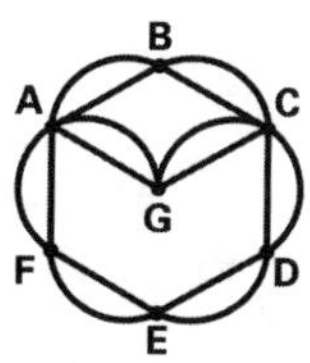

2.

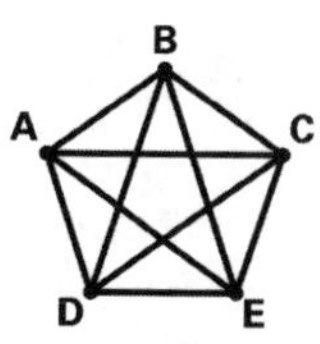

Objective 3 – Model real-life situations using graphs.

Key Terms
none

Summary
Graphs are particularly useful when modeling real life situations. You may be determining a route to get to different stores on Black Friday or the best route to take to pick up a group of friends and bring them back to your apartment. One of the most famous applications in graph theory is the Koenigsberg bridge problem, and it originated in the 18th century. The map below shows the map the Pregel River in Koenigsberg, Prussia. Note that the river divides the city into four distinct sections, labeled A, B, C and D.

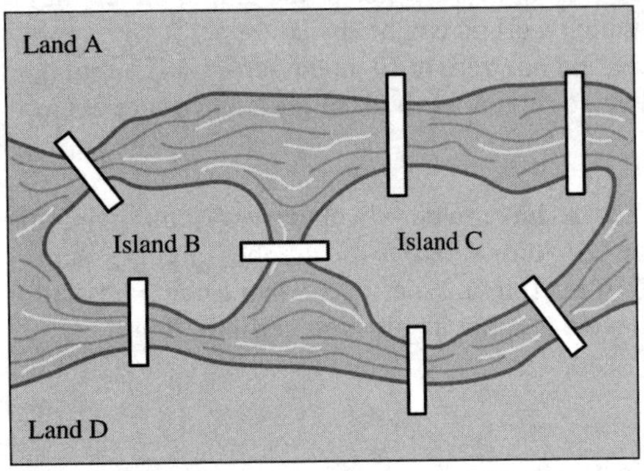

Seven bridges connected the four portions of Koenigsberg. Residents of Koenigsberg would try to visit all portions of the city and return to their starting point traveling over each bridge exactly once. This problem can be addressed using graph theory. The graph below provides a representation of the Koenigsberg bridge problem.

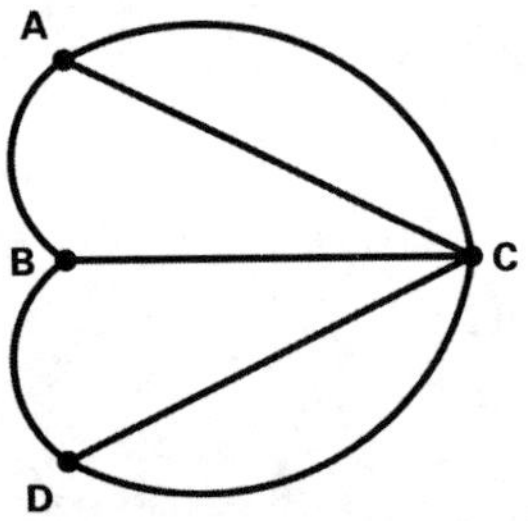

All of the vertices in this graph are odd; using Euler's theorem, we conclude that this graph cannot be traced.

Notes:

Objective 4 – Use Fleury's theorem to find Euler circuits.

Key Terms
Fleury's algorithm
Eulerize

Summary
Euler's theorem tells us if a graph has a circuit, but not how to find it. We could use trial and error, but that is not going to be efficient for any but the simplest graphs. Fleury's algorithm provides a systematic technique for finding a circuit for a graph with all even vertices.

Fleury's Algorithm: If a connected graph has all even vertices, begin at any vertex and follow the procedure below:

1. After you have traveled over an edge, erase it. If all edges for a particular vertex have been erased ,then erase that vertex also.
2. Travel over an edge that is a bridge only if there is no alternative.

We will use Fleury's algorithm to find a circuit for the graph below:

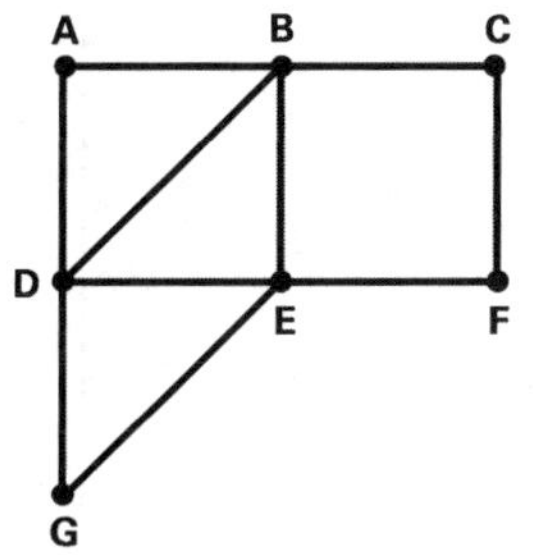

Since all of the vertices are even, we can start at any vertex and use Fleury's algorithm. Let's start at A and travel to B, C, F, then E. Note that we can then erase edges AB, BC, CF, EF, and vertices C and F yielding the graph below:

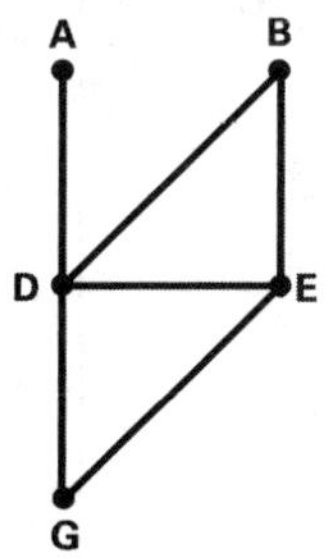

We can now travel from E to G and then D. We cannot go from D to A because it is a bridge. From D we could go to E or B, let's travel to E then B, back to D, and finally return to A. This circuit would be written ABCFEGDEBDA.

Problem Solving; When we use Fleury's algorithm, we are using the *Be systematic* problem solving strategy.

If a graph is not Eulerian, it is possible to duplicate edges of the graph so that all vertices are even. This process is called Eulerizing the graph. We do not connect vertices that were not originally connected, but rather duplicate edges.

Notes:

Notes:

Guided Example

Determine an Euler circuit for the graph below. If the graph is not Eulerian, first Eulerize it.

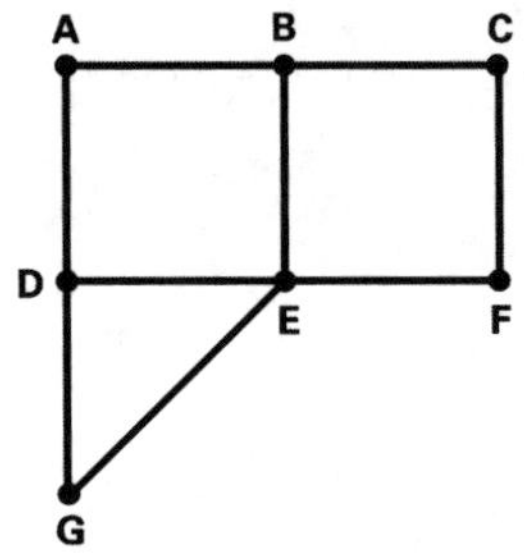

Note that this is similar to our graph above, but edge BD is removed. Vertices B and D are now odd. We Eulerize it by duplicating edges BE and DE.

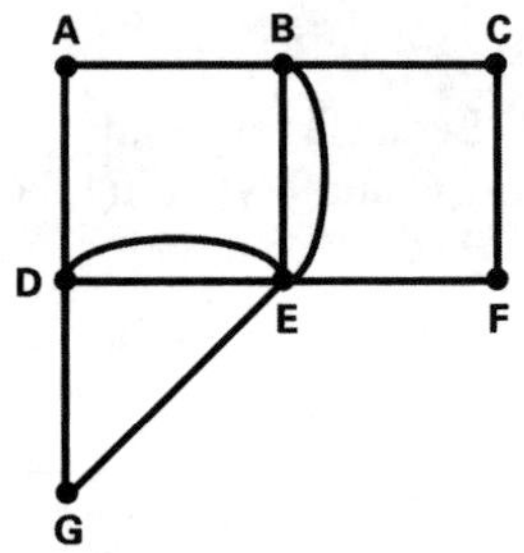

To find the circuit, we will again start at A and move to B, C, F, and E. We are left with the graph below.

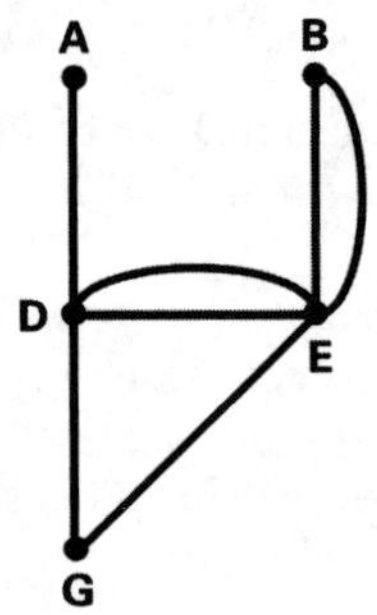

We have many options at E, but let's go to B, back to E, to G, D, back to E, then D and finally return to A.
Our path would read ABCFEBEGDEDA.

Practice (Exercises 13 & 15 p.157)

If the graphs below are Eulerian, find an Euler circuit. If the graph is not Eulerian, first Eulerize it and then find an Euler circuit. Write your answer as a series of vertices.

1.

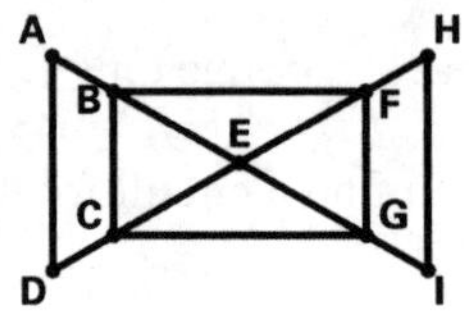

2.

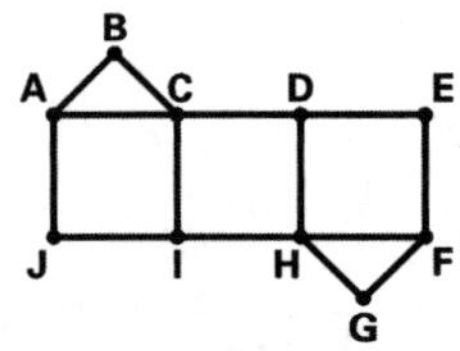

Objective 5 – Utilize graph coloring to simplify a problem.

Key Terms
Four color problem

Summary
In 1852 math student Francis Guthrie posed a question to his professor, Augustus DeMorgan. Guthrie's question is known as the Four-color problem. Using at most four colors, is it always possible to color a map so that any two regions sharing a common border receive different colors? DeMorgan was not able to answer this, and it was not until 1976 that "proof" was proposed showing that it was, in fact possible. This proof was controversial, however, as it was done using a computer program rather than traditional methods for validation

Example: We will address Guthrie's question using the map below. Use a vertex to represent each state, and put an edge between states only if they share a border.

We obtain the graph below and will color it in the guided example.

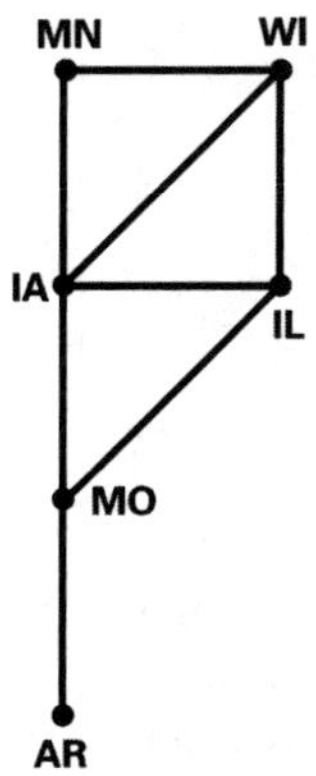

Problem Solving; *Draw Pictures* in mind.

Notes:

Guided Example	***Practice*** (Exercises 37 & 41 p.158)
Using the graph obtained above, "color" the vertices so that no two vertices of the same edge have the same color. Rather than use actual color, we will use R for red, B for blue Y for yellow, and G for green. 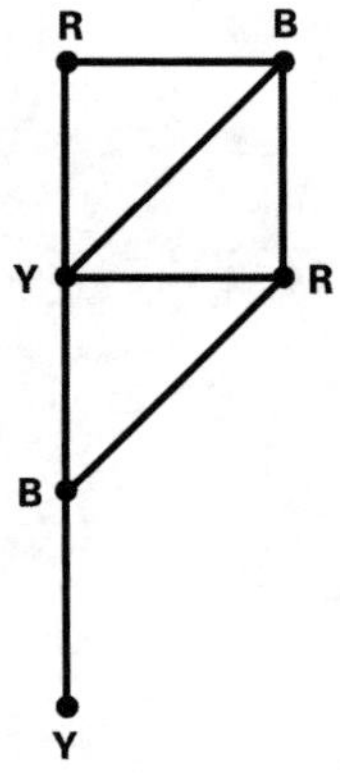 We start out coloring MN with (R)ed. Now we assign (B)lue to WI and (Y)ellow to IA. We can't use B or Y for IL so we use R. We don't assign a new color until we are forced to. We can use B for MO; when we go to AR we can use either R or Y. We have colored the map with three colors.	1. Represent the map below by a graph, where edges are joined if the states they represent share a common border. 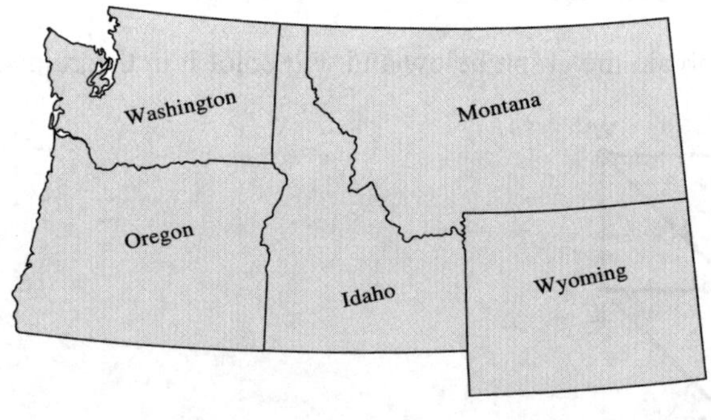 2. Using the map above, find the smallest number of colors that can be used to color the map so that any two states sharing a stretch of common border are not colored with the same color.

Section 4.2 The Traveling Salesperson Problem

Objective 1 – Use Hamilton circuits to solve the traveling salesperson problem.
Objective 2 – Determine all Hamilton circuits in a complete graph.
Objective 3 – Solve the traveling sales person problem using the brute force algorithm.
Objective 4 – Solve the traveling salesperson problem using the nearest neighbor algorithm.
Objective 5 – Solve the traveling salesperson problem using the best edge algorithm.

Objective 1 – Use Hamilton circuits to solve the traveling salesperson problem.

Key Terms
Traveling salesperson problem
Hamilton path
Hamilton circuit
Hamiltonian

Summary
The traveling salesperson problem (TSP) is a simple to understand, but potentially difficult problem to solve. Conceptually, the problem involves starting at a location, traveling to a number of other locations, and returning to the location you started at. We want to do this at a minimum cost. Translating this into a graph theory problem, we start at a vertex, travel along edges to all other vertices, and return to the initial vertex. Simple, right? Be careful – this seemingly simple problem can be difficult to solve! We add a new dimension to the graph, weights on the edges. You could think of these as the cost to travel from one vertex to another. Can we minimize the weight?

Before we get too far into the traveling salesperson problem, we need to introduce some new terminology from graph theory.

A path that passes through all of the vertices of a graph exactly once is called a Hamilton path. If a Hamilton path begins and ends at the same vertex, then it is called a Hamilton circuit. If a graph has a Hamilton circuit, we say the graph is Hamiltonian.

We can now restate the traveling salesperson problem in terms of Hamilton circuits. Given the representation of the TSP in a graph, can we find the optimal Hamilton circuit for our salesperson? Before we solve this problem, we will practice finding Hamiltonian circuits and then address minimizing the costs.

Problem Solving; *The Splitting-Hairs Principle* The definition of Hamilton path may sound a lot like the Euler path we discussed in an earlier section, but they are different. In a Hamilton path, you do not have to trace every edge, as you do with an Euler path.

Notes:

Guided Example	***Practice*** (Exercise 3 p.169)
Find a Hamilton circuit in the graph below that begins with the edge EB. 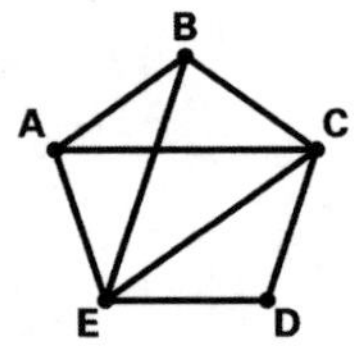 Starting at E and traveling along edge EB, we appear to have a choice of where to go next. If we go to C, however, we will not be able to get to both A and D and return to E. We will leave B and go to A, then to C, D and return to E. Following this path allows us to visit each vertex exactly once and return to E. The path would be written EBACDE.	1. Find a Hamilton circuit in the graph below that begins with the specified edge. a. AD b. EA 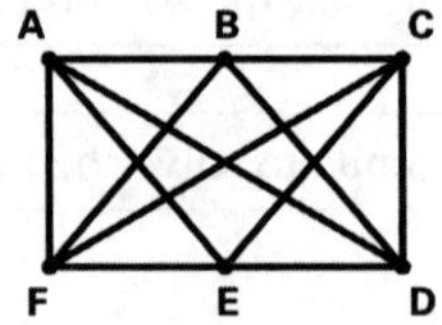

Objective 2 – Determine all Hamilton circuits in a complete graph.

Key Terms
Complete graph
Tree diagram

Summary
Recall that a path that passes through all of the vertices of a graph exactly once is called a Hamilton path. If a Hamilton path begins and ends at the same vertex, then it is called a Hamilton circuit. Not all graphs have Hamilton circuits, but we will find that there is a particular type of graph, called a complete graph, that not only has a Hamilton circuit, but actually has many of them.

We start by defining a complete graph on n vertices, represented by K_n. In a complete graph, every pair of vertices is joined by an edge.

The graph below is a graph of K_4. Note that every pair of vertices is connected with an edge.

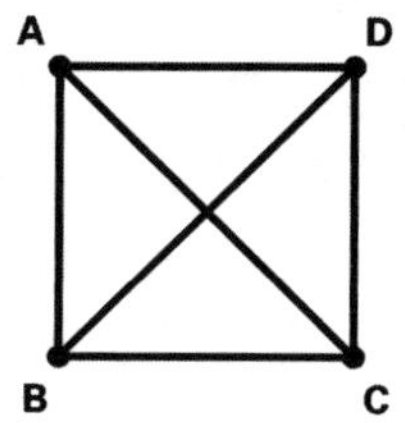

K_4

We could name several Hamilton circuits for this graph; ADCBA is one, BDCAB is another, and we could go on. If we wanted to obtain a listing of all possible Hamilton circuits it would be tedious without a good strategy.

Problem Solving; To obtain a listing of all Hamilton circuits, we will use the *Be Systematic* strategy. Recall tree diagrams from Chapter 1. The tree diagram below will help us to obtain a list of all Hamilton circuits for K_4.

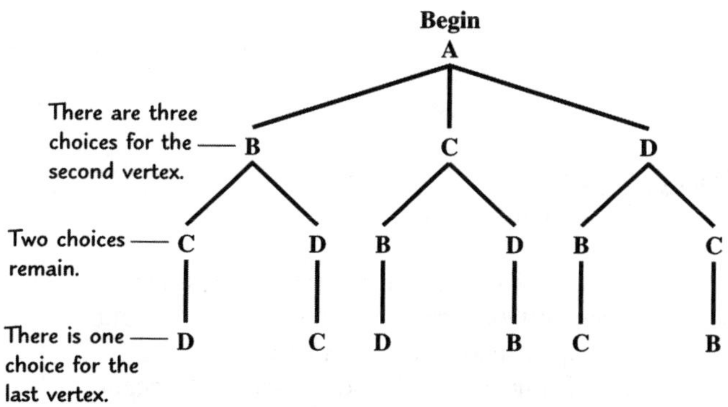

Tracing through the six branches of the tree, we see that K_4 has six Hamilton circuits that start at A: ABCDA, ABDCA ACBDA, ACDBA, ADBCA, ADCBA. You may think that we need to do a similar exercise starting at B, but notice that we could take the first path above ABCDA and start at B, BCDAB. This is considered the same circuit.

If we need only the number of Hamilton circuits on a complete graph, but not a list of them, we use the result that the number of Hamilton circuits in K_n is $(n-1)! = (n-1)\times(n-2)\times(n-3)\times\ldots\times3\times2\times1$. The exclamation point is called factorial, so we would read $(n-1)!$ as $(n-1)$ factorial. In our example above, $n = 4$ and there were $(4-1)! = 3! = 3\times2\times1 = 6$ Hamilton circuits.

Notes:

Guided Example	***Practice*** (Exercises 9 & 31 pp.169-170)
Your sibling has received offers from 5 colleges and plans to spend the upcoming vacation week visiting each The family will leave from home, visit each campus, and return home. Assume that it is possible to travel from each campus to any other. How many circuits are possible? There are 5 colleges plus your home, so if we were to represent this problem as a graph there would be 6 vertices. We assume that this is a complete graph, as we assumed travel from each campus to any other is possible. The number of circuits would be $(6-1)!=5!=5\times4\times3\times2\times1=120$ .possibilities.	1. How many Hamilton circuits are in K_7? 2. Foodandwine.com states that the best ice cream cities in the United States are Columbus (OH), Scottsdale (AZ), Austin (TX), Cambridge (MA), Honolulu (HI), Minneapolis (MN), New Orleans (LA), and San Francisco (CA). In how many ways can Jamie, who lives in Cleveland, OH, tour these cities and then return home?

Objective 3 – Solve the traveling sales person problem using the brute force algorithm.

Key Terms
Weighted graph
Brute force

Summary
Returning to the traveling salesperson problem, now let's consider the costs to travel from one place to the next. If we have a complete graph with n vertices, we know that there are $(n-1)!$ Hamilton circuits. Which is the cheapest one? Again, a seemingly simple question, but one that is not easy to solve.

Before we start, we will define a weighted graph, where we assign numerical weights to the edges of the graph. The weight of a path is the sum of the weights of the edges of the path. Solving the TSP then becomes finding the Hamilton circuit with the smallest weight.

Recall that the second step in Polya's problem solving strategy is to devise a plan. The plan we will use right now for solving the TSP is one called Brute force. The strategy isn't hard to understand, but it does involve a lot of work! We list all Hamilton circuits find the weight of each, and select the circuit with the smallest weight. This circuit is the solution to the traveling salesperson problem. We will discuss other strategies for solving the problem, but the brute force method is the only one that is guaranteed to find the minimum weight.

Notes:

Guided Example	***Practice*** (Exercise 13 p.169)
You are planning to tour three locations while visiting Washington DC over break. You plan to leave your hotel, labeled vertex A, and visit attractions at B, C, and D. The graph below shows the subway fares you anticipate for each of the trips. Use brute force to find the travel itinerary with the lowest cost. 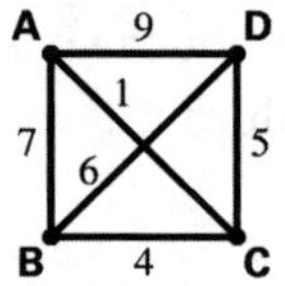 From our work in a previous section, we know the circuits for K_4 are ABCDA, ABDCA ACBDA, ACDBA, ADBCA, ADCBA. The cost for each is below: ABCDA = $7+4+5+9=25$ ABDCA = $7+6+5+1=19$ ACBDA = $1+4+6+9=20$ ACDBA = $1+5+6+7=19$ ADBCA = $9+6+4+1=20$ ADCBA = $9+5+4+7=25$ It is not a coincidence that each weight appears twice. The circuit ABCDA and ADCBA traverses the same edges, in different directions! Going forward, we would only need to look at half of the circuits. Clearly our best bet is ABDCA (or in reverse ACDBA).	1. Use the brute force algorithm to find a Hamilton circuit that has minimal weight. 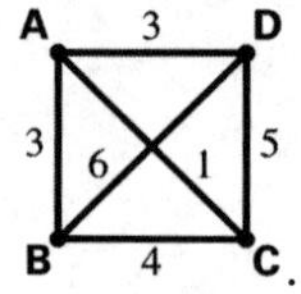

Objective 4 – Solve the traveling salesperson problem using the nearest neighbor algorithm.

Key Terms
Nearest neighbor

Summary
It would be nice if there was an algorithm that would solve the traveling salesperson problem faster than brute force. Unfortunately there isn't one, and at present it is uncertain if one exists. The strategy we will show here is one that will give a pretty good (i.e. low weight) solution, but will not always find the best. It is called the nearest neighbor algorithm.

Nearest Neighbor Algorithm:

1. Start at any vertex, X.
2. Of all edges connected to X, choose any one that has the smallest weight (there may be more than one, and just select one of them).
3. Choose subsequent new vertices as you did in step 2 (but to not return to a vertex you have visited until the end).
4. After all vertices have been chosen, return to your starting point.

Notes:

Guided Example

Use the nearest neighbor algorithm to find a Hamilton circuit for the graph below, starting at vertex A.

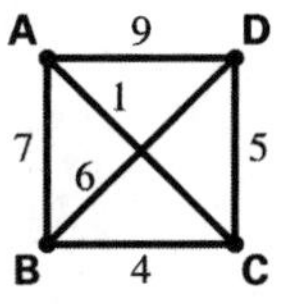

Leaving vertex A, the smallest weight is on the edge to C so we will take that. When we leave C, the best choice is to go to B. from B, we go to D then return to A. The circuit is ACBDA and has weight $1+4+6+9=20$. Note that this is not the best path that we found using the brute force method.

Practice (Exercise 17 p.170)

1. Use the nearest neighbor algorithm to find a Hamilton circuit that begins at vertex A in the graph below.

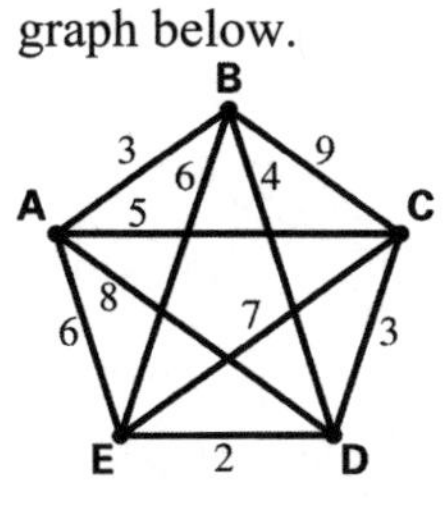

Objective 5 – Solve the traveling salesperson problem using the best edge algorithm.

Key Terms
Best edge algorithm

Summary
In the nearest neighbor algorithm, we traveled on the best edge "at the moment". What we didn't look at was the fact that we might be forced to take a very heavily weighted (perhaps expensive!) edge towards the end of our circuit. The best edge algorithm tries to address this by looking at the overall picture, and trying to ensure that we use the edges that have the smaller weights on them.

To use the Best Edge Algorithm:

1. Begin by choosing any edge with the smallest weight.
2. Choose any remaining edge in the graph with the smallest weight.
3. Keep repeating step 2; however do not allow a circuit to form until all vertices have been used. Also, because the final Hamilton circuit cannot have three edges joined to the same vertex, never allow this to happen during the construction of the circuit.

Notes:

Guided Example

Using the graph below, use the best edge algorithm to find a Hamilton circuit. List the circuit beginning at vertex A

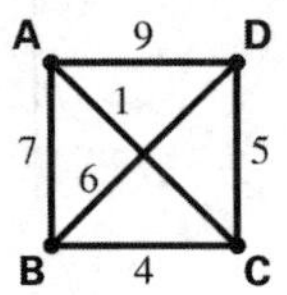

Looking at the weights, we first select AC, with weight 1. The next best weight is BC at 4. We would then select DC with 5, but that would have three edges at vertex C which is not allowed. We thus bypass DC and go to BD with a weight of 6. We can't pick BA, so are forced to use DA. This circuit is ACBDA and has weight of 20.

Practice (Exercise 23 p.170)

Using the graph below, use the best edge algorithm to find a Hamilton circuit. List the circuit beginning at vertex A.

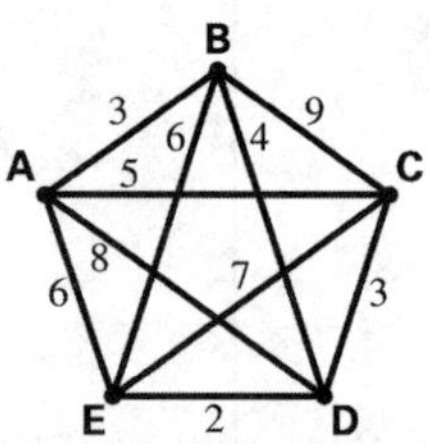

Section 4.3 Directed Graphs

Objective 1 – Use directed graphs to model relationships that go in only one direction.
Objective 2 – Use directed graphs to model influence.
Objective 3 – Model the spread of a disease by a directed graph.

Objective 1 – Use directed graphs to model relationships that go in only one direction.

Key Terms
Directed edge
Directed graph
Directed path
Length

Summary
When an edge has a direction it is called a directed edge. A graph in which all edges are directed is called a directed graph. If X and Y are vertices in a directed graph, and it is possible to begin at X, follow a sequence of edges in the directions indicated and end at Y, then we call the sequences of edges encountered a directed path from X to Y. We will denote a directed path by the sequence of vertices encountered along the path. The length of a directed path is the number of edges along that path (keep in mind that you are not allowed to repeat edges on a path).

Problem Solving; As with many graph theory questions, we will use the *Draw Pictures* strategy.

Notes:

<u>Guided Example</u>	***<u>Practice</u>*** (Exercise 3 p. 177)
Using the graph below, find the following: 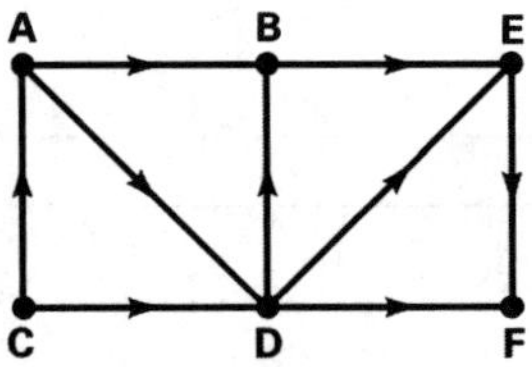 1. A directed path from C to E. We can move from C to A to B to E, so a directed path is CABE with length 3. 2. A directed path of length greater than 3 from C to E. We need a directed path with one more edge than that used in #1. We can go from C to A to D to B to E. This is a directed path CADBE with length 4. 3. A directed path of length 5 from C to F. We can obtain this path by going from C to A to D then to B next to E and finally to F. The path is CADBEF.	Find the requested items, if possible. If it is not possible, explain why not. 1. Using the graph 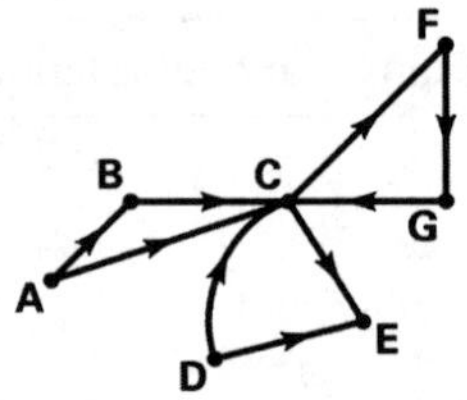 a. A directed path from B to E. b. A directed path from A to G. c. A directed path of length 5 from A to E. d. A directed path from F to D. e. A directed path of length 5 from A to C.

Objective 2 – Use directed graphs to model influence.

Key Terms
None

Summary
An interesting application of directed graphs is to model how members of a group influence each other. Because influence is not necessarily exerted equally in both directions, we use directed graphs. Our goal is to see who in the set has the most influence on others in the set. We also look at the number of ways a person can have two stage influence, that is influence on someone who then influences another. This is best seen in terms of an example.

Problem Solving; When working with modeling influence, you will need to look at each vertex of the graph and how many ways you can get from that vertex to another, and also paths of length 2 to other vertices. This can easily become confusing! Use the *Be systematic* strategy.

Notes:

Guided Example

The following influence has been observed among five faculty members:

Use a directed graph to model this situation and determine a ranking order for the committee members using one- and two-stage influence.

Member	Pratt	Quigley	Rich	Stickle	Thums
Influences	Rich	Stickle Thums	Stickle	Pratt Thums	Pratt Rich

A quick look at the table shows that Quigley, Stickle and Thums each have one-stage (direct) influence over the same number of people. Using a graph will help us to visualize who among them has the most influence if we allow two-stage influence as well.

We start out by construct a directed graph, where an arrow from P to R indicates that Pratt has influence over Rich. Verify the graph below models the influence as indicated in the table above.

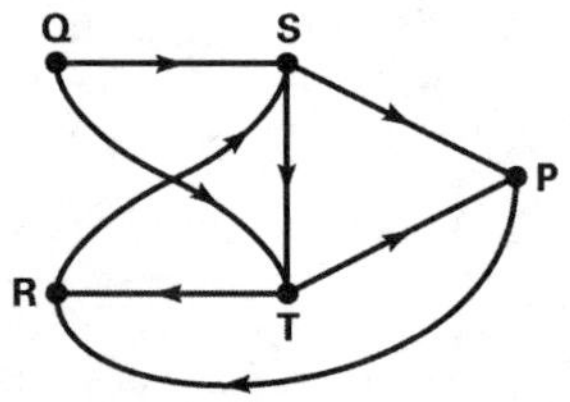

We now look for the paths of length one or two from each vertex. Let's look at Q. We can go from Q to S, Q to T in a path of length one. Looking at paths of length two, we can go from Q to P (using two different paths), from Q to T and from Q to R. This gives Quigley total influence of 6. In a similar fashion, T has one-stage influence over R and P and two-stage influence over R and S. In the table below, we model the number of ways a faculty member can influence another (using a path of length one or two). Summing over the rows, we find that Quigley has the most one- and two-stage influence over the other faculty members. Stickle and Thums are tied for the next position, and the list of influence ends with Rich and Pratt.

	P	Q	R	S	T	Total Direct and Two-Stage Influence
P	0	0	1	1	0	2
Q	2	0	1	1	2	6
R	1	0	0	1	1	3
S	1	0	2	0	1	4
T	1	0	2	1	0	4

Practice (Exercise 17 p.179)

A designer who wishes to obtain the contract for creating a new advertising campaign for Center City Community College wants to determine the most influential members of the selection committee. The table below lists the influences that have been observed among committee members. Model this committee with a directed graph and use one- and two-state influence to determine the most influential committee members.

Member	Lee	Feinstein	Johnson	Cordaro	Murciano
Influences	No one	Cordaro, Murciano	Lee, Cordaro	Lee	Lee, Johnson

Objective 3 – Model the spread of a disease by a directed graph.

Key Terms
none

Summary
We now look at another interesting application of directed graphs. In this case, we look to see if there are any vertices where you can start and have a path to each other vertex. If the answer is yes, it means you can leave that vertex and travel to each other vertex in the graph. We can use this to track the path of a disease. Who in the graph has the capability to spread a disease to everyone? In the guided example we will look at this in terms of spreading false information.

Problem Solving; We use the *Draw Pictures* strategy in this example.

Notes:

Guided Example

Consider the graph from the previous guided example, where the vertices represent five faculty members; Pratt, Quigley, Rich, Stickle, and Thums. A rumor has been circulating that the president of the university is going to resign. Using the directed graph, identify who (if anyone) could have initiated the rumor, given that all five have heard it. Assume that a directed path from Pratt to Quigley indicates that Pratt could have initiated the rumor and Quigley heard it. The length of the path is not considered important.

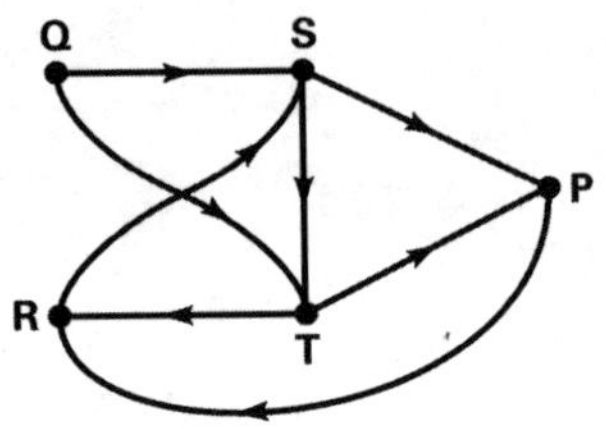

We will consider the vertices in alphabetical order.
If Pratt initiated the rumor, R, S and T would know, but there is not path from P to Q.
If Q initiated the rumor, P, R, S, and T would know. Q could have initiated the rumor.
If R initiated the rumor, P, S and T would know, but not Q.
If S initiated the rumor, P, R , and T would know, but not Q.
If T initiated the rumor, P, R, and S would know, but not Q.

It is clear that Q must have initiated the rumor. Note that no directed paths lead to Q, so the only way Q could know is if the rumor started with Quigley!

Practice (Exercise 9 p.179)

1. Ryan, Dwight, Pam, Jim, Angela, and Kevin work in the same office. A rumor has been circulating that their company is being taken over by a Singapore based firm in a hostile takeover. The given directed graph shows how rumors travel among these six employees.

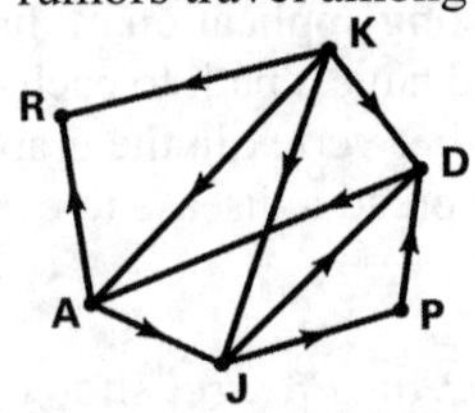

   a. Are there any people among these six who could have initiated the rumor?
   b. Change the direction of as few edges as possible to make Pam the person who started the rumor.

Section 4.4 Scheduling Projects Using PERT

Objective 1 – Use the language of PERT diagrams to model projects.
Objective 2 – Use PERT diagrams to schedule projects efficiently.

Objective 1 – Use the language of PERT diagrams to model projects.

Key Terms
PERT

Summary
The PERT organizational technique (Program Evaluation and Review Technique) is helpful when planning and organizing large projects. To employ this technique, we draw a PERT diagram, which is a special type of directed graph. We represent each task by a vertex in the graph containing the number of time units which are required to perform the task. If a task X immediately precedes a task Y, we draw a directed edge from vertex X to vertex Y in the graph.

Example: Construct a PERT diagram to model the following tasks in a project.

Task	Days Required to Perform Task	Preceding Tasks
A	1	None
B	2	None
C	3	A
D	5	A, B, C
E	4	A, B, C
F	2	B
G	6	A, B, C, D, E, F

We draw a directed edge from A to C because A immediately precedes C. Since B and C both precede D, we draw an edge from each to D. We do the same for E., B directly precedes "F, so we draw an edge to represent this relationship. Finally, D, E, and F precede G, so we draw an edge from D, E, and F to G. The vertices represent the number of days to perform each of the tasks.

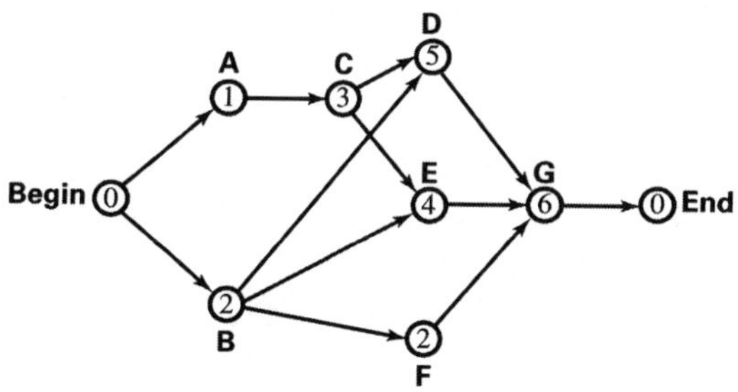

We can use the diagram to model the sequences of tasks that must be performed before the job can be completed. The numbers on the vertices help us determine how long it will take to complete the job.

Notes:

Objective 2 – Use PERT diagrams to schedule projects efficiently.

Key Terms
Critical path

Summary
Now that we have the PERT diagram, we will use it to schedule the tasks. In order to do this efficiently, we need to determine critical paths. Suppose T is a task in a PERT diagram. Let us consider all directed paths from "Begin" to T. If we add the time along each of these paths, any path requiring the most time to complete is called a critical path for task T. The idea of a critical path is important; it tells us the amount of time it will take before task T can be started.

Example: Use the PERT diagram we created in the previous objective, copied below. Find a critical path to vertex D.

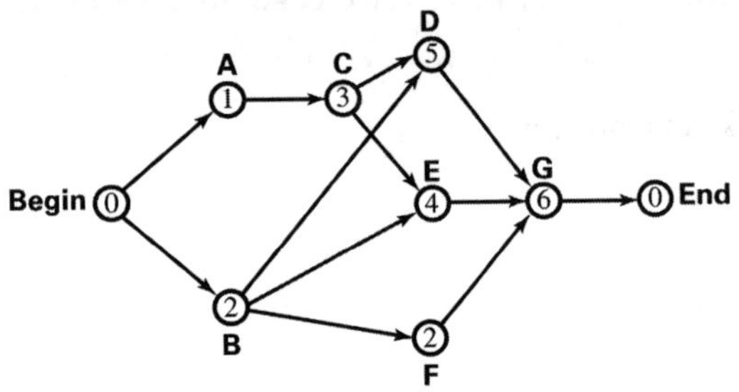

There are two paths to D; one is Begin, A,C,D the other is Begin, B, D. Begin, A, C, D is the critical path because the sum of the vertices A and C is 4, which is larger than the number on vertex B.

To schedule the tasks in a project, determine the critical path for each vertex; this will tell you when the task associated with that vertex can be started.

Problem Solving; Be sure to keep the *Order Principle* in mind when working with PERT diagrams.

Notes:

Guided Example

Using the PERT diagram above, find the critical paths for each vertex. Then use these to determine the time to complete the project and create a schedule.

Since A and B have no preceding tasks, they can be started immediately.

The critical path for C is Begin, A, C, which takes 1 day to complete.

The critical path for D was found above

To determine the critical path for E, consider the paths Begin, A, C, E and Begin, B, E. The critical path is Begin, A, C, E which will take 4 days.
There is only one path to F (Begin, B, F) so it is the critical path and takes 2 days.

.
To determine the critical path to G, we need to find the paths to G and then find the one with the largest sum of days along the way.

The paths to G, with time to complete them:
Begin, A, C, D, G 9
Begin, A, C, E, G 8
Begin, B, D, G 7
Begin, B, E, G 6
Begin, B, F, G 4

The critical path to G is Begin, A, C, D, G. We can start task G on day 10, and it will take 15 days to complete the project.

To schedule the project efficiently, we use the information obtained on the critical paths for each vertex, and identify when we can start each task:

Task	Day Task Begins
A	1
B	1
C	2
D	5
E	5
F	3
G	10

Practice (Exercise 9 p. 187)

The Earth Day Committee has begun planning for an Earth Day festival. We list the subtasks for this project and the dependencies among them in the following table. Draw a PERT diagram for this project and construct a schedule for the tasks.

Task	Preceding Tasks	Time Required (weeks)
1. Get funding	none	2
2. Get permits	none	1
3. Decide on program	1	2
4. Rent tents	1, 2, 3	2
5. Arrange for speakers, entertainers, etc.	1, 2, 3, 4	4
6. Advertise	1, 3, 5	2
7. Set up tents, booths, etc.	1, 2, 3, 4	1
8. Set up festival	1, 2, 3, 4, 5, 6, 7	1

Chapter 4 Answers

Section 4.1

Objective 1 1) connected, all vertices are even 2)

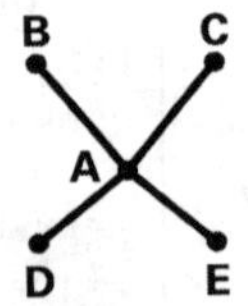

Objective 2 1) yes 2) yes

Objective 4 1) One possible answer is ABFHIGCBEFGECDA 2) Duplicate edges AJ, IJ, DE, EF; ABCDEFEDHFGHICAJIJA

Objective 5 1) 2) 3

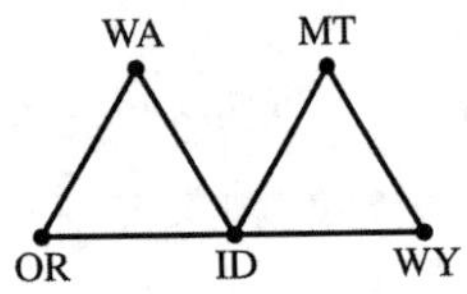

Section 4.2

Objective 1 1) a. possible answer ADCBFEA b. possible answer EADCBFE

Objective 2 1) $(7-1)!=6!=720$ 2) $(9-1)!=8!=40,320$

Objective 3 1) ACBDA has weight 14.

Objective 4 1) ABDECA

Objective 5 1) ABEDCA

Section 4.3

Objective 1 1) a. BCE b. ABCFG c. ACFGCE d. not possible e. ABCFGC

Objective 2 1) Feinstein (5), Murciano (4), Johnson (3), Cordaro (1), Lee (0)

Objective 3 1) a. Kevin b. One way: reverse the edge on KD

Section 4.4

Objectives 2 1)

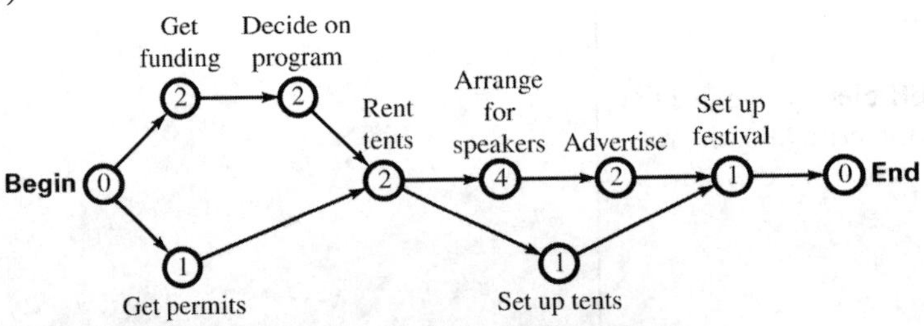

Task	Week Task Can Begin
Get funding	1
Get permits	1
Decide on program	3
Rent tents	5
Arrange for speakers, etc.	7
Advertise	11
Set up tents, booths, etc.	7
Set up festival	13

Chapter 5: Numeration Systems

Section 5.1 The Evolution of Numeration Systems.

Objective 1 – Represent numbers and compute in the Egyptian system.
Objective 2 – Calculate in the Roman numeration system.
Objective 3 – Make conversions in the Chinese numeration system.

Objective 1 – Represent numbers and compute in the Egyptian system.

Key Terms
Number
Numeral
Simple grouping

Summary
A number tells us how many objects we are counting; a numeral is a symbol that represents a number. We will look at the way various cultures represented numbers and the way various numeration systems work.

The Egyptian numeration system uses hieroglyphics and is a simple grouping system. In this system we form numerals by combining the symbols that represent various powers of 10. The table shows the symbols used and the number the symbol represents. We call this a simple grouping system because the order in which we write the symbols doesn't impact the number that the numeral represents. The table below contains Egyptian numeration symbols.

Number	Symbol	Name
1	𓏺	Stroke
10	𓎆	Heel bone
100	𓍢	Scroll
1,000	𓆼	Lotus flower
10,000	𓂭	Pointing finger
100,000	𓆐	Fish or tadpole
1,000,000	𓁨	Astonished person

Example: Convert 𓍢𓆼𓍢𓍢 𓁨𓁨𓏺𓏺𓏺𓏺𓏺 to the modern notation we use today; Hindu-Arabic notation.

Note that this would be the same as 𓁨𓁨𓆼𓍢𓍢𓍢𓏺𓏺𓏺𓏺𓏺. We have 2 millions, 1 thousand, 3 hundreds and 5 ones, so this numeral represents 2,001,305.

When Adding and subtracting in the Egyptian system, we group symbols together, and where necessary, regroup (for example, if our answer had ten heel bones, we would write one scroll).

Problem Solving: Use the *Analogies Principle* when performing addition and subtraction in the Egyptian system.

Notes:

Guided Example	***Practice*** (Exercises 3, 5, & 11 p.201)
Subtract ∩∩∩∩∩∩∩∩| from 99∩.|||||. We can subtract the one stroke from the five, but we can't subtract the eight heel bones from one. As with traditional subtraction, we borrow; that is we rewrite one of the scrolls as ten heel bones and then proceed. 9∩∩∩∩∩∩∩∩∩∩∩||||| - ∩∩∩∩∩∩∩∩| 9∩∩∩||||	1. Write the Egyptian numeral  using Hindu-Arabic numerals. 2. Write 3,245 using Egyptian numerals. 3. Perform the addition 

Objective 2 – Calculate in the Roman numeration system.

Key Terms
Subtraction principle

Summary
The Roman numeration system uses letters of the alphabet as numerals to represent certain numbers. The table below has a listing of the numerals and their associated numerical value. A great benefit of the roman system is the subtraction principle. We note that X represents the number 10, IX represents 10 – 1 or 9. Similarly, XC represents 100 – 10 or 90. There are two restrictions on the use of the subtraction principle.

1. We can only subtract the numerals I, X, C, and M. We would not write VL to represent 45.
2. We can only subtract numerals from the next two higher numerals. We only subtract I from V and X; we would only subtract C from D and M.

Number	Roman Numeral
1	I
5	V
10	X
50	L
100	C
500	D
1,000	M

Example: Represent the number 2,499 using Roman numerals.

We will use MM to represent 2,000. The subtraction principle allows us to represent 400 with the numerals CD. The most efficient way to write 99 is XCIX. Putting this together we have MMCDXCIX.

When adding two Roman numerals, you must group like numerals and then, when possible, replace numerals with a simpler representation. If we were to add VI and VIII we might think that the answer is VVIIII, but we would replace VV with X and IIII with IV. The answer would be XIV.

Problem Solving: When computing Roman sums, it is a good idea to use the *Draw Pictures* strategy. A counting board like the one used in the guided example is a good choice.

Notes:

Guided Example

Find the Roman sum CDXXXVIII + CXIV. Use the counting board below.

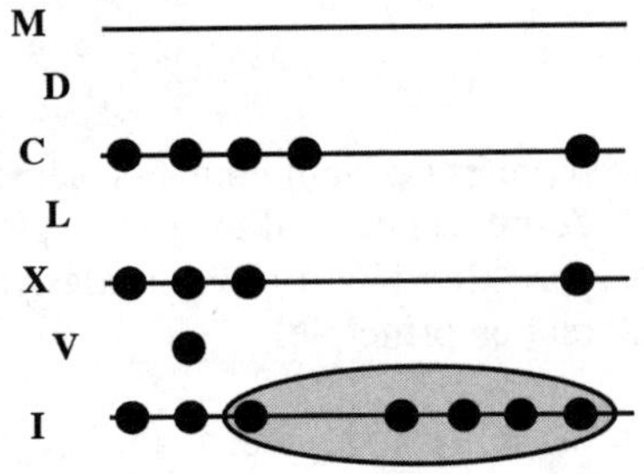

The five ones (circled) can be replaced by a V (five); we combine that V with the one already on the board to obtain an X (ten). We can then take the five tens and obtain a single L (fifty). The five hundreds can be replaced by a single D (five-hundred). The final answer is DLII.

Practice (Exercises 21, 33 & 67 pp.201-202)

1. Write DLXIV using Hindu-Arabic numerals.

2. Write 444 in Roman notation.

3. Find the sum CCCLXXXVII + CCCXVII. You may want to use the counting board below.

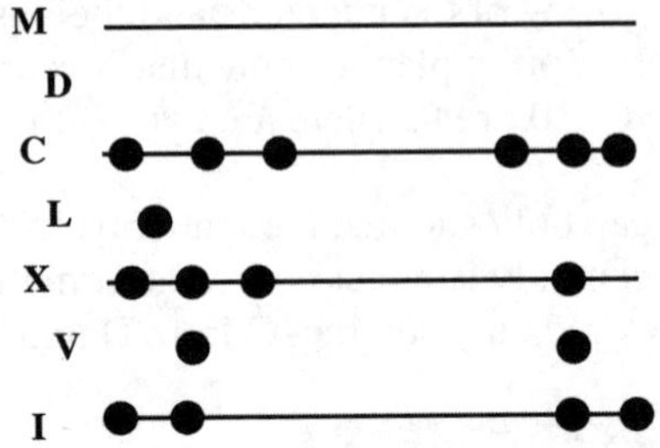

Objective 3 – Make conversions in the Chinese numeration system.

Key Terms
Multiplicative system

Summary
The traditional Chinese numeration system uses multiples of powers of 10 to represent numbers and is based on the notation shown in the table below.

Chinese Numeral	Value
一	1
二	2
三	3
四	4
五	5
六	6
七	7
八	8
九	9
十	10
百	100
千	1,000

This system is an example of a multiplicative system because we form numerals by writing products of integers between 1 and 9, inclusive, and powers of 10. While ancient Chinese numerals were written vertically, we will follow the current practice which is to write them horizontally.

Notes:

Guided Example	***Practice*** (Exercises 43 & 47 p.201)
1. Write 4,537 using Chinese numerals. To write this numeral, we need 4 thousands, 5 hundreds, 3 tens, and 7 units. Think of this as $(4\times1{,}000)+(5\times100)+(3\times10)+7$. Using the Chinese numerals in the table above, we have 四千五百三十七. 2. Write the Chinese numeral as a Hindu-Arabic numeral. 九千五十六 The largest power of 10 is 1,000 and just to the left of that character is a 9, so we have 9 thousands or 9,000. There are no hundreds (because there is no 百). We have 5 tens, and then 6 units. Our number is thus 9,056.	1. Write as a Hindu-Arabic numeral (recall that we are writing Chinese numerals horizontally). 九千九百九十九 2. Write the numeral using a Chinese numeral. 2,805

Section 5.2 Place Value Systems.

Objective 1 – Represent numbers and make conversions using the Babylonian system.
Objective 2 – Represent numbers and make conversions using the Mayan system.
Objective 3 – Perform operations using expanded notation in the Hindu-Arabic system.
Objective 4 – Use galleys and Napier's rods to calculate products.

Objective 1 – Represent numbers and make conversions using the Babylonian system.

Key Terms
Place value system
Positional system

Summary
In a place value system, also called a positional system, the placement of the symbols in a numeral determines the value of the symbols. For example, in the numeral 87, we have 8 tens and 7 ones. If we had written 78 that would have been very different.

The ancient Babylonian system is an example of a positional system. The system had only two symbols;  which represented 1 and  which represented 10. We could write 12 as . To write larger numbers, the Babylonians used several groups of these symbols, separated by spaces, and multiplied the value of these groups by increasing powers of 60. A system such as this one, based on powers of 60, is called a sexagesimal system.

Example: Convert  to Hindu Arabic notation.

There are three groups; the first represents 2 and is multiplied by 60^2, the second, 11, is multiplied by 60^1 and the right most group 21, is multiplied by 60^0, which is the units. Thus our number is
$2\times60^2+11\times60+21=2\times3,600+11\times60+23=7,883$.

The symbol  was used to indicate subtraction. The numeral  represents $30-2=28$.

Converting from Hindu-Arabic to Babylonian notation requires division by successive powers of 60, starting with the largest power of 60 less than or equal to the given number.

Example: Convert 75 to Babylonian notation.

The highest power of 60 which is less than or equal to 75 is 60^1 so we divide 75 by 60.

$$\begin{array}{r} 1 \\ 60\overline{)75} \\ \underline{60} \\ 15 \end{array}$$

We see that the quotient is 1 and the remainder is 15. We thus have one power of 60 and 15 units. We would write .

Notes:

Guided Example

Write 4,112 using Babylonian notation.

We start with division by 60^2 or 3,600 since this is the largest power of 60 which is less than or equal to 4,112; then divide the remainder by the next power, 60^1.

$$\begin{array}{r} 1 \\ 3600\overline{)4112} \\ \underline{3600} \\ 512 \end{array} \qquad \begin{array}{r} 8 \\ 60\overline{)512} \\ \underline{480} \\ 32 \end{array}$$

. We have 1 term of 60^2, 8 of 60^1 and 32 units. We write

 which, with fewer symbols, could be written

Practice (Exercises 3 & 9 p.212)

1. Write the Babylonian numeral  as a Hindu-Arabic numeral.

2. Write 123,485 using Babylonian notation.

Objective 2 – Represent numbers and make conversions using the Mayan system.

Key Terms

Mayan numerals

Summary

The Maya Indians had a number system based on the number 20. It used patterns of dots and bars to count, as shown below. To count beyond 19, the Mayans positioned symbols vertically, with the lowest position representing units, the next higher position representing 20s, the next higher position representing 20×18, the next representing $20 \times 18 \times 20$, and so on. The symbol represented zero.

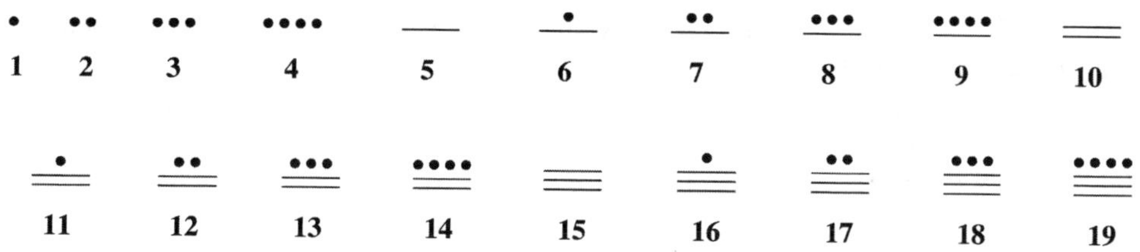

Example: Translate the Mayan numeral  to Hindu-Arabic notation.

We have 4 units, 0 20s, and 7 360s (remember the row above 20s is for 20×18). Thus we obtain $4 + 0 \times 20 + 7 \times 20 \times 18 = 4 + 2,520 = 2,524$.

We convert from Hindu-Arabic numbers to Mayan in a similar way to that with Babylonian numerals, but in this case our divisors are 20, 360, 7200, and so on.

Problem Solving: We *Relate a New Problem to an Older One* and do the conversion for Mayan numerals with the same process as the Babylonian numerals, but with new divisors.

Notes:

Guided Example

Write the number 8,003 using Mayan notation.

The largest divisor less than or equal to 8,003 is 7,200 so that is our first division. We then take the remainder, 803, and divide by 360. Lastly, the remainder 83 is divided by 20.

$$\begin{array}{r} 1 \\ 7200\overline{)8003} \\ \underline{7200} \\ 803 \end{array} \quad \begin{array}{r} 2 \\ 360\overline{)803} \\ \underline{720} \\ 83 \end{array} \quad \begin{array}{r} 4 \\ 20\overline{)83} \\ \underline{80} \\ 3 \end{array}$$

We have 1 term of 7200, 2 terms of 360, 4 20s, and 3 units.

Our Mayan numeral would be

•
••
••••
•••

.

Practice (Exercises 13 & 19 p.212)

1. Translate the Mayan numeral below to Hindu-Arabic notation.

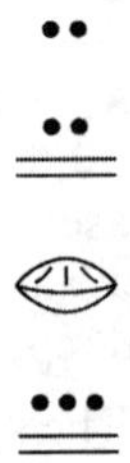

2. Write the number 2,173 using Mayan notation.

Objective 3 – Perform operations using expanded notation in the Hindu-Arabic system.

Key Terms
None

Summary
The Hindu-Arabic number system we are familiar with is a place value system based on the number 10. If we write the Hindu-Arabic numbers in expanded form, based on their place value, it helps to explain the algorithms for addition and subtraction that we learned when young.

Example: Write the numbers 476 and 1,018 in expanded form.

$476 = 4\times10^2 + 7\times10^1 + 6\times10^0$; $1,018 = 1\times10^3 + 0\times10^2 + 1\times10^1 + 8\times10^0$.

Notes:

Guided Example

Calculate 1,018 +476 using expanded notation.

We have written the two numbers in expanded notation; to perform the sum

$$\begin{array}{lrrrr} 1,018 = 1\times10^2 + & 0\times10^2 & +1\times10^1 & + 8\times10^0 \\ +\underline{476} = & + \underline{4\times10^2} & \underline{+7\times10^1} & \underline{+ 6\times10^0} \\ 1,494 = 1\times10^3 + & 4\times10^2 & + 8\times10^1 & + 14\times10^0 \end{array}$$

We take 10 of the 14 unit values and add to the tens yielding 9 tens; the sum is 1,494.

Practice (Exercises 37 & 41 pp. 212- 213)

Perform the following additions and subtractions using expanded notation.

1. $2,863 + 425$
2. $926 - 784$

Objective 4 – Use galleys and Napier's rods to calculate products.

Key Terms

Galleys

Napier's Rods

Summary

Although many of us use a calculator (or phone!) to perform multiplication, during elementary school you no doubt learned a method for doing it by hand. Another method, popular in Italy in the 15th century, was the galley method, which uses a rectangle divided into triangles to organize the computations.

Example: Use galleys to perform the multiplication 736×52.

We create a rectangle with three columns and two rows (because we have a 3 digit numeral multiplied by a 2 digit numeral). Subdivide the 6 squares to make 12 triangles.

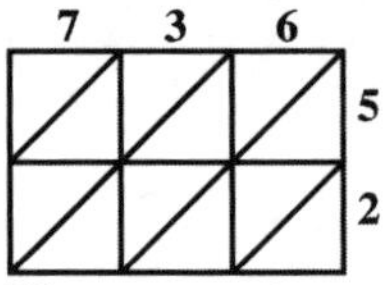

Now compute partial products by multiplying the numbers in each column and row; for example $7 \times 5 = 35$ so we put a 3 and a 5 in the triangles in the top left square.

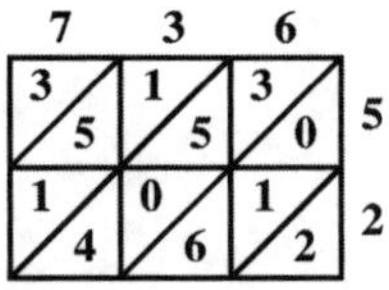

The next step is to sum the diagonals to find the digits in the product. As we do in our more modern method, if the digits sum to more than 9, we carry to the next diagonal.

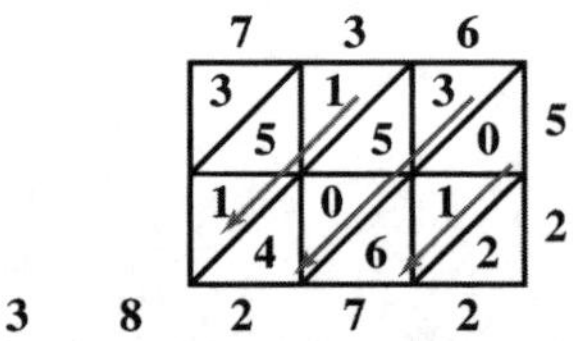

The answer to our problem is 38,272.

In the 17th century, English mathematician John Napier developed a device called Napier's rods for doing multiplication. The device consists of a series of strips, each labeled at the top with a digit. The remainder of the strip contains multiples of that digit. There is also a strip called the index.

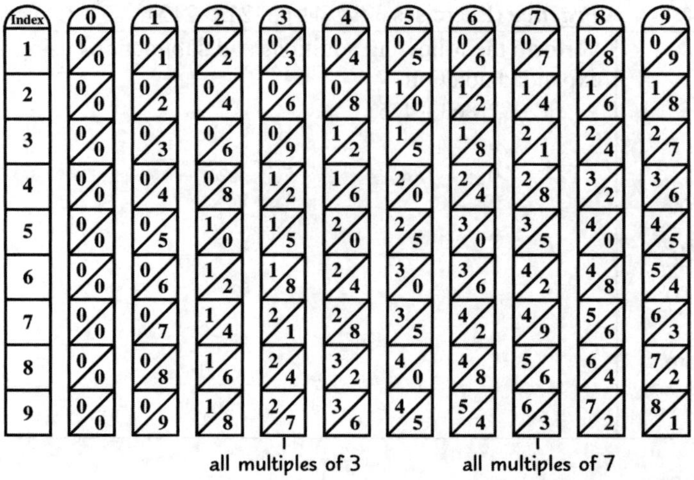

To calculate a product using Napier's rods, you select the strips for the digits you need and place them next to the index; the resulting galleys are used to determine the product. We perform a multiplication using Napier's rods in the guided example below.

Notes:

Guided Example	***Practice*** (Exercises 49 & 57 p.213)
Use Napier's rods to perform the multiplication 3×672. We need the index strip and the strips for 6, 7, and 2. 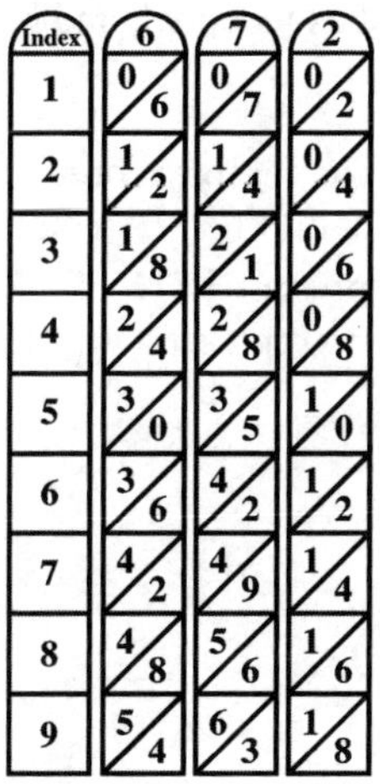 We use the galley in the row associated with the index 3. The galley that we use to find the product is 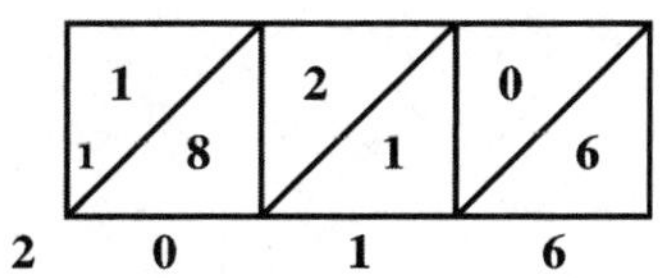 We find $3 \times 672 = 2{,}016$.	1. Use the galley method to perform the multiplication 23×876. 2. Use Napier's rods to perform the multiplication 6×924

Section 5.3 Calculating in Other Bases

Objective 1 – Count in non-base-10 systems.

Objective 2 – State similarities between a base-5 and a base-10 system.

Objective 3 – Perform arithmetic operations in a non-base-10 system.

Objective 4 – Make conversions among binary, octal, and hexadecimal systems.

Objective 1 – Count in non-base-10 systems.

Key Terms
Octal
Hexadecimal
Binary

Summary
In order to understand principles of other bases, we first need to consider what it means to count in base 10. When we count, we begin 1, 2, 3, etc., until we reach 9. Instead of writing a single symbol for ten, we write "one zero", this represents 1×10^1 and 0×10^0. Extending this to base b, we would write 10_b represents $1\times b^1$ and $0\times b^0$.

Example: Express 10_5 as a base-10 numeral.

10_5 would be equal to $1\times5^1 + 0\times5^0 = 5$. Note that 11_5 would equal 6 base 10.

Now let's try counting in base 5. We would start out 1, 2, 3, 4; but rather than writing 5 we would write 10_5 then 11_5, 12_5, 13_5, 14_5. The next number would be 20_5. Checking, this would be $2\times5^1 + 0\times5^0$.

Some other common bases are base-8, which is called octal; base-16, which is called hexadecimal, and base-2 which is the binary system.

The hexadecimal system has a unique feature; we use letters in addition to numbers. The reason is pretty simple; if we were to start counting, 1, 2, 3, …7, 8, 9, we don't have a single representation for 10. You might ask why we can't write 10_{16}, but realize that given our definition above, this would represent $1\times16^1 + 0\times16^0$ which is decimal 16, not 10! To get around this problem, we use A, B, C, D, E, and F to represent 10, 11, 12, 13, 14, and 15 (in that order). In the hexadecimal system,
$2F1_{16} = 2\times16^2 + 15\times16^1 + 1\times16^0 = 512 + 240 + 1 = 753$.

Notes:

Guided Example	***Practice*** (Exercises 1, 3, & 5 p.224)
In base 10, if we start at 23, we know the next three numbers are 24, 25, and 26. Use the starting point below and the given base and write the next three numbers. 1. 32_5 The first two numbers are fairly straightforward, 33_5, 34_5. What is next? We can't write a 5, so we use the fact that 5 is 10_5 to add one and obtain 40_5. 2. 100_2 The next number would be 101_2; the following number is 110_2 and then 111_2. Verify on your own that, in base 10, these numbers would be 4, 5, 6, and 7.	List the numbers that precede and follow each of the given numbers in the give base. 1. 24_5 2. 1011_2 3. EF_{16}

Objective 2 – State similarities between a base-5 and a base-10 system.

Key Terms
None

Summary
The base 5 system behaves as the base 10 system does; but rather than looking at powers of 10, we look at powers of 5. Recall that when counting in base 10, we begin 1, 2, 3, etc., until we reach 9. Instead of writing a single symbol for ten, we write "one zero", this represents 1×10^1 and 0×10^0. Comparing this to base 5, we would write 10_5 represents 1×5^1 and 0×5^0.

Example: Express 132_5 as a base-10 numeral.

132_5 would be equal to $1 \times 5^2 + 3 \times 5^1 + 2 \times 5^0 = 25 + 15 + 2 = 42$.

To convert from base 10 to base 5, we perform successive divisions by 5 until the quotient is zero, and note the remainders. Reading the remainders in the reverse order of the divisions gives us the base 5 numeral.

Notes:

Guided Example

Convert the decimal number 271 to base 5.

We start by dividing 271 by 5 and determining the quotient and remainder. We then divide the quotient by 5 and continue until the quotient is 0.

54	10	2	0
5)271	5)54	5)10	5)2
270	50	10	0
1	4	0	2

To determine the base 5 numeral, we read the remainders in reverse order, obtaining $2{,}041_5$.

Practice (Exercises 7 & 23 p.224)

1. Write the number 432_5 as a base-10 numeral.

2. Convert the decimal number 334 to base 5.

Objective 3 – Perform arithmetic operations in a non-base-10 system.

Key Terms
None

Summary
We will practice adding, subtracting, and multiplying in base 5. Prior to doing this, you might want to review these procedures in base 10. Recall that when doing addition and subtraction in base 10, anytime you carried or borrowed, you used a unit of 10 (that is you carried a 1 when you had a group of 10, and when you borrowed 1 you used a group of 10). In a similar way, when adding in base 5, when you obtain a group of 5 you carry 1, and when subtracting you borrow 1 and obtain a group of 5. Let's try a few examples.

Example 1: Perform the addition $241_5 + 132_5$.

It will help to line up the digits as we do in column addition. $\begin{array}{r} 241_5 \\ \underline{+132_5} \end{array}$.

We start with the units, 2 + 1 = 3. $\begin{array}{r} 241_5 \\ \underline{+132_5} \\ 3_5 \end{array}$ Next, we add the numbers in the fives place, which would seem to be a 7, but we can't write that so we think of it as 12_5 so we write a 2 and carry a group of 5, which we represent as a 1 when we carry; $\begin{array}{r} \overset{1}{2}41_5 \\ \underline{+132_5} \\ 23_5 \end{array}$ We then add the final column which is the five-squared place. 2 + 1 + 1 = 4. $\begin{array}{r} \overset{1}{2}41_5 \\ \underline{+132_5} \\ 423_5 \end{array}$

So we have 423_5.

When subtracting, when we borrow we must think about the place from which we are borrowing (a group of 5).

Example 2: Perform the subtraction $241_5 - 132_5$

We will again line up digits to make our work easier. We have $\begin{array}{r} 241_5 \\ \underline{-132_5} \end{array}$. To start, we can't subtract 2 from 1 so we borrow. When we do this, we obtain one group of 5, so we have 11_5 or decimal 6 fives. Our first step is to subtract 2 from these so we obtain 4. The remaining columns don't require borrowing. We have $\begin{array}{r} 2\overset{3}{\cancel{4}}\,{}^{1}1_5 \\ \underline{-1\;3\;\;2_5} \\ 1\;0\;\;4_5 \end{array}$.

We will multiply in base 5 in the guided example; the table of multiplication facts below will be helpful.

×	0	1	2	3	4
0	0	0	0	0	0
1	0	1	2	3	4
2	0	2	4	11_5	13_5
3	0	3	11_5	14_5	22_5
4	0	4	13_5	22_5	31_5

$3 \times 4 = 22_5$

Notes:

Guided Example	***Practice*** (Exercises 37, 47, & 55 p.224-225)
Multiply $141_5 \times 32_5$ To start out, we write this in columns and then work on the partial product, multiplying 141_5 by 2_5. First we multiply 2 times 1 and get 2 units. Then we multiply 2 times 4 fives to get 8 fives, which we would write as 13_5 (check the table to see that $2\times4=13_5$). We write the 3 and carry the 1 to the five-squared column. Finally, 2 units times 1 five-squared plus the 1 we carried equals 3, as shown. $\begin{array}{r} \scriptstyle 1 \\ 141_5 \\ \times\ 32_5 \\ \hline 332_5 \end{array}$ The next partial product involves multiplying the three in the fives place times 141_5. We multiply 3 fives times 1 unit to get 3; 3 fives times 4 fives to get 22_5 fives. We write the 2 and carry the 2 into the twenty-fives column. The final calculation at this stage is to multiply 3 fives times 1 twenty fives, then adding 2 to get 10_5. $\begin{array}{r} \scriptstyle 2 \\ 141_5 \\ \times\ 32_5 \\ \hline 332_5 \\ 1023\ \end{array}$. To complete the problem, we add, obtaining $11{,}112_5$.	Perform the indicated operation. 1. $3{,}412_5 + 231_5$ 2. $2{,}412_5 - 321_5$ 3. $41_5 \times 23_5$

Objective 4 – Make conversions among binary, octal, and hexadecimal systems.

Key Terms
None

Summary
In a previous worksheet, we defined several important systems; base-8, which is called octal; base-16, which is called hexadecimal, and base-2, binary. We will now look at ways to convert between the binary and octal system and the binary and hexadecimal system.

When converting from binary to octal, we will look at digits in groups of three from the right to the left and convert each group of 3 binary digits to one octal digit. To convert from binary to hexadecimal, we will consider groups of 4 binary digits, again starting from the right, and will convert them to hexadecimal. Recall that in the hexadecimal system, we incorporate the letters use A, B, C, D, E, and F to represent 10, 11, 12, 13, 14, and 15 (in that order).

The tables below will help to see the relationship between the systems.

Binary		**Octal**
0 0 0	=	0
0 0 1	=	1
0 1 0	=	2
0 1 1	=	3
1 0 0	=	4
1 0 1	=	5
1 1 0	=	6
1 1 1	=	7

Binary		**Hexadecimal**
0 0 0 0	=	0
0 0 0 1	=	1
0 0 1 0	=	2
0 0 1 1	=	3
·		·
·		·
·		·
1 1 1 0	=	E
1 1 1 1	=	F

Notes:

Guided Examples

Write the binary number 110111010001_2 as an octal number and then as a hexadecimal number.

To write 110111010001_2 in octal notation, we group the digits in threes, as shown below, and then convert using the table provided.

Binary →110 111 010 001
Octal → 6 7 2 1
So we have $110111010001_2 = 6721_8$.

Next we convert 110111010001_2 to a hexadecimal number by looking at groups of four digits.

Binary →1101 1101 0001
Hexadecimal → D D 1
We conclude that $110111010001_2 = DD1_{16}$.

Practice (Exercise 63 p.225)

1. Write the binary number 1011101101_2 as an octal number and as a hexadecimal number.

Section 5.4 Modular Systems

Objective 1 – Identify congruent numbers in modular systems.

Objective 2 – Perform operations in modular systems.

Objective 3 – Solve congruences in modular systems.

Objective 4 –Apply modular systems to solve real-world problems.

Objective 1 – Identify congruent numbers in modular systems.

Key Terms
Modulo-*m* system
Congruent

Summary
Even though you may not realize it, you use modular systems in your daily life; the best example being the 12 hour clock. What time is it one hour after noon? It's one o'clock! What time is it 15 hours after midnight? It's three o'clock. Clock arithmetic is an example of a modulo-12 system.

If *m* is an integer greater than 1, then a modulo-*m* system consists of the numbers 0, 1, 2, …, *m*-1. Counting and arithmetic operations are performed in a manner corresponding to movements in a *m*-hour clock. The number *m* is called the modulus of the system.

We say that a is congruent to b modulo m, written $a \equiv b(\bmod m)$ provided that *m* evenly divides $a - b$ (or $b - a$).

Problem Solving: Remember the *Splitting-Hairs Principle*; the lines ≡ look somewhat like the equal sign but do not mean exactly the same thing. When we write $a \equiv b(\bmod m)$, we are not saying that a and b are equal in the usual sense, but that in terms of a *m*-hour clock, we can consider them to be the same.

Notes:

Guided Example	***Practice*** (Exercises 5 & 9 p.232)
Determine if the statement $7 \equiv 21(\bmod 4)$ is true or false. If it is false, change the statement to make a true statement. In order for the statement $a \equiv b(\bmod m)$ to be true, m must evenly divide $a-b$ or $b-a$. Since $b-a$ = 14, which is not evenly divisible by 4, the statement is not true. To make it true, let's determine what $21(\bmod 4)$ would be equal to. The easiest way to do this is to divide 21 by 4, and the remainder, 1, is equal to $21(\bmod 4)$. Use the statement above (in order for the statement $a \equiv b(\bmod m)$ to be true, m must evenly divide $a-b$ or $b-a$) to verify that $1 \equiv 21(\bmod 4)$. It is important to note that this is not the only true statement we could have made. Verify that $5 \equiv 21(\bmod 4)$ is also true.	Determine which of the statements are true. 1. $35 \equiv 59(\bmod 12)$ 2. $46 \equiv 78(\bmod 10)$

Objective 2 – Perform operations in modular systems

Key Terms
None

Summary
It is fairly straightforward to add, subtract, and multiply in a modulo-*m* system. It will help to think of a modulus-*m* clock when doing so! To add, subtract, or multiply in a modulo-*m* system, perform the operation as usual, then replace the result with one of the numbers 0, 1, 2, … *m*-1 that is congruent to the result.

Notes:

Guided Example	***Practice*** (Exercises 11, 13, & 15 p.232)
Find the following:. 1. $10+11 \pmod{9}$ To find this sum, we add 10 and 11, obtaining 21, and then convert to modulo-9 by dividing by 9 and finding a remainder of 3. Thus $10+11 \equiv 3(\text{mod}\,9)$. 2. $10-11 \pmod{9}$ In this case, we find the difference, and express this as a number between 0 and 8 that is congruent. Ten minus 11 is -1; think of a clock with 9 at the top, one to the left of 9 is 8. Verify that $10-11 \equiv 8\,\text{mod}(9)$. 3. $10\times11 \pmod{9}$ Since 10 times 11 is 110, we can divide 110 by 9 to find the remainder; $10\times11 \equiv 2(\text{mod}\,9)$.	Perform the indicated operations. 1. $3+4 \pmod{5}$ 2. $4-9 \pmod{12}$ 3. $4\times9 \pmod{12}$

Objective 3 – Solve congruences in modular systems.

Key Terms

None

Summary

To solve a congruence modulo-m, we can use trial and error, considering m different possibilities. There may be more than one solution, so you can't stop once you find an answer that works. It is also possible that there are no solutions.

Example: Solve the congruence $2-x \equiv 5 \pmod{8}$

We will use trial and error with $x = 0, 1, 2, 3, 4, 5, 6, 7$.

$2-0 \equiv 5 \pmod{8}$ is not true since $5-2$ is not divisible by 8.

$2-1 \equiv 5 \pmod{8}$ is not true since $5-1$ is not divisible by 8.

$2-2 \equiv 5 \pmod{8}$ is not true since $5-0$ is not divisible by 8.

$2-3 \equiv 5 \pmod{8}$ is not true since $5+1$ is not divisible by 8.

$2-4 \equiv 5 \pmod{8}$ is not true since $5+2$ is not divisible by 8

$2-5 \equiv 5 \pmod{8}$ is true! $-3 \equiv 5 \pmod{8}$ is true because $5-(-3)=8$ which is divisible by 8.

$2-6 \equiv 5 \pmod{8}$ is not true since $5+4$ is not divisible by 8.

$2-7 \equiv 5 \pmod{8}$ is not true since $5+5$ is not divisible by 8.

The solution to the congruence is 5.

Notes:

Guided Example	***Practice*** (Exercises 21 & 29 p.232)
Solve the congruence $6x \equiv 4 \pmod{4}$. We will check to see if any of the values of x between 0 and 3 make this congruence true. $6 \cdot 0 \equiv 4 \pmod{4}$ is true, so 0 is a solution. $6 \cdot 1 \equiv 4 \pmod{4}$ is not true $6 \cdot 2 \equiv 4 \pmod{4}$ is true so 2 is a solution $6 \cdot 3 \equiv 4 \pmod{4}$ is not true There are two solution to the congruence, $x = 0$ and $x = 2$.	Solve the following congruences. 1. $6+x \equiv 4 \pmod{8}$ 2. $4x \equiv 2 \pmod{10}$

Objective 4 – Apply modular systems to solve real-world problems.

Key Terms
None

Summary
There are many applications where modular systems are used. We will focus on an application where we need to solve a pair of congruences simultaneously to find a solution that satisfies both. Our method in solving these problems will be to establish a sequence of numbers that would satisfy each congruence, and then look for the solution that meets the required criteria (in our case the smallest number that satisfies both).

Problem Solving: When solving a pair of congruences in this manner, we are using the *Look for Patterns* problem solving strategy.

Notes:

Guided Example	***Practice*** (Exercise 37 p.232)
Bob is packing hockey pucks in a storage container. If he makes stacks of 7, there are 3 left over, and if he makes stacks of 5 there are 4 left over. What is the smallest number of pucks he has to store? Let's call the number of pucks *p*. To say that we have 3 left over if stacks contain 7 is the same as saying $p \equiv 3(\text{mod}\,7)$ and to have 4 left over when stacking in groups of 5, we have $p \equiv 4(\text{mod}\,5)$. Let's consider the possible values of *p* in each case. For $p \equiv 3(\text{mod}\,7)$, *p* must be one of the numbers in the sequence 3, 10, 17, 24, 31, … (think of a 7 hour clock, we start at the 3, make a complete revolution and are at 10, and so on). For $p \equiv 4(\text{mod}\,5)$ *p* must be a number in the sequence 4, 9, 14, 19, 24, 29, … We look for the smallest number which is common to both, which is 24. Bob must have 24 pucks (so you might ask, why not stack them in groups of 6?)	1. A package designer for a chocolate manufacturer is designing boxes to hold a certain number of candies. When arranging the pieces six in a row, there will be four left over. When arranging the pieces five in a row, there will be three left over. What is the smallest number of pieces that the manufacturer might be considering to place in the box?

Chapter 5 Answers

Section 5.1

Objective 1	1) 2,121,210 2)  3) 
Objective 2	1) 564 2) CDXLIV 3) DCCIIII
Objective 3	1) 9,999 2) 二千入百五

Section 5.2

Objective 1	1) 8,187 2) 
Objective 2	1) 18,733 2) 
Objective 3	1) 3,288 2) 142
Objective 4	1) 20,148 2) 5,544

Section 5.3

Objective 1	1) 23_5, 30_5 2) 1010_2, 1100_2 3) EE_{16}, $F0_{16}$
Objective 2	1) 117 2) $2{,}314_5$
Objective 3	1) $4{,}143_5$ 2) $2{,}041_5$ 3) $2{,}043_5$
Objective 4	1) $1{,}355_8$; $2ED_{16}$

Section 5.4

Objective 1	1) true 2) false
Objective 2	1) 2 2) 7 3) 0
Objective 3	1) 6 2) 3,8
Objective 4	1) 28

Chapter 6: Number Theory and the Real Number System

Section 6.1 Number Theory

Objective 1 – Determine when numbers are prime.
Objective 2 – Apply divisibility tests to factor numbers into products of prime factors.
Objective 3 – Find the GCD and LCM of two natural numbers.
Objective 4 – Use the GCD and LCM to solve applied problems.

Objective 1 – Determine when numbers are prime.

Key Terms

Natural numbers
Counting numbers
Number theory
Divides
Divisor
Factor
Multiple
Prime number
Composite number

Summary

In chapter 2, we defined several important sets of numbers. One of the sets was the natural numbers, or counting numbers, which is defined $\{1, 2, 3, \ldots, \}$. The natural numbers possess interesting patterns and properties that have engaged mathematicians over time; the study of the natural numbers and their properties is called number theory.

One important characteristic of numbers is divisibility. If a and b are natural numbers, we say that a divides b, written $a|b$, provided there is a natural number q such that $b=aq$. We say that a is a divisor of b, a is a factor of b, or b is a multiple of a. A way of testing to see if a divides b is to actually divide b by a and see if you get a zero remainder.

A natural number greater than 1 that has only itself and 1 as factors is called a prime number. A natural number greater than 1 that is not prime is called composite.

If you are trying to determine if a number b is prime, check for divisors which are less than or equal to $\sqrt{b}$. If you don't find a divisor by that point, you can stop and conclude b is a prime.

Notes:

Guided Example	***Practice*** (Exercises 25 & 27 p.247)
Determine if 23 is a prime number. To check to see if 23 is prime, we must see if any prime numbers less than or equal to $\sqrt{23}$ is a divisor of 23. The primes less than or equal to $\sqrt{23}$ are 2 & 3, neither of which is a divisor of 23. We conclude 23 is prime.	What is the largest prime divisor you must try as a divisor of the given number n before you know whether or not n is prime? 1. 179 2. 331

Objective 2 – Apply divisibility tests to factor numbers into products of prime factors.

Key Terms
Factor tree

Summary
If you need to check larger numbers to determine if they are prime, it is helpful to have some quick tests for divisibility rather than doing lengthy division. The table below provides some of the divisibility tests you could use:

Number Is Divisible by	Test	Example
2	The last digit of the number is divisible by 2.	2 divides 13,578 because 2 divides 8.
3	The sum of the digits is divisible by 3.	3 divides 21,021 because 3 divides $2 + 1 + 0 + 2 + 1 = 6$.
4	The number formed by the last two digits is divisible by 4.	102,736 is divisible by 4 because 4 divides 36.
5	The last digit is 0 or 5.	607,895 is divisible by 5.
6	The number is divisible by both 2 and 3.	802,674 is divisible by both 2 and 3, so it is divisible by 6.
8	The number formed by the last three digits is divisible by 8.	8 divides 230,264 because 8 divides 264.
9	The sum of the digits is divisible by 9.	2,081,763 is divisible by 9 because 9 divides $2 + 0 + 8 + 1 + 7 + 6 + 3 = 27$.
10	The number ends in 0.	12,865,890 is divisible by 10.

The Fundamental Theorem of Arithmetic tells us that every natural number greater than 1 is a unique product of prime numbers. One way to determine the prime factors of a number is to build a factor tree. When constructing such a tree, you start by expressing the number as a product of smaller simpler numbers. Then try to re-express the numbers you have found as simpler numbers. Continue until you end up with prime factors. This is one case where the order of the steps is not important; you and a friend could build a factor tree with different first steps!

Notes:

Guided Example

1. Check the number 960 for divisibility by 2, 3, 4, 5, 6, 8, 9, 10. Then factor 960

Using the divisibility tests above, we determine that 960 is divisible by

2, because the last digit, 0, is divisible by 2.
3, because the sum of the digits is 15, which is divisible by 3.
4, because the last two digits, 60, is divisible by 4.
5, because the last digit is a 0.
6, because the number is divisible by 2 and 3.
8 because 960 is divisible by 8.
10 because the number ends in 0.

Note that 960 is not divisible by 9, because the sum of the digits is 15, which is not divisible by 9.

To find the prime factorization, we can start with any of the divisors above! We might start with 10 since it is largest, and obtain the tree.

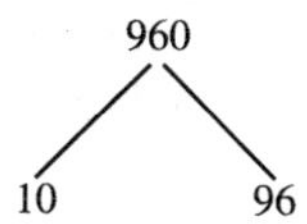

We then break both 10 and 96 into factors, and continue until no more prime factors can be found, as we see below

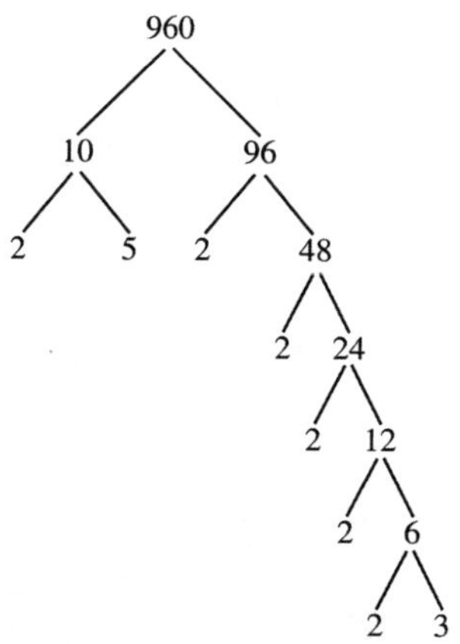

We write $960 = 2^6 \times 3 \times 5$.

Practice (Exercises 9, 11, 41 pp.247-248)

For exercise 1 & 2, check the numbers for divisibility by each of these numbers: 2, 3, 4, 5, 6, 8, 9, 10. State the number that divides the given number.

1. 141,270

2. 47,385

For question 3, factor the number. If the number is prime, state so.

3. 621

Objective 3 – Find the GCD and LCM of two natural numbers.

Key Terms
Greatest common divisor (GCD)
Least common multiple (LCM)
Euclidean Algorithm

Summary
The greatest common divisor of two natural numbers is the largest natural number that divides both numbers. The least common multiple of two natural numbers is the smallest natural number that is a multiple of both numbers. In order to find both the GCD and LCM, follow the steps below:

1. Factor both numbers, and write each as a product of powers of primes.
2. To calculate the GCD, multiply the smallest powers of any primes that are common to both numbers.
3. To calculate the LCM, multiply the largest powers of all primes that occur in either number.

Example: Find the GCD and LCM of 68 and 110.

Writing 68 as a power of primes, we obtain $68 = 2^2 \times 17$; doing the same with 110 yields $110 = 2 \times 5 \times 11$. To find the GCD, we take the smallest powers of common primes which is 2. Checking, we see that both numbers are divisible by 2.
To find the LCM, we take the largest powers of primes, yielding $2^2 \times 5 \times 11 \times 17 = 3740$.

We state a property which links the LCM and the GCD; If a and b are two numbers, then $GCD(a,b) \times LCM(a,b) = a \times b$. Let's check this with the result of our example above.
$GCD(68,110) \times LCM(68,110) = 2 \times 3740 = 7480 = 68 \times 110$.

We can also use the Euclidean Algorithm to find the GCD of two numbers. To follow the steps of this algorithm, start by dividing the smaller number into the larger number. Divide the remainder obtained into the divisor of the previous step. Continue until the remainder is 0. The remainder at the next to the last division is the GCD of the two numbers.

Example: Use the Euclidean Algorithm to find the GCD of 68 and 110.

We start by dividing 110 by 68, which yields the remainder 42 as seen below. The next step is to divide the remainder, 42, into the previous divisor, 68. We continue this process until we find a remainder of 0.

$$\begin{array}{r} 1 \\ 68\overline{)110} \\ \underline{-68} \\ 42 \end{array} \quad \begin{array}{r} 1 \\ 42\overline{)68} \\ \underline{-42} \\ 26 \end{array} \quad \begin{array}{r} 1 \\ 26\overline{)42} \\ \underline{-26} \\ 16 \end{array} \quad \begin{array}{r} 1 \\ 16\overline{)26} \\ \underline{-16} \\ 10 \end{array} \quad \begin{array}{r} 1 \\ 10\overline{)16} \\ \underline{-10} \\ 6 \end{array} \quad \begin{array}{r} 1 \\ 6\overline{)10} \\ \underline{-6} \\ 4 \end{array} \quad \begin{array}{r} 1 \\ 4\overline{)6} \\ \underline{-4} \\ 2 \end{array} \quad \begin{array}{r} 2 \\ 2\overline{)4} \\ \underline{-4} \\ 0 \end{array}$$

The remainder at the next to last division (2) is the GCD of 110 and 68.

Notes:

Guided Example

Use prime factorization and the Euclidean algorithm to find the GCD of 121 and 110.

Factoring each, we find $121 = 11^2$ and $110 = 2 \times 5 \times 11$. The GCD is 11.

Using the Euclidean algorithm, we find

$$\begin{array}{r} 1 \\ 110\overline{)121} \\ \underline{-110} \\ 11 \end{array} \qquad \begin{array}{r} 10 \\ 11\overline{)110} \\ \underline{-11} \\ 00 \\ \underline{-00} \\ 0 \end{array}$$

The remainder at the next to last step is 11, which is the GCD.

Practice (Exercises 47 & 57 p.248)

1. Use prime factorization to find the GCD and LCM of 56 & 70.

2. Use the Euclidean algorithm to find the GCD of the numbers 280 and 588; then use your results to find the LCM.

Objective 4 – Use the GCD and LCM to solve applied problems

Key Terms
None

Summary
Finding the GCD or LCM can be useful in solving application problems. In the case of the GCD, a carpenter may want to take two boards of different lengths and cut them into smaller, boards of equal size. We see in the example below that we can use the LCM determine when two species with different periodic hatching cycles will emerge at the same time.

Example: A builder has left over 2x4 pieces measuring 6 and 10 feet. She wishes to cut them into as many similar length pieces as possible. What length board should be used?

To answer this, we find the GCD of 6 and 10 which is 2. If she cuts them into 2 foot segments, there will be 3 pieces cut from the 6 foot board, and 5 from the 10 foot board, and no waste.

Notes:

Guided Example	***Practice*** (Exercise 69 p.248)
In addition to avoiding their predators, another advantage of the 13- and 17- year emergence pattern for cicadas is that the two species rarely emerge during the same year to compete for food. Suppose that type A cicada has a 13- year emergence cycle and type B cicada has a 17- year cycle. If both type A and type B cicadas emerged during 2010, when is the next year that both cicadas will emerge during the same year? To determine the solution, we need to find the LCM of 13 and 17. Since both numbers are prime, the LCM is equal to $13 \times 17 = 221$. Type A and B cicadas will appear 221 years after 2010, or 2231.	1. Assume that Jet Blue flights for Miami leave every 35 minutes and fights for Dallas leave every 20 minutes. If flights to Miami and Dallas have just departed, how many minutes will it be before this will happen again?

Section 6.2 The Integers

Objective 1 – Model the rules for adding and subtracting integers.
Objective 2 – Use the rules for multiplying and dividing signed numbers to perform arithmetic operations.
Objective 3 – Perform computations using integers..

Objective 1 – Model the rules for adding and subtracting integers.

Key Terms
Whole numbers
Integers
Opposites

Summary
Recall from chapter 2 that the set of whole numbers $W = \{0, 1, 2, 3, \ldots\}$ is the set of natural numbers with one addition, the number 0. If we take the whole numbers and add the additive inverses (opposites) of the natural numbers, we get the set of Integers. This set is frequently represented by I, $I = \{\ldots, -3, -2, -1, 0, 1, 2, 3, \ldots\}$. We show below a way to add and subtract integers with the use of a number line.

Problem Solving; When we look to the number line for guidance, we are using the *Draw Pictures* strategy.

When looking at the difference between two integers, we use the definition $a - b = a + (-b)$.

Example: Rewrite as an addition problem and solve $11 - 5$.
We can re-write $11 - 5$ as $11 + (-5)$. The number line below reminds us that addition moves to the right, and subtraction to the left. $11 + (-5) = 6$.

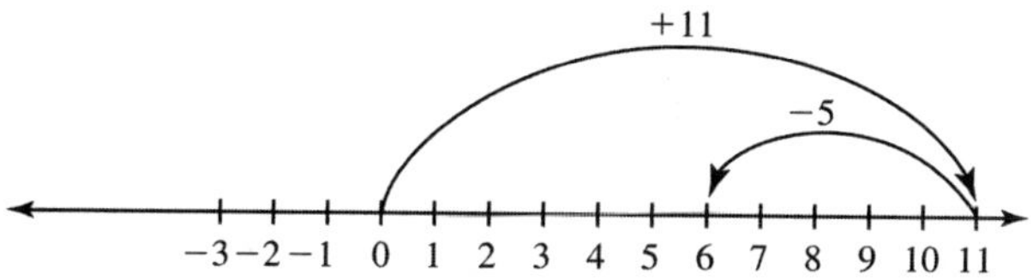

Notes:

Guided Example

Rewrite as an addition problem and solve:

$(-2) - 4$

We re-write the subtraction as addition:

$(-2) - (4) = (-2) + (-4) = -6$

Practice (Exercises 7, 9 & 11 p.255)

Rewrite the following subtraction problems as addition problems and then find the answers:

1. $(+18) - (-5)$
2. $(+6) - (+19)$
3. $(-28) - (+37)$

Objective 2 – Use the rules for multiplying and dividing signed numbers to perform arithmetic operations.

Key Terms
None

Summary
We often have to multiply and divide integers. There are important rules to keep in mind. If a and b have the same sign, then $a \cdot b$ is positive. If a and b have opposite signs, then $a \cdot b$ is negative. Division is a form of multiplication ; if a, b, and c are integers, then $a/b = c$ means $a = b \cdot c$ The rules for signs hold with division; if a and b have the same sign, then $\frac{a}{b}$ is positive. If a and b have opposite signs, then $\frac{a}{b}$ is negative.

Notes:

Guided Examples	***Practice*** (Exercises 17, 23, & 31 p.255)
Perform the following: 1. $(-6)(-5)$ Since the terms have the same sign, the answer is positive. $(-6)(-5) = 30$. 2. $\frac{-121}{11}$ Since the terms have opposite signs, the answer is negative. $\frac{-121}{11} = -11$	Find the following: 1. $(-7)(+8)$ 2. $(-8)(-9)$ 3. $\frac{+15}{+3}$

Objective 3 – Perform computations using integers.

Key Terms
None

Summary
Most calculations involve more than just one computation. We will now turn to more complex expressions, and use our rules for working with signed integers to simplify them.

Problem Solving; We need to keep the *Order Principle* in mind when performing calculations with signed integers. Work inside parentheses must be done first, then multiplication and division, and finally addition and subtraction.

Notes:

Guided Example	***Practice*** (Exercises 39 & 41 p.255)
Simplify $\left(\frac{-1-(5)}{4-5}\right)\cdot\left(\frac{-4+7}{-5-(-2)}\right)$ Our first step will be to work inside the parentheses, changing all subtraction to addition: $\left(\frac{-1+(-5)}{4+(-5)}\right)\cdot\left(\frac{-4+7}{-5+(+2)}\right)$. Next we will calculate the sums inside the parentheses: $\left(\frac{-6}{-1}\right)\cdot\left(\frac{3}{-3}\right)$. Performing the divisions, being careful to have the correct signs, we find $(6)\cdot(-1)=-6$.	Simplify each of the following expressions. 1. $(-6-(-14))\cdot\left(\frac{-12+(-6)}{7-4}\right)$ 2. $\left(\frac{-1-(-5)}{4-6}\right)\cdot\left(\frac{-4+(-2)}{-5+3}\right)$

Section 6.3 The Rational Numbers

Objective 1 – Determine when two rational numbers are equal.

Objective 2 – Add and subtract rational numbers.

Objective 3 – Multiply and divide rational numbers.

Objective 4 – Perform operations using mixed numbers.

Objective 5 – Represent rational numbers as repeating decimals.

Objective 1 – Determine when two rational numbers are equal

Key Terms

Rational numbers
Numerator
Denominator
Equal
Cross multiplying
Cross product
Reducing the rational number
Lowest terms

Summary

We define rational numbers to be the set of all numbers that can be written in the form $\frac{a}{b}$ where a and b are integers and $b \neq 0$. The top number, a, is called the numerator, and the bottom number, b, is called the denominator. Two rational numbers $\frac{a}{b}$ and $\frac{c}{d}$ are equal if and only if $ad = bc$. Calculating the products ad and bc is referred to as cross multiplying, or determining the cross product.

It is often desirable to write a rational number in lowest terms. We do this by canceling any factors which are common to the numerator and the denominator. This is also called reducing the rational number. We use the fact that $\frac{a \cdot c}{b \cdot c} = \frac{a \cdot \not{c}}{b \cdot \not{c}} = \frac{a}{b}$

Notes:

Guided Example

Check to see if the rational numbers below are equal. If they are, express them in lowest terms.

$\frac{14}{18}, \frac{42}{54}$

To test for equality, we will calculate the cross product:

$\frac{14}{18} \times \frac{42}{54}$

We multiply and find $14 \cdot 54 = 756$ and $18 \cdot 42 = 756$. This shows that $\frac{14}{18} = \frac{42}{54}$. We can reduce either one to lowest terms, so let's consider $\frac{14}{18}$. We notice that both numerator and denominator are divisible by 2, so canceling 2 from both numerator and denominator yields

$\frac{14}{18} = \frac{\cancel{2} \cdot 7}{\cancel{2} \cdot 9} = \frac{7}{9}$.

Practice (Exercises 3 & 7 p.267)

Which of the following pairs of rational numbers are equal?

1. $\frac{12}{14}, \frac{14}{16}$

2. $\frac{5}{14}, \frac{15}{42}$

Objective 2 – Add and subtract rational numbers.

Key Terms
Common denominator

Summary
In order to add or subtract rational numbers, they must have the same denominator. As long as this is true, we simply add or subtract the numerators. If we wish to add or subtract two rational numbers with different denominators, we must first rewrite the rational numbers into equal ones with the same, common, denominator. While there are many common denominators you could use, the product of the denominators will always work.

For addition of two rational numbers: $\frac{a}{b}+\frac{c}{d}=\frac{a\cdot d+b\cdot c}{b\cdot d}$ To subtract, use $\frac{a}{b}-\frac{c}{d}=\frac{a\cdot d-b\cdot c}{b\cdot d}$.

Example:

To add $\frac{2}{3}+\frac{4}{5}$ we will use the common denominator 15 and express the sum in the following form:

$\frac{2}{3}+\frac{4}{5}=\frac{2\cdot 5+4\cdot 3}{3\cdot 5}$ yielding the answer $\frac{22}{15}$. As a final step, always check to see that your answer is in lowest terms (ours is).

Notes:

Guided Example	***Practice*** (Exercises 19 & 23 p.268)
Find $\frac{2}{3}+\frac{4}{5}-\frac{1}{2}$ express your answer in reduced form. Our strategy will be to first add $\frac{2}{3}+\frac{4}{5}$ and then subtract $\frac{1}{2}$. The common denominator for the sum is 15. $\frac{2}{3}+\frac{4}{5}-\frac{1}{2}=\frac{2\cdot 5+4\cdot 3}{3\cdot 5}-\frac{1}{2}=\frac{22}{15}-\frac{1}{2}$ Now we subtract using common denominator 30. $\frac{22}{15}-\frac{1}{2}=\frac{22\cdot 2-1\cdot 15}{30}=\frac{29}{30}$ A final step is to check that the answer is in lowest terms, which it is. There are no common factors to cancel.	Perform the following operations. Express your answer as a positive or negative quotient of two integers in reduced form. 1. $\frac{7}{16}-\frac{1}{3}$ 2. $\frac{1}{8}-\frac{2}{3}+\frac{1}{2}$

Objective 3 – Multiply and divide rational numbers.

Key Terms
None

Summary
When we multiply two rational numbers $\frac{a}{b}$ and $\frac{c}{d}$, you multiply numerators and denominators, as indicated in the equation $\frac{a}{b}\cdot\frac{c}{d}=\frac{a\cdot c}{b\cdot d}$. It is important that your final answer be in lowest terms. Since you will do this by canceling common factors, one time saving strategy is to cancel any common factors before you multiply $\frac{a\cdot c}{b\cdot d}$.
To divide rational numbers, you should re-write the division problem as a multiplication problem. The rule to follow is to invert the second term and multiply as indicated in the following formula: $\frac{a}{b}\div\frac{c}{d}=\frac{a}{b}\cdot\frac{d}{c}$. You may also see the quotient of $\frac{a}{b}$ and $\frac{c}{d}$ written as a compound fraction; $\frac{\frac{a}{b}}{\frac{c}{d}}=\frac{a}{b}\div\frac{c}{d}=\frac{a}{b}\cdot\frac{d}{c}$.

Notes:

Guided Examples	***Practice*** (Exercises 25, 29, & 33 p.268)
Find the following and express answers in simplest form. 1. $\frac{3}{25}\cdot\left(-\frac{10}{12}\right)$ We know the answer will be negative, so multiply without regard to signs and then express your answer as a negative. $\frac{3}{25}\cdot\left(\frac{10}{12}\right)=\frac{\cancel{3}^{1}}{\cancel{25}_{5}}\cdot\frac{\cancel{10}^{2}}{\cancel{12}_{4}}=\frac{1}{5}\cdot\frac{\cancel{2}^{1}}{\cancel{4}_{2}}=\frac{1}{5}\cdot\frac{1}{2}=\frac{1}{10}$ We conclude $\frac{3}{25}\cdot\left(-\frac{10}{12}\right)=-\frac{1}{10}$. 2. $\frac{5}{16}\div\frac{5}{8}$ We will invert and multiply: $\frac{5}{16}\div\frac{5}{8}=\frac{\cancel{5}^{1}}{\cancel{16}_{2}}\cdot\frac{\cancel{8}^{1}}{\cancel{5}_{1}}=\frac{1}{2}\cdot\frac{1}{1}=\frac{1}{2}$	Perform the following operations. Express your answers as a positive or negative quotient of two integers in reduced form. 1. $\frac{2}{3}\cdot\frac{1}{2}$ 2. $\frac{7}{8}\div\left(-\frac{5}{24}\right)$ 3. $\left(\frac{11}{30}\div\left(-\frac{1}{6}\right)\right)\cdot\left(\frac{15}{4}\right)$

Objective 4 – Perform operations using mixed numbers.

Key Terms
Improper fraction

Summary
A rational number with numerator greater than the denominator is called an improper fraction. To convert the improper fraction $\frac{a}{b}$ to a mixed number, perform the division $b\overline{)a}$ with quotient q and remainder r, and rewrite $\frac{a}{b}$ as $q\frac{r}{b}$.

To change a mixed number $q\frac{r}{b}$ to an improper fraction, the numerator is found by multiplying q times b and adding r, the denominator is b. More simply, $q\frac{r}{b}=\frac{q\cdot b+r}{b}$

Notes:

Guided Example	***Practice*** (Exercises 43 & 47 p.268)
1. Convert $\frac{37}{5}$ to a mixed number. We divide 37 by 5, and find a remainder of 2 $5\overline{)37}$ gives quotient 7, 35 subtracted, remainder 2, so the mixed number is $7\frac{2}{5}$. 2. Check your work by converting your answer to an improper fraction. $7\frac{2}{5}=\frac{7\cdot 5+2}{5}=\frac{37}{5}$	3. Convert $\frac{121}{15}$ to a mixed number. 4. Convert $9\frac{1}{6}$ to an improper fraction.

Objective 5 – Represent rational numbers as repeating decimals.

Key Terms
Decimal form

Summary

We have defined rational numbers to be the set of all numbers that can be written in the form $\frac{a}{b}$ where a and b are integers and $b \neq 0$. Another way to define the rational numbers is that they are numbers in the form $\frac{a}{b}$ which can be expressed as repeating decimals. We consider the decimal 0.75 to be a repeating decimal where the place after the 5 is a repeating 0. For a given rational number, how do we find this decimal expansion? We do this by dividing the denominator into the numerator and obtaining the quotient.

Notes:

Guided Example	***Practice*** (Exercises 51 & 55 p.268)
Express the following rational numbers as repeating decimals. 1. $\frac{3}{8}$ We divide 8 into the numerator 3, and find 0.375. 2. $\frac{17}{33}$ If you use your calculator to divide 33 into 17, you will see .5151515152 but this is not quite correct. When we do this by hand, we see a pattern in the quotient which will continue. $$\begin{array}{r} 0.5151 \\ 33\overline{)17.0000} \\ \underline{165} \\ 50 \\ \underline{33} \\ 170 \\ \underline{165} \\ 50 \\ \underline{33} \\ 17 \end{array}$$ We write $\frac{17}{33}$ as the repeating decimal $0.\overline{51}$ to indicate that the pattern continues infinitely.	Write each rational number as a terminating or repeating decimal. 1. $\frac{3}{16}$ 2. $\frac{4}{13}$

Section 6.4 The Real Number System

Objective 1 –Represent irrational numbers as nonrepeating decimal expansions.
Objective 2 – Compute using radicals.
Objective 3 – Illustrate the properties of common number systems.

Objective 1 – Represent irrational numbers as nonrepeating decimal expansions.

Key Terms
Irrational number

Summary
Irrational numbers are numbers which are not rational; they do not have a repeating decimal expansion and cannot be written as the ratio of two integers. We can thank Pythagoras for discovering the irrational number $\sqrt{2}$ around 500 BC. There are actually more irrational numbers than there are rational numbers; any number in the form $\sqrt{n}$, where n is not a perfect square, is irrational. When looking at a decimal approximation of an irrational number, we may see a pattern, but if the decimal values do not repeat exactly, the number is not considered a repeating decimal and thus is not considered rational. An example would be the decimal 0.21221222122221…. This is not a repeating decimal because the number of times 2 appears between each 1 is increasing. This is an irrational number.

There are some well-known irrational numbers; notably π, φ, and e. The number π is the ratio of the circumference to the diameter of a circle. While many people say that π is approximately 3.14, it is actually a non-repeating decimal. The number φ (phi) is defined by the equation $\varphi = \frac{\sqrt{5}+1}{2}$ and is called the golden ratio as it describes the ratio of the length to the width of a beautiful rectangle. The number $e = 2.7182818284...$, named after Leonhard Euler, has many important applications in business, science, and engineering.

Notes:

Guided Example	***Practice*** (Exercises 3 & 7 p.279)
Give the decimal expansion for $\sqrt{5}$ and explain why it is irrational. Using a calculator, you will likely find $\sqrt{5}$ =2.236067977. You could (incorrectly) assume that this is a repeating decimal. As it turns out, $\sqrt{5}$ =2.236067977499789696… and never establishes a repeating pattern. We conclude that it is irrational. You can find the first million digits of $\sqrt{5}$ with a quick web search and see that it doesn't repeat in that range!	Which of the following are rational and which are irrational? 1. 3.1416 2. 0.101101111 0…

Objective 2 – Compute using radicals.

Key Terms
Radical

Summary
Square roots, cube roots, fourth roots, and so on are examples of radical expressions. The root symbol $\sqrt{\ }$ is called the radical sign. Performing calculations involving radicals involves tedious manipulations. For our purposes we will estimate radical expressions using a calculator or app. It is important that we state how accurate our results are by stating where rounding has been done.

Notes:

Guided Example	***Practice*** (Exercises 15 & 23 p.279)
Express $\varphi = \dfrac{\sqrt{5}+1}{2}$ as a decimal rounded to the hundredths place (second place to the right of the decimal). Using a calculator or app, we approximate $\sqrt{5}$ to be 2.236; substituting this into the equation for phi, we obtain $\varphi = \dfrac{\sqrt{5}+1}{2} \approx \dfrac{2.236+1}{2} = 1.618$. Rounding as instructed yields an approximation of 1.62.	Use a calculator to perform the operations and round your answers to two decimal places. 1. $\sqrt{20}+6\sqrt{5}$ 2. $\dfrac{\sqrt{32}}{\sqrt{18}}$

Objective 3 – Illustrate the properties of common number systems

Key Terms
Closure property
Commutative property for addition and multiplication
Associative property
Identity element for addition
Identity element for multiplication
Additive inverse
Multiplicative inverse
Distributive Property

Summary
If we combine the set of rational numbers with the set of irrational numbers, we get the set of real numbers. The real numbers, together with the operations of addition and multiplication, form a mathematical system that has the properties listed below:

Property	Definition	Examples
Closure property for addition and multiplication	If a and b are real numbers, then $a + b$ and $a \cdot b$ are also real numbers.	3 and π are both real numbers, so $3 + \pi$ is a real number as well. $\frac{1}{2}$ and $\sqrt{3}$ are real numbers, so $\frac{1}{2} \cdot \sqrt{3}$ is also a real number.
Commutative property for addition and multiplication	If a and b are real numbers, then $a + b = b + a$ and $a \cdot b = b \cdot a$. (The *order* in which we combine the numbers is not important.)	$3 + \sqrt{2} = \sqrt{2} + 3$ and $(-5) \cdot \pi = \pi \cdot (-5)$
Associative property for addition and multiplication	If a, b, and c are real numbers, then $(a + b) + c = a + (b + c)$ and $(a \cdot b) \cdot c = a \cdot (b \cdot c)$. (The way we *group* the numbers is not important.)	$(3 + \sqrt{5}) + \sqrt{7} = 3 + (\sqrt{5} + \sqrt{7})$ and $\left((-6) \cdot \frac{1}{2}\right) \cdot \pi = (-6) \cdot \left(\frac{1}{2} \cdot \pi\right)$
Zero is an **identity element for addition,** and 1 is an **identity element for multiplication.**	If a is any real number, then $0 + a = a + 0 = a$ and $1 \cdot a = a \cdot 1 = a$.	$0 + 5 = 5 + 0 = 5$ and $1 \cdot \sqrt{10} = \sqrt{10} \cdot 1 = \sqrt{10}$
If x is a real number, then $-x$ is an **additive inverse** for x. If $x \neq 0$, then $\frac{1}{x}$ is a **multiplicative inverse** for x.	For any real number x, $x + (-x) = (-x) + x = 0$ and $x \cdot \left(\frac{1}{x}\right) = \left(\frac{1}{x}\right) \cdot x = 1$.	$3 + (-3) = (-3) + 3 = 0$ and $\pi \cdot \frac{1}{\pi} = \frac{1}{\pi} \cdot \pi = 1$
Multiplication **distributes** over addition.	If a, b, and c are real numbers, then $a \cdot (b + c) = (a \cdot b) + (a \cdot c)$.	$\frac{1}{2} \cdot (4 + 6) = \left(\frac{1}{2} \cdot 4\right) + \left(\frac{1}{2} \cdot 6\right) = 5$

Problem Solving; The *Always principle* reminds us that if a property holds for real numbers, it always works – no exceptions!

Notes:

Guided Example	***Practice*** (Exercises 43, 45, & 47 p.279)
State the property of the real numbers that is illustrated below in the following examples. 1. $11+(-11)=(-11)+11=0$ This example illustrates the additive inverse property. For any real number x, $-x$ is the additive inverse of x; $x+(-x)=(-x)+x=0$ 2. φ and 2 are real numbers, so $\varphi+2$ is a real number and $2\cdot\varphi$ is a real number. These examples illustrate that the real numbers are closed for addition and multiplication. When you add or multiply two real numbers, the result is a real number.	State which property of the real numbers we are illustrating in the following exercises. 1. $3(4+5)=3\cdot4+3\cdot5$ 2. $3+(6+8)=(6+8)+3$ 3. $3+(6+8)=(3+6)+8$

Section 6.5 Exponents and Scientific Notation

Objective 1 – Use the basic rules of exponents to do calculations.
Objective 2 – Perform operations using scientific notation.
Objective 3 – Solve problems using scientific notation.

Objective 1 – Use the basic rules of exponents to do calculations.

Key Terms
Base
Exponent

Summary
Using exponential notation allows us to write very large or very small numbers efficiently and easily. If a is any real number and n is a counting number, then $a^n = a \cdot a \cdot a \cdot a \cdot \ldots \cdot a$, a product of n a's. The number a is called the base, and n is the exponent.

There are several rules which guide or work with exponents.
Product Rule: if x is a real number and m and n are natural numbers, then $x^m x^n = x^{m+n}$.
Power Rule: if x is a real number and m and n are natural numbers, then $\left(x^m\right)^n = x^{m \cdot n}$.

Quotient Rule: if x is a nonzero real number and m and n are natural numbers, then $\frac{x^m}{x^n} = x^{m-n}$.

Definition of negative exponent: If $a \neq 0$, then $a^0 = 1$, and if $a \neq 0$ and n is a natural number, $a^{-n} = \frac{1}{a^n}$.

Problem Solving; Be sure to keep the *Order principle* in mind when working with exponents. You may want to review the order of operations, located in the Appendix A of your textbook.

Notes:

Guided Example	***Practice*** (Exercises 11, 17 & 23 p.289)
Rewrite the following, indicating which properties are being used. 1. $\left((-3)^2\right)^4$ Using the power rule, we rewrite the expression as $(-3)^8$ which is equal to 6,561. 2. $\frac{4^{-1}}{4^2}$ Using the quotient rule, we write the expression as $4^{-1-2} = 4^{-3}$. Now using the definition of negative exponents, we have $4^{-3} = \frac{1}{4^3} = \frac{1}{64}$.	Use the rules for exponents to first rewrite each of the following expressions and then evaluate the new expression. 1. $3^2 \cdot 3^4$ 2. $\left(3^2\right)^{-3}$ 3. $\frac{6^{-2}}{6^{-4}}$

Objective 2 – Perform operations using scientific notation.

Key Terms
Scientific notation

Summary
Working with very large or very small numbers is much easier if we first write the numbers in scientific notation. A number is written in scientific notation if it is of the form $a \times 10^n$ where $1 \le a < 10$ and n is any integer. To convert a decimal number to scientific notation use the rules below.

If the number x is very large

a) Make the number smaller by moving the decimal point to the left until the number is equal to or greater than 1 and less than 10.

b) Restore the size of x by multiplying the number you obtained in part a) by 10 raised to the power, which is the number of places you moved the decimal point to the left.

If the number x is very small,

c) Make the number larger by moving the decimal point to the right until the number is equal to or greater than 1 and less than 10.

d) Restore the size of x by multiplying the number you obtained in part c) by 10 raised to the power, which is the negative of the number of places that you moved the decimal point to the right.

Example: Convert 0.0052 to scientific notation.

Since 0.0052 is small, we make the number larger by moving the decimal place 3 places to the right to obtain 5.2. To restore the size of the original number, we multiply by 10^{-3} yielding 5.2×10^{-3}.

To perform operations with scientific notation, we regroup the numbers and perform the calculations using the rules for exponents.

Problem Solving; Notice that in multiplying and dividing numbers written in scientific notation, you are not doing anything new. You are simply applying properties such as associativity and commutativity, as well as the rules for working with rational numbers that we have used earlier. We are applying the *Analogies Principle.*

Notes:

Guided Example	***Practice*** (Exercises 33, 37 & 51 p.289)
Perform the following division by regrouping terms and using rules of exponents. Leave your answer in scientific notation form. $\left(6\times10^{-3}\right)\div\left(2\times10^{2}\right)=\dfrac{\left(6\times10^{-3}\right)}{\left(2\times10^{2}\right)}$ We regroup the terms and divide $\dfrac{6}{2}\times\dfrac{10^{-3}}{10^{2}}$ $=3\times10^{-5}$. As a final step, check to see that our solution is written in proper scientific notation form (it is).	1. Rewrite 0.0024 in scientific notation. 2. Rewrite 3.25×10^{4} in standard notation. Use scientific notation to perform the following operations and leave your answer in scientific notation form. 3. $\left(1.2\times10^{-3}\right)\left(3\times10^{5}\right)$

Objective 3 – Solve problems using scientific notation.

Key Terms
None

Summary
There are many problem solving situations which involve very large or small numbers. In these cases, it is helpful to first express the numbers in scientific notation and then perform the mathematical calculations. Applications involving population sizes, distances, and the national debt are particularly suited to this method.

Notes:

Guided Example	***Practice*** (Exercises 83 & 87 pp.290-291)
In 2010, the U.S. national debt was \$13,561 billion and the population was approximately 308 million. Calculate the amount of national debt owed by each person in the united states in 2010. First we will express each quantity in scientific notation. Since \$1 billion is \$1,000,000,000 we express the national debt as $13,561\times10^{9}=1.3561\times10^{13}$ and the population was $308\times10^{6}=3.08\times10^{9}$. To determine the debt per person we divide the two numbers. $\dfrac{\left(1.3561\times10^{13}\right)}{\left(3.08\times10^{8}\right)}$ $=\dfrac{1.3561}{3.08}\times10^{(13-8)}$ $\approx0.44\times10^{5}$ $=4.4\times10^{4}$ The debt is approximately \$44,000 per person.	1. In 2007, the U.S. population was 301 million and the country's defense budget was \$439.3 billion. How much was spent on defense for each person in the United States? 2. By 2020, it is estimated that health care spending will be \$4.4 trillion, or \$13,850 per resident. Use this information to calculate the estimated U.S. population in 2020.

Section 6.6 Sequences

Objective 1 – Calculate the terms of arithmetic and geometric sequences.
Objective 2 – Explain the differences in behavior between arithmetic and geometric sequences.
Objective 3 – Recognizes applications of the Fibonacci sequence.

Objective 1 – Calculate the terms of arithmetic and geometric sequences.

Key Terms

Sequence
Terms
Arithmetic sequence
Common difference
Geometric sequence
Common ratio

Summary

A sequence is a list of numbers that follows some rule or pattern. The numbers in the list are called the terms of the sequence. We represent the terms of a sequence with letters, and indicate the term number with a subscript. As an example, a_3 is the third term is the sequence for a.

Two common types of sequences are arithmetic and geometric. In an arithmetic sequence, terms are obtained by adding a fixed amount to the previous term. This amount is called the common difference of the sequence. The common difference may be negative, in which case the terms in the sequence are obtained by subtraction. A geometric sequence is a sequence in which each term after the first is a nonzero constant multiple of the preceding term. The constant by which we are multiplying is called the common ratio.

Notes:

Guided Examples	***Practice*** (Exercises 5 & 11 p.300)
1. Given the arithmetic sequence $-1, 3, 7, 11, \ldots$, identify the common difference and write the next two terms of the sequence. The common difference is 4. We can find this by subtracting any two consecutive terms. The next two terms will be 15 and 19. 2. Consider the geometric sequence $2, -4, 8, -16, \ldots$ identify the common ratio and list the next two terms of the sequence. The common ratio is -2. This can be obtained by dividing any term in the sequence by the one prior to it. The next two terms of the sequence are 32 and -64.	Identify the sequence as either arithmetic or geometric. List the next two terms of each sequence. 1. $1, \frac{1}{2}, \frac{1}{4}, \frac{1}{8}, \ldots$ 2. 1.5, 2.0, 2.5, 3.0, …

Objective 2 – Explain the differences in behavior between arithmetic and geometric sequences.

Key Terms
*n*th term
Sum of first *n* terms of arithmetic sequence.

Summary
As we have learned, an arithmetic sequence is one in which each term after the first differs from the preceding term by a fixed constant, called the common difference. We can determine the *n*th term of the sequence, a_n, if we know a_{n-1}. But what if we want to know a_n and don't know the term before it? The following formula gives us a way to find a_n provided we know the first term and the common difference.

The *n*th term of an arithmetic sequence can be obtained using $a_n = a_1 + (n-1)d$ where a_1 is the first term in the sequence and *d* is the common difference.

Example: An arithmetic sequence has first term 4 and common difference 3. Find the 10th term in the sequence.

We know a_1=4 and $d = 3$. Using the formula, we find $a_{10} = a_1 + (10-1)d = 4 + (9)3 = 31$.

What if we wanted to find the sum of the first 10 terms in this sequence? We know a_1 and a_{10} and could certainly determine the other 8 terms. We can save some time, however, by using the fact that the sum of the first *n* terms of an arithmetic sequence is $\frac{n(a_1 + a_n)}{2}$. For our example above, the sum of the first 10 terms would be $\frac{10(4+31)}{2} = 175$. You should verify that this is the correct sum.
In the case of the geometric sequence, the *n*th term can be obtained using the formula $a_n = a_1 \cdot r^{n-1}$. We use this in the guided example, below.

Notes:

Guided Example	***Practice*** (Exercises 15 & 23 p.300)
Given the geometric sequence 2, -4, 8, -16, … find a_9. We are given the first four terms, and could certainly continue the pattern and find the next five. We can save time, however, by using the formula $a_n = a_1 \cdot r^{n-1}$. We substitute $a_1 = 2$ and $r = -2$ obtaining: $a_9 = 2 \cdot (-2)^{9-1} = 2 \cdot (-2)^8 = 512$.	1. For the arithmetic sequence 2, 8, 14, 20, …find a_{15} and then find the sum from a_1 to a_{15}. 2. For the geometric sequence $1, \frac{1}{2}, \frac{1}{4}, \frac{1}{8}, \ldots$ find a_7.

Objective 3 –. Recognizes applications of the Fibonacci sequence

Key Terms
Fibonacci sequence

Summary
In 1202, Leonardo of Pisa, better known as Fibonacci, posed a problem which led to the well known Fibonacci sequence. The problem he posed had to do with the breeding patterns of rabbits, but the sequence which came from this problem is 1, 1, 2, 3, 5, 8, 13, 32, 34, … where terms from the third on are determined by summing the two prior terms. Rather than using a_n to describe the *n*th term, we use F_n. In general $F_1 = 1$, $F_2 = 1$, and $F_{n-1} + F_n = F_{n+1}$. The Fibonacci sequence occurs in nature in many different ways. The seeds of daisies and sunflowers are arranged in spirals going in two different directions. If you count the number of spirals going in one direction and then count the number of spirals going in the other direction, you will often find that the numbers you get are a pair of consecutive numbers in the Fibonacci sequence.

One interesting fact about the Fibonacci sequence is that if you go through the sequence dividing each term by the term preceding it, you get the numbers listed in the table below. It seems as though these numbers are settling down and going toward some fixed number. In fact, they are!

$\frac{1}{1} = 1$
$\frac{2}{1} = 2$
$\frac{3}{2} = 1.5$
$\frac{5}{3} = 1.667$
$\frac{8}{5} = 1.6$
$\frac{13}{8} = 1.625$
$\frac{21}{13} = 1.615$
$\frac{34}{21} = 1.619$
$\frac{55}{34} = 1.618$
$\frac{89}{55} = 1.618$
$\frac{144}{89} = 1.618$
$\frac{233}{144} = 1.618$

It was proved over 200 years ago that the number these quotients are approaching is the golden ratio, φ ,which we have studied. Recall that $\varphi = \dfrac{\sqrt{5}+1}{2} \approx \dfrac{2.236+1}{2} = 1.618$

Notes:

Guided Example	***Practice*** (Exercises 27 & 29 p.300)
Use the information to find the indicated terms of the Fibonacci sequence. $F_{11} = 89$ and $F_{10} = 55$. Find F_9. We know that $F_9 + F_{10} = F_{11}$. Substituting, we have $F_9 + 55 = 89;\ F_9 = 89 - 55 = 34$.	Given the terms of the Fibonacci sequence, find the indicated term. 1. $F_{11} = 89$ and $F_{13} = 233$. Find F_{12}. 2. $F_{13} = 233$ and $F_{15} = 610$. Find F_{14}.

Chapter 6 Answers

<u>Section 6.1</u>

Objective 1 1)13 2)17

Objective 2 1) 2, 3, 5, 6, 10 2) 3, 5, 9 3) $3^3 \times 23$

Objective 3 1) 14; 280. 2) 28; 5880

Objective 4 1) 140

<u>Section 6.2</u>

Objective 1 1) +23 2) -13 3) -65

Objective 2 1) -56) +72 3) +5

Objective 3 1) -48 2) -6

<u>Section 6.3</u>

Objective 1 1) not equal 2) equal

Objective 2 1) $\frac{5}{48}$ 2) $-\frac{1}{24}$

Objective 3 1) $\frac{1}{3}$ 2) $-\frac{21}{5}$ 3) $-\frac{33}{4}$

Objective 4 1) $8\frac{1}{15}$ 2) $\frac{55}{6}$

Objective 5 1) 0.1875 2) $0.\overline{307692}$

<u>Section 6.4</u>

Objective 1 1) rational 2) irrational

Objective 2 1) 17.89 2) 1.33

Objective 3 1) distributive 2) commutative (addition) 3) associative (addition)

<u>Section 6.5</u>

Objective 1 1) 729 2) $\frac{1}{729}$ 3) 36

Objective 2 1) 2.4×10^{-3} 2) 32,500 3) 3.6×10^{2}

Objective 3 1) 1.459×10^{3} 2) 3.18×10^{2} million.

<u>Section 6.6</u>

Objective 1 1) geometric, $\frac{1}{16},\frac{1}{32}$ 2) arithmetic, 3.5, 4.0

Objective 2 1) 86; 660 2) $\frac{1}{64}$

Objective 3 1) 144 2) 377

Chapter 7: Algebraic Models

Section 7.1 Linear Equations.

Objective 1 – Solve Linear Equations.

Objective 2 –Use Intercepts to graph linear equations.

Objective 3 –Apply the slope-intercept form of a line in solving problems.

Objective 1 – Solve Linear Equations.

Key Terms

Linear equation
Standard form
Solution
Ordered pair
Equivalent

Summary

A linear equation in two variables is an equation that can be written in the form $Ax + By = C$, where *A, B*, and *C* are real numbers and *A* and *B* are both not zero. When written this way, the equation is in standard form.

A solution for a linear equation is an ordered pair of numbers which makes the equation true.
We call equations equivalent if they have the same solutions.

We may rewrite equations in an equivalent form by

1) Adding or subtracting the same expression from both sides of an equation
2) Multiply or divide both sides of the equation by the same nonzero expression.

Notes:

Guided Example	***Practice*** (Exercises 9 & 13 p.316)
Given the linear equation $7x = 21 + 35y$ 1. Rewrite the equation in standard form. In order to obtain an equation in standard form, we need to subtract the term $35y$ from both sides, which gives us $7x - 35y = 21$. We simplify the equation by dividing both sides of the equation by 7, $\frac{7x - 35y}{7} = \frac{21}{7}$ $\frac{7x}{7} - \frac{35y}{7} = 3$ $x - 5y = 3.$ 2. Show that the ordered pair $(3,0)$ is a solution of the equation. We substitute 3 for *x* and 0 for *y* then check to see if the equation is a true statement. $x - 5y = 3$ $3 - 5(0) = 3$ $3 = 3$ Since we obtain a true statement, $(3,0)$ is a solution to the equation.	Solve each equation for the stated variable. 1. $P = 2l + 2w$; solve for *w*. 2. $A = P(1 + rt)$; solve for *r*.

Objective 2 – Use Intercepts to graph linear equations.

Key Terms
Graph
x-intercept
y-intercept

Summary
When we graph an equation, we plot the points that are solutions to the equation, that is, they make the equation true. If the equation is a linear equation, the graph is always a straight line.

The *y*-intercept of a graph is the point where the graph crosses the *y*-axis. It is found by setting $x = 0$ in the equation and solving for *y*. The *x*-intercept is the point where the graph crosses the *x*-axis. To determine this value, we set $y = 0$ in the equation and solve for *x*. We can then graph the linear equation by plotting the two intercepts and drawing a straight line.

Notes:

Guided Example

Graph the equation $5x - y = 10$ by first finding the *x*- and *y*-intercepts.

We will start with the *x*- intercept which is found when $y = 0$. Substituting, we find $5x - 0 = 10$, so $5x = 10$ and $x = 2$. The *x*-intercept is the point $(2,0)$.

For the y-intercept, we set $x = 0$ and solve for *y*. We obtain $5(0) - y = 10$, so $y = -10$. The *y*-intercept is $(0,-10)$.

We plot these two points and obtain the graph below.

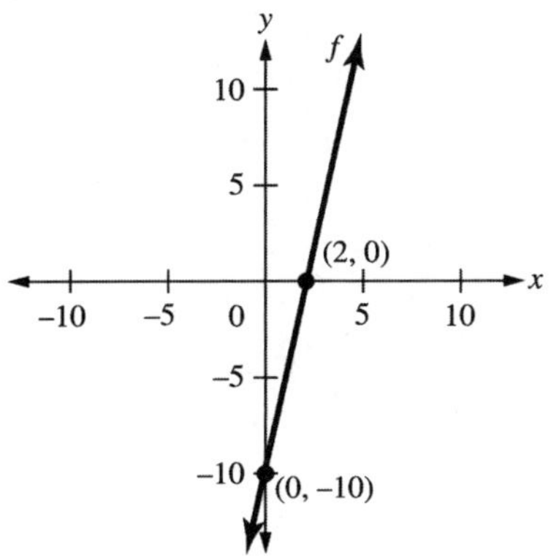

Practice (Exercises 23 & 25 p.316)

Graph each equation by first finding the *x*- and *y*-intercepts.

1. $4x - 3y = 16$

2. $\frac{1}{3}x + \frac{1}{2}y = 3$

Objective 3 – Apply the slope-intercept form of a line in solving problems.

Key Terms
Slope
Slope intercept form

Summary

The slope of a linear equation is defined as $m = \frac{\text{rise}}{\text{run}} = \frac{\text{change in } y}{\text{change in } x} = \frac{y_2 - y_1}{x_2 - x_1}$. The slope tells us how steep the graph of the linear equation appears. If the slope is zero, the line is horizontal. If the slope is negative, the line falls as x moves from left to right; if the slope is positive the line rises as x moves from left to right.

Problem Solving; The *Analogies Principle* can help us understand and remember the concepts of rise and run. Think of the rise of an air balloon (vertical change) and the run of a train (horizontal change).

A linear equation is in the slope-intercept form if it is written in the form $y = mx + b$. The number m is the slope of the line that is the graph of the equation, and $(0, b)$ is the y-intercept.

Notes:

Guided Example	***Practice*** (Exercises 57 & 59 p.317)
In 2015, 70 percent of college students preferred print textbooks over electronic versions. Suppose that in 2017, the percentage had dropped to 65. 1. Use this information to build a linear model in slope intercept form where x = years after 2015 and y = percent of college students who prefer a print textbook. We first use the points (0, 70) and (2,65) to find the slope. $m = \frac{\text{change in } y}{\text{change in } x} = \frac{65-70}{2-0} = \frac{-5}{2} = -2.5$ Now we use the slope and the y-intercept (0, 70) to obtain the slope intercept form $y = -2.5x + 70$. 2. Use your model to predict the percentage of college students who will prefer a print textbook in 2020. 2020 is $x = 5$ years from 2015. Using this in our model yields $y = -2.5(5) + 70 = 57.5$.	According to the U.S Census Bureau, there were 80 people per square mile in the United States in 2000. By 2015, the number of people per square mile had grown to 91. Use this information to develop a linear equation in slope-intercept form to solve the exercises below. In developing your equation, think of 2000 as year zero and round the slope to two decimal places. 1. Write the equation you are using to model this information (in slope-intercept form). 2. What do you expect the population density to be in 2025?

Section 7.2 Modeling with Linear Equations

Objective 1 – Build a linear model using a point and the slope.
Objective 2 – Use two points to build a linear model.
Objective 3 – Describe how to use the line of best fit to model real data.

Objective 1 – Build a linear model using a point and the slope.

Key Terms
None

Summary
We can find a linear equation if we know the slope of its graph and a point on the graph. We first substitute the slope for *m* in the slope-intercept form of the line, $y = mx + b$. The next step is to substitute a point (x, y) into the equation in order to find the *y*-intercept, *b*. Sometimes the point we are given is in the form $(0, b)$, so we can immediately put that value in for *b* in the equation $y = mx + b$.

Notes:

Guided Example	***Practice*** (Exercise 17 p.325)
Construct a linear model in the slope intercept form for an equation with slope 2 passing through the point $(1.-3)$. Since the line has slope 2, we know that $m = 2$ and thus the equation has form $y = 2x + b$. Now we use the point $(1.-3)$ to find the *y*-intercept, *b*. We substitute $x = 1$ and $y = -3$ yielding $-3 = 2(1) + b$ and determine that $b = -5$. The linear equation is $y = 2x - 5$.	1. According to the Centers for Disease Control and Prevention, the life expectancy of a female born in the United States in 2009 was 80.6 years and increasing at a rate of 0.1 year per year. Assume that this rate of increase remains constant. Model this situation by a linear equation; use this equation to estimate the life expectancy of a female born in the United States in 2030.

Objective 2 – Use two points to build a linear model.

Key Terms
none

Summary
Sometimes we are not given the slope of a linear equation, but are given two points. In such situations, we can use the two points to find the slope, and then use the slope and one of the points to build the model as we did in the previous worksheet.

Problem Solving: When using two points to build a linear model, we *Relate a New Problem to an Older One*

Notes:

Guided Example	***Practice*** (Exercises 9 & 15 p.325)
Find a linear equation whose graph passes through the points $(-6,-5)$ and $(0,-1)$. We use the two points to find the slope: $m=\frac{\text{change in } y}{\text{change in } x}=\frac{-1-(-5)}{0-(-6)}=\frac{4}{6}=\frac{2}{3}$ We can then use either of the given points to find *b*. Let's pick $(-6,-5)$. Substituting this point for (*x*, *y*) we obtain $-5=\frac{2}{3}(-6)+b$ which simplifies to $-5=-4+b$ thus $b=-1$. Our equation is $y=\frac{2}{3}x-1$ We could have saved ourselves work if we noticed that the point we were given, $(0,-1)$, was the *y*-intercept. We could have then substituted $b=-1$ in the equation $y=\frac{2}{3}x+b$.	Find a linear equation whose graph passes through the given points. Write the equation in slope-intercept form. 1. $(2,3)$ and $(5,9)$. 2. $(-6,-8)$ and $(-4,-1)$.

Objective 3 – Describe how to use the line of best fit to model real data.

Key Terms
Line of best fit
Linear regression
Linear correlation coefficient

Summary
When we work with real data, the points often do not all fall on a straight line. In order to get an idea of the relationship between the variables, we determine the line that best fits the data points.

There are several ways we can determine this line of best fit. We can draw the line by eye, that is look at the data points and try to draw in the line that visually appears to model the trend in the data. In the guided example below, we model selecting two of the data points and using the two points as we have done before to determine the equation of the line. Statisticians use linear regression, a sophisticated method with detailed calculations, to find the slope and y-intercept of the line which is deemed to best fit all of the data points by minimizing distances between the data points and the line. This topic is addressed in chapter 14 of the text. In that chapter, you will also learn about the linear correlation coefficient, which measures the strength of the linear relationship between the x and y variables.

Notes:

Guided Example

The table below gives the median sales price for a two bedroom house in a major city between 2011 and 2014.

Year	Selling price (thousands of dollars)
2011	360
2012	419
2013	539
2014	520

a. What is the average yearly increase between 2011 and 2014?
We will let 2011 represent year 0. Use the points (0, 360) and (3, 520) to find the slope.

$$m = \frac{\text{change in } y}{\text{change in } x} = \frac{520-360}{3-0} = \frac{160}{3} \approx 53.33$$

b. Use the result from part (a) and the data point which represents 2011 to write an equation that describes the sales price in terms of year.
We will use $m = 53.3$ and the point (0, 360) to obtain the equation. Note that $b = 360$, so $y = 53.33x + 360$.

c. What does your equation predict as a sales price in 2020?
In 2020, $x = 9$, so we obtain $y = 53.33(9) + 360$ or \$839.97 thousand dollars.

Practice (Exercise 29 p.327)

1. The graph below shows the increase in the civilian labor force from 2003 to 2006, according to the U.S. Department of Labor. Assume that the rate of increase continues for the next several years and can be modeled by a linear equation of the form $l = mt + b$.

U.S. Civilian Labor Force

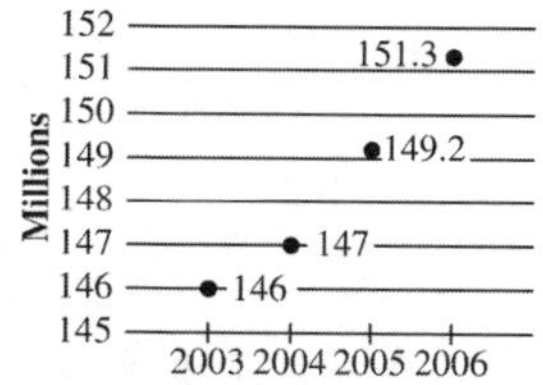

Source: U.S. Bureau of Labor Statistics

a. What is the average yearly increase from 2003 to 2006?
b. Use the result from part (a) as the slope and the point that represents the 2003 data to write an equation that describes this increase in the labor force.
c. What does your equation in part (b) predict the size of the labor force to be in 2009?
d. Use the equation you found in part (b) to estimate in what year the size of the labor force will reach 200 million.

Section 7.3 Modeling with Quadratic Equations

Objective 1 – Use the quadratic formula to solve equations.
Objective 2 – Graph a quadratic equation.
Objective 3 – Use a quadratic equation to model data.

Objective 1 – Use the quadratic formula to solve equations.

Key Terms

Quadratic equation
Quadratic formula
Discriminant
Vertex

Summary

Equations in the form $y = ax^2 + bx + c$ are called quadratic equations. In order to find the *x*-intercepts of a quadratic equation, we need to solve the equation $0 = ax^2 + bx + c$. This equation might or might not have solutions, but if they exist, they can be found using the quadratic formula. We use the coefficients *a, b*, and *c* from the equation and substitute into the formula $x = \frac{-b \pm \sqrt{b^2 - 4ac}}{2a}$. The expression under the square root symbol, $b^2 - 4ac$, is called the discriminant. If the discriminant is negative, there are no real solutions. If the discriminant is 0, there is exactly one solution $\left(\text{at the vertex, } x = \frac{-b}{2a}\right)$. If the discriminant is positive, there are two real solutions.

Notes:

Guided Example	***Practice*** (Exercises 1 & 7 p.335)
Solve the quadratic equation $2x^2+3x-2=0$ In this equation, $a=2$, $b=3$, and $c=-2$. We substitute these values into the quadratic formula to get the solutions: $\frac{-b\pm\sqrt{b^2-4ac}}{2a}=\frac{-3\pm\sqrt{3^2-4(2)(-2)}}{2(2)}$ $=\frac{-3\pm\sqrt{9+16}}{4}=\frac{-3\pm5}{4}$ So our solutions are $\frac{-3-5}{4}=-2$ and $\frac{-3+5}{4}=\frac{2}{4}=\frac{1}{2}$. Note that in our example, the discriminant, b^2-4ac, was equal to 25. Since this is a positive number, we obtain two real solutions to the equation.	Solve each quadratic equation. 1. $x^2-10x+16=0$ 2. $2x^2+11x+15=0$

Objective 2 – Graph a quadratic equation.

Key Terms
Vertex
Parabola

Summary
The graph of the quadratic equation $y = ax^2 + bx + c$ is a parabola. In order to sketch the graph, we first address the following questions and then use the results to construct the graph.

1. Is the parabola opening up or down? We answer this by looking at a. If $a > 0$ the parabola opens up, if $a < 0$ the parabola opens down.
2. Find the vertex of the parabola. The vertex occurs when $x = \frac{-b}{2a}$.
3. Determine the value of the discriminant, $b^2 - 4ac$, and use this to determine the number of solutions to the equation $0 = ax^2 + bx + c$. This will also tell us how many x-intercepts there are (if any).
4. Find the x- and y-intercepts.

Notes:

Guided Example	***Practice*** (Exercises 13 & 17 p.335)
Sketch the graph of $y = 2x^2 + 3x - 2$ We will address the four questions above and then sketch the graph. 1. The parabola will open up because $a = 2$ which is greater than 0. 2. The vertex of the parabola occurs when $x = \frac{-b}{2a} = \frac{-3}{2 \cdot 2} = -\frac{3}{4}$. The y-coordinate of the vertex is $y = 2\left(-\frac{3}{4}\right)^2 + 3\left(-\frac{3}{4}\right) - 2$ $= 2\left(\frac{9}{16}\right) - \frac{9}{4} - 2 = \frac{-25}{8}$ 3. There are two solutions to the equation $2x^2 + 3x - 2 = 0$. The discriminant is equal to $b^2 - 4ac = 3^2 - 4(2)(-2) = 25$. 4. The x-intercepts are -2 and $\frac{1}{2}$ (see previous worksheet). The y-intercept is found by substituting $x = 0$ and solving for y. The ordered pair is $(0, -2)$ The graph of the parabola is below: 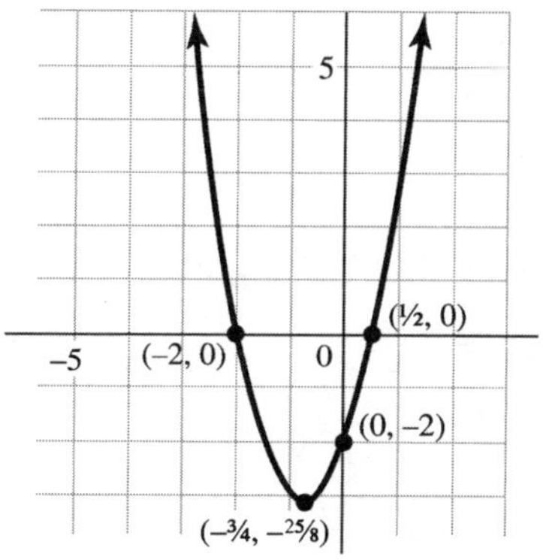	Answer each of the following questions and then draw the graph for each quadratic equation. Is the equation's graph opening up or down? What is the vertex of the graph? What are the x-intercepts? What is the y-intercept? 1. $y = 4x^2 - 4x - 2$ 2. $y = -x^2 + 7x - 12$

Objective 3 – Use a quadratic equation to model data.

Key Terms
none

Summary
Quadratic models are used to model situations of two general types. If we have a quadratic model which, when graphed, would be a parabola opening down, we have a period of increase followed by a period of decrease. If the model, when graphed, would be a parabola opening up, we have a period of decrease followed by a period of increase. We will see in the example and exercises below that once we identify the type of model, it is often of interest to find the vertex and the intercepts.

Problem Solving: We use the Draw Pictures Strategy when using quadratic equations as models.

Notes:

Guided Examples	***Practice*** (Exercise 23 p.336)
A fountain in a city park shoots water from a nozzle. The height of the water t seconds after leaving the nozzle is given by $H=-16t^2+72t+40$. 1. Determine the values of t that make sense for this model. Since t represents time after water leaves the nozzle, we need $t \geq 0$. 2. Determine the time at which the water in the fountain reaches the maximum height. We need to find the t-coordinate of the vertex, which is $t=-\frac{b}{2a}=-\frac{72}{2\cdot(-16)}=2.25$ seconds. 3. Determine the height of the water at the maximum point. Substitute 2.25 for t, $H=-16t^2+72t+40$ $=-16(2.25)^2+72(2.25)+40$ $=121$ The maximum height is 121 feet. 4. Find the time at which the water will hit the ground. To answer this we need to find the t-intercepts. Using the quadratic formula with $a=-16$, $b=72$ and $c=40$ we find $\frac{-b\pm\sqrt{b^2-4ac}}{2a}$ $=\frac{-72\pm\sqrt{(72)^2-4(-16)(40)}}{2(-16)}$ $=\frac{-72\pm\sqrt{7744}}{-32}$ $=\frac{-72\pm 88}{-32}$ So $t=\frac{-72-88}{-32}=5$ or $t=\frac{-72+88}{-32}=-0.5$ Since we required $t \geq 0$ the only reasonable answer is $t=5$. The water hits the ground 5 seconds after leaving the nozzle.	1. A plane is dropping emergency food supplies to relieve famine in a less-developed country. Crates are dropped, and the height of the create above the ground at time t is given by the equation $H=160-16t^2$. a. Graph this equation. b. Are there any values for t that it would not make sense to use for this equation because of the physical characteristics of a falling crate? c. Find the time at which the crate will strike the ground.

Section 7.4 Exponential Equations and Growth

Objective 1 – Explain the differences among linear, quadratic, and exponential growth models.
Objective 2 – Use exponential equations to model growth.
Objective 3 – Solve exponential equations using the log function
Objective 4 – Use logistic models to describe growth.

Objective 1 – Explain the differences among linear, quadratic, and exponential growth models.

Key Terms
Exponential equation

Summary
When we use equations to model behavior, it is important to think about the type of equation that would best fit the situation. If growth is constant, then a linear model would be appropriate. A quadratic model is used when you are modeling a quantity which switches from increasing to decreasing (or decreasing to increasing). An exponential model is the best choice when the rate of growth is proportional to the amount present at any given time. The general form of an exponential model is $y = a \cdot b^x$. Note that in an exponential model, the variable x is an exponent.

Notes:

Guided Example	***Practice*** (Exercise 63 p.349)
In each of the following examples, suggest the model that would be best to use for modeling the situation. 1. Model sales tax, y, as related to the total cost of purchases. In this example, a linear model would be best. Depending on the tax rate in your state, the amount of tax paid is a percentage of the total purchase amount. It doesn't matter if your purchase is large or small, the rate is constant. 2. Model the growth of an invasive aquatic plant in a lake by your home. In this case, the exponential model would be best. Growth of populations, including plants, would be proportional to the amount in existence at a given time, which would make the appropriate model choice the exponential model.	1. Determine the key difference between a linear model and an exponential model.

Objective 2 – Use exponential equations to model growth.

Key Terms
Compound interest formula

Summary
A good example of a place where an exponential equation is a good fit is in finance, specifically when computing compound interest. The amount of interest that you earn at any given time is proportional to the amount of money that you have in the account! The compound interest formula, is $A = P(1+r)^n$ where P is the principal, the amount invested, r is the annual interest rate, and n is the amount of time (in years) the money is invested.

We will find that this formula can also be used to model population growth.

Notes:

Guided Example	***Practice*** (Exercises 7 & 13 p.347)
The population of Oman was 3.4 million in 2016 and growing at an annual rate of 2.05%. Assuming this rate continues, predict the population of Oman in 2025. We will use $A = P(1+r)^n$ where P is the population of Oman in 2016, $r = 0.0205$ and $n = 2025 - 2016 = 9$. $A = P(1+r)^n = 3.4(1+.0205)^9 = 4.08$ We predict the population of Oman to be 4.08 million in 2025.	1. You have $4,000 in a bank account at the beginning of a year and the interest is compounded annually at an 8% interest rate. Assuming that no withdrawals are made, use the compound interest formula to compute the amount in the account after 2 years. 2. The population of Bangladesh was 169 million in 2015, with a growth rate of 1.6%. Predict the population of Bangladesh in 2025.

Objective 3 – Solve exponential equations using the log function.

Key Terms
Logarithm

Summary
A common practice when working with equations is to solve for an unknown. In exponential equations, if the unknown is the exponent, we can make use of the logarithm function to solve for an exponent.

By definition, $\log_b y = x$ if and only if $y = b^x$. Simply put, $\log_b y$ asks for the exponent you would raise b in order to obtain y. The logarithm function with base $b = 10$ is called the common logarithm. As an example, $\log_{10} 1,000 = 3$, since $10^3 = 1,000$. If you write $\log x$ without specifying the base, the assumption is that it is log base 10. The exponent property of the logarithm function makes them particularly useful when solving exponential equations. This property states $\log y^x = x \log y$

Notes:

Guided Example	***Practice*** (Exercises 23 & 41 p.348)
The population of Oman was 3.4 million in 2016 and growing at an annual rate of 2.05%. Assuming this rate continues, when will the population of Oman double? Using the formula $A = P(1+r)^n$ with $P = 3.4, A = 6.8,$ and $r = 0.0205$ we find $6.8 = 3.4(1+0.0205)^n$. Our first step is to divide both sides of the equation by 3. 4, yielding $2 = (1.0205)^n$. We take the log base 10 of each side of the equation and use the exponent property of logarithms to write $\log 2 = \log(1.0205)^n$ $\log 2 = n\log(1.0205)$ Dividing both sides of the equation by log(1.0205) we obtain $n = \dfrac{\log 2}{\log 1.0205}$. Use of a calculator or app yields $n \approx \dfrac{0.3010}{0.0088}$ so in about 34.2 years the population in Oman will double. We anticipate this happening during 2050.	1. Use the log function to solve the equation $(3.4)^x = 6.85$ for x. 2. In 2011, the population of the United States was 307 million and the growth rate was 0.7%. If the growth rate remains the same, in what year will it be double what it was in 2011?

Objective 4 – Use logistic models to describe growth.

Key Terms
Logistic growth model
Rate reduction factor

Summary

In our study of exponential models, we assumed that the growth would continue at the indicated rate indefinitely. In the presence of scare resources, space, food, and conditions, such growth is not reasonable as the environment can sustain only a limited population. As the population grows, the percentage of this capacity that has been used up influences the rate of growth. In a logistic model, it is common to represent the maximum capacity that an environment can support by 1, or 100%. We define P_n as the percentage of the maximum capacity that has been attained by the population in year n. Assuming that the population is growing originally at rate r, P_n, the percentage of the maximum capacity that has been obtained by year n satisfies the equation $P_{n+1} = [1 + r(1 - P_n)]P_n$. The collection of equations for $n = 0, 1, 2, 3, \ldots$ is called a logistic model. The term $(1 - P_n)$ is called the rate reduction factor, as it represents the percentage of the maximum capacity that has <u>not</u> been attained, and is thus subject to the growth rate, r.

Notes:

Guided Example	***Practice*** (Exercises 29 & 31 p.348)
Consider a population which has growth rate 2%, but we know the rate of growth will change due to limited resources and other climate conditions. Given that $P_3 = 0.50$, determine P_4 and P_5. Describe your results. We use the formula $P_{n+1} = [1 + r(1 - P_n)]P_n$ Our first task is to find P_4 Substituting $r = 0.02$ and $P_3 = 0.50$ we obtain $P_4 = [1 + 0.02(1 - .5)].5$ which is equal to $\left[1 + 0.02(0.5)\right]0.5 = 0.505$. Note that our answer represents the percent of maximum capacity at time period 4. Our answer represents less than a 2% increase (which would have been 0.51) because we used the rate reduction factor. What about time period 5? We return to the formula $P_{n+1} = [1 + r(1 - P_n)]P_n$ now substituting $r = 0.02$ and $P_4 = 0.505$. $P_5 = [1 + 0.02(1 - .505)].505 \approx 0.51$	Assume that a population is growing initially at the specified rate. You are given a value for P_n and are to use the logistic growth model to compute the value of P_{n+1}. Explain what your calculations tell you. 1. Rate = 3%; $P_4 = 0.36$. 2. Rate = 4.5%; $P_8 = 0.72$.

Section 7.5 Proportions and Variation

Objective 1 – Solve problems using ratios and proportions.

Objective 2 – Apply the capture-recapture method for estimating populations.

Objective 3 – Use direct and inverse variation to solve problems.

Objective 1 – Solve problems using ratios and proportions.

Key Terms

Ratio
Proportion
Cross products

Summary

Many problems involve solving a problem involving a ratio. A ratio is a quotient of two numbers. We write the ratio of *a* to *b* as $a:b$ or $\frac{a}{b}$.

Example: Do you look at the unit price when shopping? You might find that your energy drink costs 8 cents per ounce. How is that computed?

The drink may cost \$1.59 and the bottle contains 20 ounces. The quotient $\frac{1.59}{20} = 0.0795$ is the cost per ounce and is an example of a ratio.

When we write $\frac{a}{b} = \frac{c}{d}$, expressing that two ratios are equal, it is called a proportion. If you want to check and see if in fact $\frac{a}{b} = \frac{c}{d}$ you can compute the cross products *ad* and *bc*. If $ad = bc$, then we know $\frac{a}{b} = \frac{c}{d}$.

Notes:

Guided Example

How much should a 30 ounce drink cost if a 20 ounce drink costs \$1.59 and you wish to keep the unit cost the same?

We set up the proportion $\frac{1.59}{20} = \frac{x}{30}$ and then cross multiply to obtain the equation $(1.59)(30) = 20x$. We divide both sides of the equation by 20 to get $\frac{(1.59)(30)}{20} = x$ or $x = 2.385$. Rounding to the nearest cent gives a cost of \$2.39.

Practice (Exercises 5 & 23 p.355)

1. Solve for x in the proportion $\frac{30}{40} = \frac{27}{x}$.
2. The dosage of a particular drug is proportional to the patient's body weight. If the dosage for a 150-pound woman is 6 milligrams, what would the dosage be for her daughter Maria, who weighs 65 pounds?

Objective 2 – Apply the capture-recapture method for estimating populations.

Key Terms
None

Summary
An interesting application of proportions occurs in estimating the size of wildlife populations. Biologists capture and tag a number of animals and release them into the population. They later capture another group of animals and observe how many are tagged. The basic idea is that ratio of tagged animals in the sample to the number of the sample will be equal to the ratio of the number of tagged animals to the number of animals in the population.

Notes:

Guided Example	***Practice*** (Exercises 29 & 31 p.356)
A rally is planned at a large university and organizers plan to use a capture-recapture approach to estimate the number in attendance. They give out 200 bracelets to students entering the rally area. Later in the event, they randomly sample 100 people at the rally and 18 have bracelets. Use this information to predict the number in attendance at the rally. Let x represent the total number at the rally. We set up a proportion $\frac{200}{x} = \frac{18}{100}$ and solve by finding cross products: $(200)(100) = 18x$ so $x = \frac{20,000}{18} \approx 1111.1$. Using this model we estimate that about 1111 students attended the rally.	1. Biologists capture, tag, and release 400 bald eagles. Several months later, of 240 bald eagles that are captured, 8 are tagged. Estimate the population of bald eagles. 2. It is estimated that there are 1,000 grizzly bears living in a certain region. Assume that biologists, capture, tag, and release 55 bears. Several months later, they capture a sample of 95 bears. How many would you expect to find tagged?

Objective 3 – Use direct and inverse variation to solve problems.

Key Terms
Varies directly
Directly proportional
Constant of variation
Constant of proportionality
Varies Inversely
Inversely Proportional

Summary
We say that y varies directly as x, or that y is directly proportional to x, if $y = kx$, where k is a nonzero constant. The constant k is called the constant of variation or the constant of proportionality. The constant of proportionality is equal to $\frac{y}{x}$.

Example: Assume that y varies directly as x. If $y = 51$ when $x = 3$, what is the value of y when $x = 8$?

We will use the given information to find the constant of proportionality. We know that $y = kx$ so $51 = k(3)$ thus $k = \frac{51}{3} = 17$. We conclude that $y = 17x$ and use this to find y when $x = 8$; $y = 17(8) = 136$.

In the case of inverse variation, we say that y is inversely proportional to x if $y = \frac{k}{x}$, where k is a nonzero constant. The constant of constant of proportionality is equal to the product xy.

Notes:

Guided Example	***Practice*** (Exercises 7 & 41 pp.355-356)
Assume that y varies inversely as the square of x. If $y = 25$ when $x = 10$, find y when $x = 4$. Our first step is to find the constant of proportionality, k. Since y varies inversely as the square of x, we know $y = \frac{k}{x^2}$. Substituting $y = 25$ and $x = 10$, we find $25 = \frac{k}{10^2}$ and $k = 2500$. Now we can find y when $x = 4$ using $y = \frac{2500}{x^2}$. Substituting $x = 4$ gives $y = \frac{2500}{4^2} = 156.25$.	1. Assume that y varies directly as x. If $y = 37.5$ when $x = 7.5$, what is the value for y when $x = 13$? 2. Assume that the amount of water that Mario uses for irrigation at his vineyard is inversely proportional to the amount of rainfall. If he uses 30,000 gallons during a month in which there is 3 inches of rain, how much water would he use in a month that has 5 inches of rain?

Section 7.6 Modeling with Systems of Linear Equations and Inequalities

Objective 1 – Represent systems of linear equations and inequalities graphically.
Objective 2 – Solve a system of equations using the elimination method.
Objective 3 – Solve a system of linear inequalities graphically.
Objective 4 – Use systems of linear equations and inequalities as models.

Objective 1 – Represent systems of linear equations and inequalities graphically.

Key Terms
System of linear equations
Solution

Summary
We have discussed methods for graphing linear equations. We now turn to a collection of linear equations, called a system of linear equations. While there can be many equations in a system, we will focus on systems of two linear equations. A solution to such a system, if it exists, would be an ordered pair.

Example: Show that $(4,-1)$ is a solution to the system of linear equations $\begin{matrix} x+2y=2 \\ 3x+4y=8 \end{matrix}$.

We take each equation and show that $(4,-1)$ makes the sentence true. For the first equation, $(4)+2(-1)=2$ and in the second equation $3(4)+4(-1)=8$. Both are true statements.

A system of equations may or may not have a solution. In a case like the one above, the solution represents the intersection point of the equations $x+2y=2$ and $3x+4y=8$. If the linear equations in the system are parallel, then there are no intersection points, and thus no solutions. In the case where the two lines are the same, we get infinitely many solutions. These cases are illustrated in the graphs below.

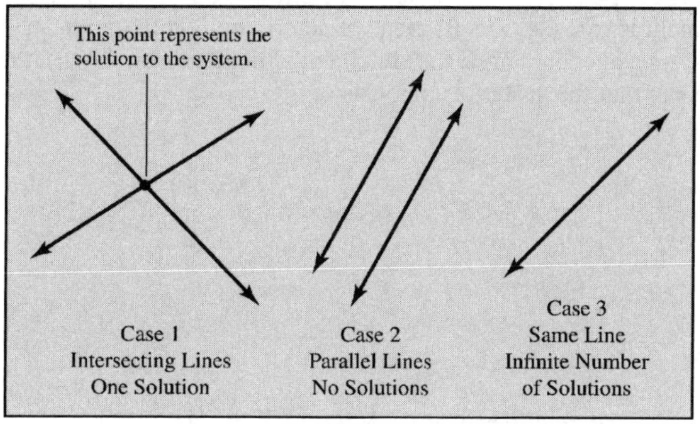

Notes:

Objective 2 – Solve a system of equations using the elimination method.

Key Terms
Elimination method
Back substitution
Inconsistent
Dependent

Summary
We need to develop a strategy to find the solution to a system of linear equations, or determine that no such solution exists. One strategy is called the elimination method. As the name suggests, we eliminate one of the variables and obtain a single equation with only one variable. This is done by multiplying both sides of the equations in the system by some constants so that one of the variables appears with opposite coefficients in the two equations.

We then add the equations together, and obtain an equation in just one variable. We solve this equation, and then go back to one of the original equations to find the value of the other variable. This step is called back substitution.

Example: Solve the system $\begin{aligned} x+2y&=2 \\ 3x+4y&=8 \end{aligned}$

Our goal is to change the system so that the coefficient of one of the variables appears with opposite signs. We can do this by multiplying both sides of the top equation by -3. The system is then $\begin{aligned} -3x-6y&=-6 \\ 3x+\ 4y&=8 \end{aligned}$. Adding the two equations together yields $0x-2y=2$ so $y=-1$. We back substitute this value back into one of the equations. For this purpose we will use $x+2y=2$; we find $x+2(-1)=2$. Adding 2 to both sides of the equation yields $x = 4$.

When we add the equations together, it is possible for both of the variables to be eliminated! If this occurs, look to see what type of equation results. If the statement is true (say $0 = 0$) then we determine that there are an infinite number of solutions. We refer to the system as dependent. If the statement is not true (say $3 = 0$), then we determine that there are no solutions and we say that the system is inconsistent.

Notes:

Guided Example	***Practice*** (Exercises 7 & 9 pp.370 - 371)
Solve the system $\begin{aligned} 3x+\ y&=6 \\ 6x+2y&=12 \end{aligned}$. We will start by multiplying the top equation by -2 so as to have opposite coefficients on y. Doing this yields $\begin{aligned} -6x-2y&=-12 \\ 6x+2y&=12 \end{aligned}$. When we add the equations together, we obtain $0x+0y=0$ or $0 = 0$. In this case, there are infinitely many ordered pairs that are solutions to the system. The system is dependent.	Use the elimination method to solve the following systems of equations. 1. $\begin{aligned} 3x-2y&=-14 \\ -4x+\ y&=\ 22 \end{aligned}$ 2. $\begin{aligned} 8x-2y&=-2 \\ -4x+y&=3 \end{aligned}$

Objective 3 – Solve a system of linear inequalities graphically.

Key Terms
Linear inequality
One point test
Half plane

Summary
The systems we have solved have been for linear equations. We now move to systems of linear inequalities. A linear inequality is one that can be written in the form $ax+by \geq c$, $ax+by > c$, $ax+by \leq c$, or $ax+by < c$. We will look at systems where there are two linear inequalities.

Before we attempt to solve the system, let's look at what the solution to a single linear inequality would look like. Our strategy will be the following:

1. Change the inequality to an equation and graph the line. If the inequality contains $\leq$ or $\geq$, draw a solid line. If the inequality contains < or >, draw a dotted line.
2. Choose a point that is clearly either above or below the line drawn in step 1.
3. If the point you selected in step 2 satisfies the inequality (in other words makes the statement true), then the side of the line contain the point contains the solutions for the inequality. Otherwise the opposite side of the line contains the solutions. We refer to this as the one-point test.
4. Use arrows or shading to indicate which side of the line contains the solution. The solution is a half-plane.

Example: Use the one point test to solve $x+2y \leq 2$.

We change the inequality to the equation $x+2y=2$ and graph. We plot the intercepts (0, 1) and (2, 0).

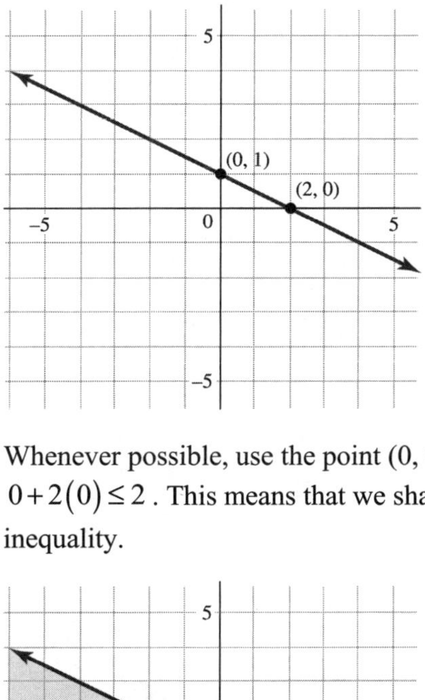

Whenever possible, use the point (0, 0) for step 2.Substituting (0, 0) into the inequality yields the true statement $0+2(0) \leq 2$. This means that we shade the side of the line where (0, 0) falls to represent the solutions to the inequality.

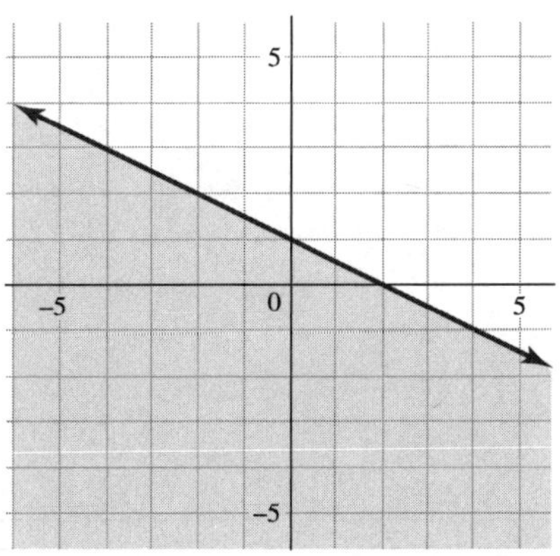

Now we can turn to a system of inequalities. Since we have two inequalities, we go through the steps outlined above for each of the inequalities.

Problem Solving; We will need to determine if the two linear inequalities intersect. In doing this, we use the strategy *Relate a New Problem to an Older One* and use the elimination method.

Notes:

Guided Example	***Practice*** (Exercises 25 & 37 p.371)
Solve the system $\begin{matrix} x+2y\le 2 \\ 3x+4y>8 \end{matrix}$. We have already graphed the top inequality, so we will turn to the second one. Changing the inequality into an equation yields $3x+4y=8$. Verify that $(0,2)$ and $(4,-1)$ are solutions to the equation. We draw a dotted line for the inequality $3x+4y>8$ on the axes with the inequality $x+2y\le 2$. You should also verify that the intersection point of the lines would be $(4,-1)$. We determined this intersection point at the beginning of this section. We do not darken this point in our graph since $(4,-1)$ would not be a solution to the inequality $3x+4y>8$. We identify this by drawing an open circle at $(4,-1)$. Our final step is to use the one-point test to see which side of the inequality $3x+4y>8$ should be shaded. We can again use the test point (0, 0), but in this case $3(0)+4(0)>8$ is not a true statement, so we shade the opposite side of the line $3x+4y=8$. The solution of the system is the set of all points that are in the double shaded section of the graph. 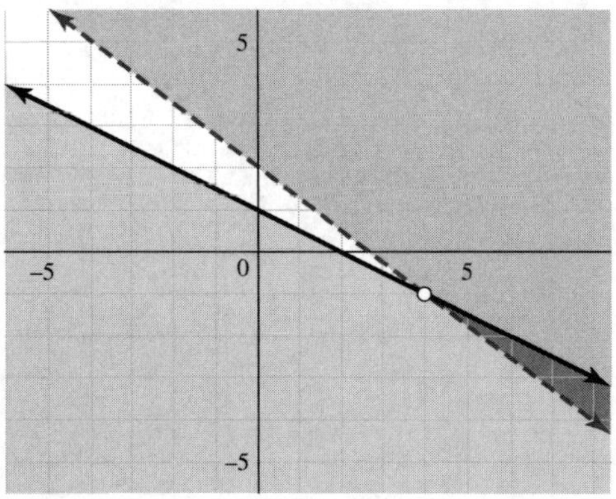 	1. Determine which points satisfy $3x+4y\ge 2$ a. $(3,5)$ b. $(1,-2)$ c. $(0,0)$ d. $(-4,6)$ 2. Solve the system $\begin{matrix} 2x-y\ge 3 \\ x-y\le -1 \end{matrix}$. Indicate all corner points and shade the solution set.

Objective 4 – Use systems of linear equations and inequalities as models.

Key Terms
None

Summary
We now apply the tools learned to solve application problems involving linear equations and inequalities. It is a good idea to first read the problem and get the big picture. Then assign variables to the unknowns and translate the problem into a system of equations or inequalities.

Problem Solving; When making models, it is always a good idea to use the *Chose Good Names strategy.*

Example: A slice of cheese pizza contains 40 g of carbohydrates and 220 mg of calcium. A 12-oz cola contains 40 g of carbohydrates and 15 mg of calcium. If Karl eats several slices of pizza and drinks cola how much of each must he eat to get a nutritional benefit of exactly 200 g of carbohydrates and 690 mg of calcium?

The two unknowns are the amount of cola (c) and the amount of pizza Karl will eat (p). Let's organize the information that we know about the carbohydrates and calcium in each food.

Nutrients	Cola	Pizza	Desired Amount
Carbohydrates	40	40	200
Calcium	15	220	690

We can now build a linear equation for the amount of carbohydrates Karl will consume. He will get 40 g from each cola and 40g from each slice of pizza and, so if he consumers c colas and p slides of pizza, he will have $40c + 40p$ grams. Since the desired amount is 200 we obtain the equation $40c + 40p = 200$. This is an equation and not an inequality because he desires exactly 200 g of carbohydrates.

We build a similar equation for calcium; with 15 mg in a cola and 220 mg in a slice of pizza, if he consumers c colas and p slides of pizza, he will have $15c + 220p$ milligrams. With a desired amount of 690 mg, we find $15c + 220p = 690$.

Now we need to solve the system; use the elimination strategy discussed earlier to verify the solution to this system is 2 colas and 3 slices of pizza.

Notes:

Guided Example	***Practice*** (Exercises 47 & 63 pp. 371-372)
How would the following differ from the example above? A 12-oz cola contains 40 g of carbohydrates and 15 mg of calcium. A slice of cheese pizza contains 40 g of carbohydrates and 220 mg of calcium. If Karl eats several slices of pizza and drinks cola how many of each could he eat to get a nutritional benefit of no more than 200 g of carbohydrates and at least 690 mg of calcium? The example has changed because we no longer want exactly 200 g of calcium, but no more than 200 g. We could change the equation to the inequality $40c + 40p \leq 200$. Since Karl wants at least 690 mg of calcium the inequality would be $15c + 220p \geq 690$ Using the strategies developed for solving inequalities, verify that the solution is 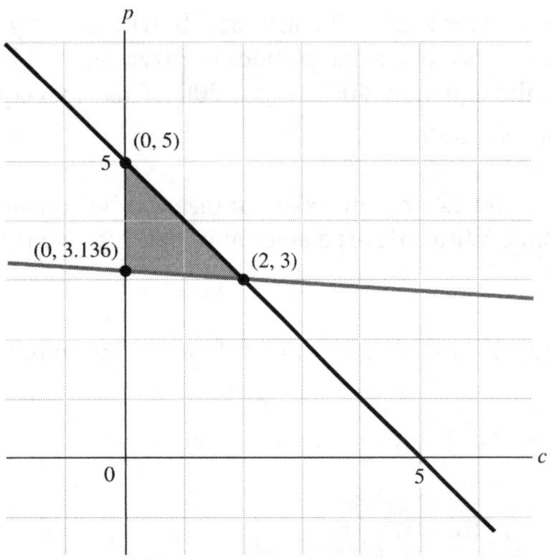 If we wanted to limit our solutions to those with whole numbers for *c* and *p*, we could look in the shaded regions for the ordered pairs (*c, p*) which have whole number values. One would be the solution we found to the system of equations, (2, 3), other choices would be (1, 4) and (0, 4) and (0, 5)	Solve the problem by modeling with a system of equations or inequalities and solving. 1. An average bagel contains 30 mg of calcium and 2 mg of iron. One ounce of cream cheese contains 25 mg of calcium and 0.4 mg of iron. If Leyla wants to eat a combination of bagels and cream cheese that contains exactly 245 mg of calcium and 10 mg of iron, how much of each should she eat? 2. Ann's personal trainer has told her to spend no more than 75 minutes on a workout, which will consist of strength and aerobic activities. Her time spent on strength training should be at least twice as much as the time spent on aerobic. Indicate the combinations of time spent on each that will meet her requirements.

Section 7.7 Dynamical Systems

Objective 1 – Apply dynamical systems as models.
Objective 2 – Identify stable and unstable dynamical systems.

Objective 1 – Apply dynamical systems as models.

Key Terms
Dynamical system

Summary
A dynamical system is a sequence of numbers A_0, A_1, A_2, A_3, … such that for each value of n, A_{n+1} = an expression involving A_n.

Example: Your friend has offered to take care of your dog while you are on vacation next month. She has asked that you pay her 5 cents the first day, and double her compensation each day. Model her compensation using a dynamical system.

Her compensation on day $n + 1$ can be computed using the dynamical system $A_{n+1} = 2A_n$ where $A_0 = 0.05$. The first few values in the sequence would be 0.05, 0.10, 0.20, 0.40. 0.80, … but the values will grow quickly! On day 20 her compensation will be $26,214.40!

Notes:

Guided Example	***Practice*** (Exercise 15 p.379)
1. You deposit $5,000 in a bank account paying 4% yearly interest that is compounded annually. Model this with a dynamical system. Since the deposit amount is 5,000 we have $A_0 = 5{,}000$. With an interest rate of 4%, the amount of money in the account grows by 4% each year, and at the end of year 1 you will have 1.04 times the amount you deposited. In general, the dynamical system is $A_{n+1} = 1.04A_n$ 2. Use the system to determine the amount in the account at the end of 2 years. We know $A_0 = 5{,}000$, and will use the system to determine A_1 and A_2. $A_1 = 1.04(5{,}000) = 5{,}200$ $A_2 = 1.04(5{,}200) = 5{,}408$	Model this situation with a dynamical system and use it to answer the question. 1. You take a 250-mg dose of an antibiotic every 4 hours. Your body eliminates 40% of the drug in a 4-hour period. How much antibiotic will be in your bloodstream after three doses?

Objective 2 – Identify stable and unstable dynamical systems.

Key Terms
Equilibrium value
Stability

Summary
When using dynamical systems, we are often interested in the long run behavior of the system. What happens in the long term? With some systems, the behavior in the long run reaches an equilibrium point, where $A_{n+1} = A_n$. If this occurs, we say the system has reach equilibrium. Not all systems reach equilibrium, but if they do we can determine the equilibrium value by following these steps:

1. Write the equation for A_{n+1} in terms of A_n.
2. Substitute *a* for both A_n and A_{n+1}.
3. Solve the equation in step 2 for *a*.

Example: Consider the dynamical system $A_{n+1} = 0.25A_n + 4$. Does this system reach equilibrium?

Assume that at some point, $A_{n+1} = A_n$, and both take on the same value *a*. Now we try to solve $A_{n+1} = A_n = a$.

We have $a = 0.25a + 4$ which leads to $0.75a = 4$ and finally $a = \frac{4}{0.75} = 5.\bar{3}$. This system does reach equilibrium.

If a system reaches equilibrium, it may or may not be stable. Stability in a dynamical system implies that small changes in the initial conditions do not impact the long term behavior. If our dynamical system is linear, in other words $A_{n+1} = mA_n + b$, the equilibrium value will be stable if $-1 < m < 1$ and unstable if $m > 1$ or $m < -1$.

Notes:

Guided Example	***Practice*** (Exercises 7 & 9 p.379)
Determine if the equilibrium value reached in the example above is stable. Our dynamical system $A_{n+1} = 0.25A_n + 4$ is in the form $A_{n+1} = mA_n + b$ where $m = 0.25$. Since this value is between and -1 and 1, the equilibrium value we obtained is stable.	Find the equilibrium value for each dynamical system. Comment on whether the value you find is stable or unstable. 1. $A_{n+1} = 2A_n + 3$ 2. $D_{n+1} = 0.40D_n + 500$

Chapter 7 Answers

<u>Section 7.1</u>

Objective 1 1) $w = \dfrac{P-2l}{2}$ 2) $r = \dfrac{A-P}{Pt}$

Objective 2 1)

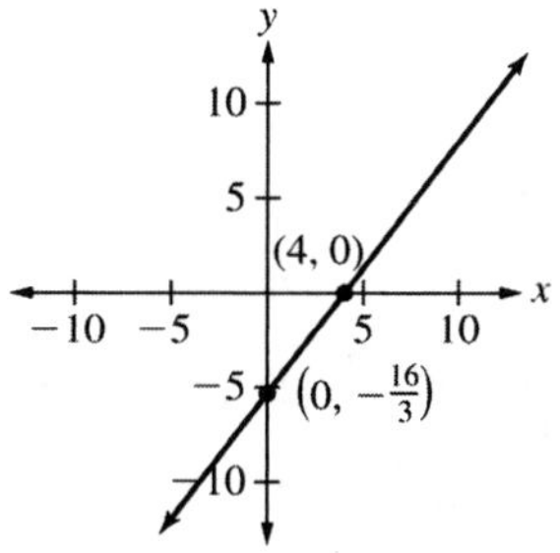

2)

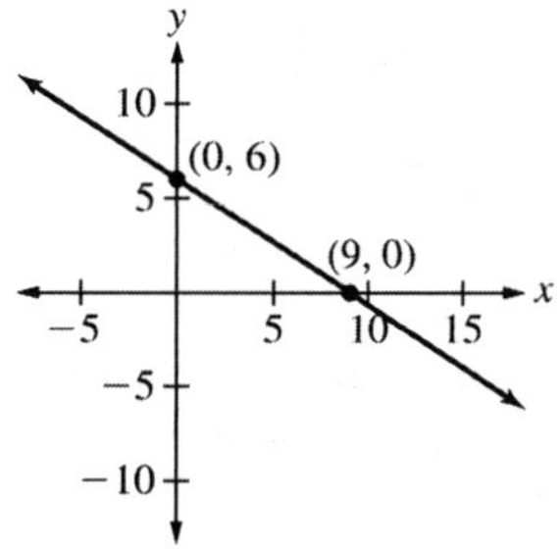

Objective 3 1) $d = 0.73t + 80$ 2) 98.25

<u>Section 7.2</u>
Objective 1 1) $y = 0.1t + 80.6$; 82.7 years.

Objective 2 1) $y = 2x - 1$ 2) $y = \dfrac{7}{2}x + 13$

Objective 3 1) a) 1.77 million b) $l = 1.77t + 146$ c) 156.62 million d) 2034

<u>Section 7.3</u>

Objective 1 1) 2,8 2) $-3, -2\dfrac{1}{2}$

Objective 2 1) opening up; vertex $\left(\dfrac{1}{2}, -3\right)$; $\left(\dfrac{1+\sqrt{3}}{2}, 0\right)$ $\left(\dfrac{1-\sqrt{3}}{2}, 0\right)$; $(0, -2)$

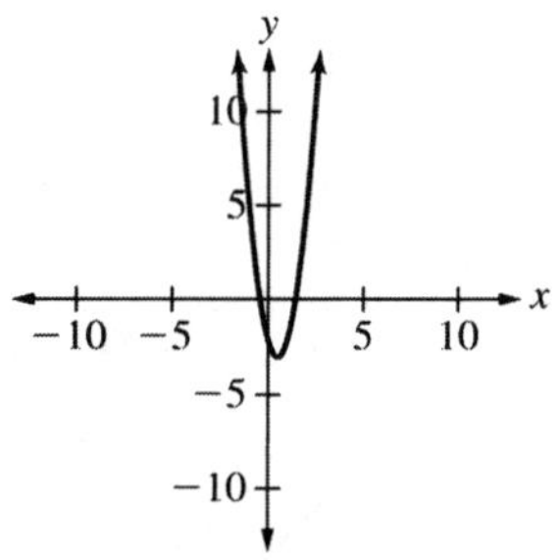

2) opening down; vertex $\left(\dfrac{7}{2}, \dfrac{1}{4}\right)$; (3, 0), (4, 0); $(0, -12)$

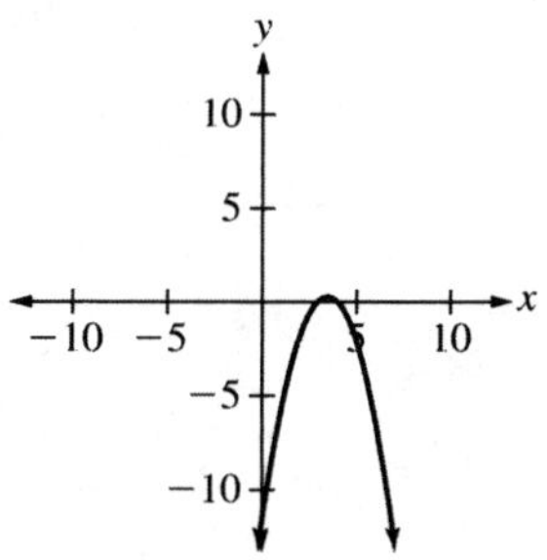

Objective 3 1) a.

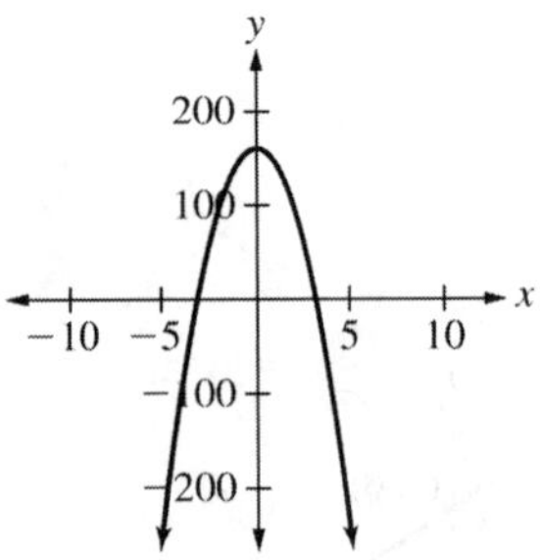

b. t (time) cannot be negative c. 3.16 seconds.

Section 7.4

Objective 1 1) In a linear model, the rate of change is constant; whereas in an exponential model the change is proportional to the amount present.

Objective 2 1) \$4665.60 2) 198.07 million

Objective 3 1) 1.57 2) 2111

Objective 4 1) 0.3669 2) 0.7291

Section 7.5

Objective 1 1) 36 2) 2.6 mg

Objective 2 1) 12,000 2) 5 or 6.

Objective 3 1) 65 2) 18,000 gallons

Section 7.6

Objective 2 1) $(-6, -2)$ 2) no solution

Objective 3 1) a, d 2)

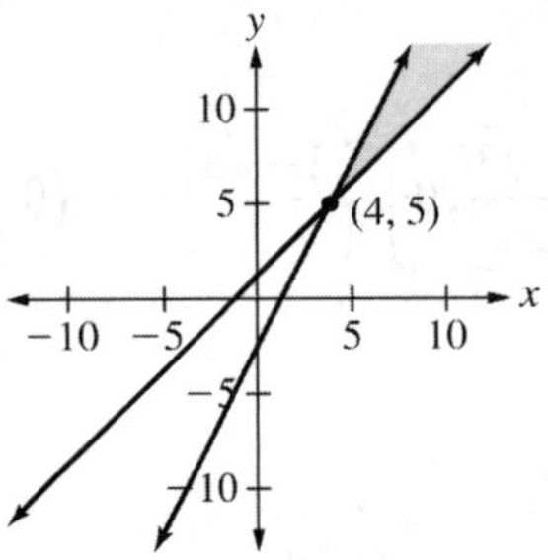

Objective 4 1) 4 bagels and 5 ounces of cream cheese 2)

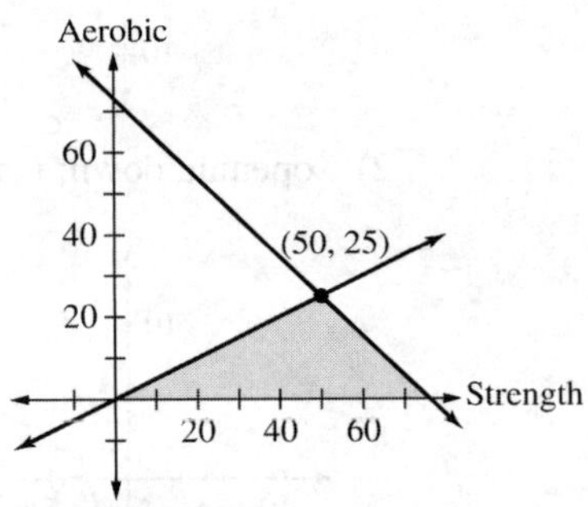

Section 7.7

Objective 1 1) $D_{n+1} = 0.6D_n + 250$, $n = 0, 1, 2, \ldots$, where $D_0 = 0$; 490 mg.

Objective 2 1) $a = -3$; unstable 2) $a = \frac{2500}{3} \approx 833.3$; stable.

Chapter 8: Consumer Mathematics

Section 8.1 Percent, Taxes, and Inflation.

Objective 1 – Perform calculations using percent.

Objective 2 – Use percents to represent change.

Objective 3 – Apply the percent equation to solve applied problems.

Objective 4 – Use percent in calculating income taxes.

Objective 5 – Use the Consumer Price Index to estimate the effect of inflation.

Objective 1 – Perform calculations using percent.

Key Terms
Percent

Summary
The word percent is derived from Latin "per centum," which means "per hundred." You can think of 25% as 25 parts of 100, so we write 25% as $\frac{25}{100}$ or 0.25. Generally speaking, when converting from percent to a decimal, drop the percent sign and divide by 100, moving the decimal two places to the left. To convert from a decimal to a percent, multiply by 100, moving the decimal to places to the right, and add a percent sign.

Problem Solving When converting from a fraction with denominator not equal to 100 to a percent, use the *Relate a New Problem to an Older One Strategy*. First convert the fraction to a decimal and then use the rule above to convert from decimal to percent.

Notes:

Guided Examples	***Practice*** (Exercises 5, 15 & 19 p.394)
1. Convert 159% to a decimal. We divide by 100, which moves the decimal two places to the left and drop the percent sign yielding 1.59. 2. Convert $\frac{7}{8}$ to a percent. We first express $\frac{7}{8}$ as the decimal 0.875 and then multiply by 100 and add the percent sign. Our final answer is 87.5%.	1. Convert 27.35% to a decimal. 2. Write 0.002 as a percent. 3. Convert $\frac{5}{16}$ to a percent.

Objective 2 – Use percents to represent change.

Key Terms
Base amount
New amount
Percent of change

Summary
Percent is often used to express a change in a quantity. When doing so, you need to be very clear about which amount is the base. The change in the quantity is compared to this base amount using the formula

$$\text{Percent of change} = \frac{\text{new amount - base amount}}{\text{base amount}}.$$

Notes:

Guided Example	***Practice*** (Exercise 53 p.395)
1. You had 248 friends on Facebook but have recently "unfriended" some of them and you now have 214 friends. What is the percent change in your number of friends? We use the percent of change formula with 248 as the base and 214 as the new amount: $\frac{214-248}{248} \approx -0.137$. The percent of change is negative because your number of friends has decreased. 2. In June of 2017, 220,000 passengers flew in and out of Cityside Airport, an increase of 2.5% over June of 2016. Use this to determine the number of passengers who flew in and out of Cityside airport in June 2016. In this problem the new amount is 220,000 but we don't know the base amount, so will represent it by x in the percent of change formula and then solve. We have $\frac{220,000-x}{x} = 0.025$. Multiplying both sides by x gives us $220,000 - x = 0.025x$. We add x to both sides and solve finding $220,000 = 1.025x$ and $x = \frac{220000}{1.025} \approx 214,634$	1. If a dealer buys a car from the manufacturer for \$19,875 and then sells it for \$21,065, what is the percent change? 2. In doing problem #1, Angela got an incorrect answer of 5.6%. What mistake did she make?

Objective 3 – Apply the percent equation to solve applied problems..

Key Terms
Percent equation

Summary
When solving applied problems involving percent, we use the basic percent equation, which states percent $\times$ base = amount. If you know the percent and the base, first convert the percent to a decimal and multiply to find the unknown amount. If the unknown is either the percent or the base, substitute the information you know into the percent equation and solve for the unknown.

Notes:

Guided Example	***Practice*** (Exercises 33 & 45 pp.394-395)
A laptop usually priced at \$750 is on sale for \$675. Express the reduction in price in percent. We will determine the amount by which the price was reduced and then find the percent using the percent equation. The price is reduced by $750-675=75$ dollars. Using the percent equation, with 75 as the amount and 750 as the base, we determine that the percent is $percent \times 750 = 75$ so the percent is $\frac{75}{750}=0.10$ or 10%.	1. In a recent year, the top-selling cookie in America was Nabisco's Chips Ahoy, with sales of \$294.6 million. The total cookie sales for that year were \$3,124 million. What percent of the total cookie sales was due to Chips Ahoy? 2. According to *The Useless Information Society*, only 6% of Beatles' autographs in circulation are estimated to be real. If there are 81 authentic autographs of the Beatles in circulation, then how many non-authentic autographs are there in circulation? (Hint: first find the total number in circulation)

Objective 4 – Use percent in calculating income taxes.

Key Terms
None

Summary
We make use of the percent equation when calculating income taxes. The examples and exercises below use the following tax table from Form 1040. It is used to compute the taxes due for a single person with the given taxable income. Be careful! You must read the table carefully and make necessary adjustments before you substitute for the base! Note that the percentages cited in the fourth column are computed based on the amount of income over that which is listed in the fifth column.

	If taxable income is over —	But not over —	The tax is	Of the amount over —
Line 1	$0	$9,275	$0 + 10%	$0
Line 2	9,275	37,650	927.50 + 15%	9,275
Line 3	37,650	91,150	5,183.75 + 25%	37,650
Line 4	91,150	190,150	18,558.75 + 28%	91,150
Line 5	190,150	413,350	46,278.75 + 33%	190,150
Line 6	413,350	415,050	119,934.75 + 35%	413,350
Line 7	415,050	—	120,529.75 + 39.6%	415,050

Notes:

Guided Example	***Practice*** (Exercises 67 & 69 p.396)
Use the tax table to determine the federal taxes due for a person with a taxable income of 1. $7,293 We use Line 1 in the table for this income. The entire income is subject to 10% tax, so the amount of tax is $0.10 \times 7,293 = 729.30$. This person owes $729.30 in tax. 2. $532,000 For this income, we go to line 7. The person has a flat amount of 120,529.75, but must also pay 39.6% of any income over 415,050. We determine the tax using the equation $120,529.75 + (0.396)(532,000 - 415,050)$ $= 120,529.75 + (0.396)(116,950)$ $= 120,529.75 + 46,312.20$ This person owes $166,841.95 in taxes.	Use the tax table above to calculate the federal income tax due for the given taxable income. 1. $148,000 2. $47,800

Objective 5 – Use the Consumer Price Index to estimate the effect of inflation.

Key Terms
Inflation
Consumer Price Index

Summary
We use percent and percent of change to measure changes in the amount that we pay for items. Inflation is the rise in the level of prices of goods and services over a period of time. One common measure of inflation is the Consumer Price Index, or CPI. The base CPI is 100, which corresponds to average prices during the years 1982 to 1984. As an example, if you were to read that the overall CPI in 2015 was 237.0, that would tell you that the prices of goods and services in 2015 were roughly 237% what they cost in 1982-84. In addition to the overall CPI, The Bureau of Labor Statistics computes the CPI for individual commodities, as you will see in the practice exercises.

Notes:

Guided Example	***Practice*** (Exercises 71 & 73 p.396)
The overall CPI for goods and services in 2010 was 218.1 and in 2015 it was 237.0. Use this to find the percent of increase in overall CPI from 2010 to 2015. Using the percent of change formula, we have (new amount – base amount)/ base amount, which is $\frac{237.0-218.1}{218.1} \approx 0.087$ or 8.7%. This represents the rate of inflation over that time period.	1. The CPI for ice cream was 179.1 in 2005 and 221.8 in 2015. Use this to determine the percent of increase in CPI for ice cream over this period. 2. The CPI for milk in 2015 was 3.310, which represents a 2.1% increase from 2005. Use this to find the CPI for milk in 2005.

Section 8.2 Interest

Objective 1 – Use the simple interest formula to solve problems.
Objective 2 – Use the compound interest formula to find a future value.
Objective 3 – Solve the compound interest formula for different unknowns, such as the present value, length, and interest rate for a loan.

Objective 1 – Use the simple interest formula to solve problems.

Key Terms
Simple interest
Principal

Summary
When we invest or borrow money, we should pay close attention to the amount of money paid or owed! Interest is the money that one person (a borrower) pays to another (a lender) to use the lender's money. Savers earn interest; borrowers pay it. Interest is expressed in percent and is paid on a base amount called the principal.

There is a simple formula that relates principal, interest earned, rate, and time. We calculate simple interest using the formula $I = Prt$ where I is the interest earned, P is the principal, r is the annual interest rate expressed as a decimal, and t is the time in years.

Example: Imagine that you invest \$1,000 into an account paying 3% simple interest for 4 years. How much interest will you earn?

Using the formula $I = Prt$ with $P = 1{,}000$, $r = 0.03$ and $t = 4$, we find $I = 1{,}000 \times 0.03 \times 4 = 120$. You have earned \$120 in interest.

In the example we just completed, the interest was \$120, but you still have your principal amount, \$1,000. At the end of the investment, you have a total of \$1,120 which we call the future value. In general, to find the future value of an account that pays simple interest, use the formula $A = P(1 + rt)$, where A is the future value and P, r, and t are defined as they were with the interest formula. In this case, P is also called the present value.

Notes:

Guided Example	***Practice*** (Exercises 1, 3, 5, & 7 p.405)
You and some friends are planning a trip to Europe in two years. How much money must you deposit now if you need to have \$2,500 as the future value and the interest rate is 5%? Assume the bank pays simple interest. We will use the formula $A = P(1+rt)$ in order to find *P*, which would be the present value (the value you need now to have the future value A = 2,500 in t = 2 years). Substituting, we find $2,500 = P(1+(0.05)(2))$, which simplifies to $2,500 = P(1.1)$. Dividing both sides of the equation by 1.1 gives us P = \$2,272.73. **Problem Solving:** *Verify your answer!* Double check that this answer makes sense. The most common mistake in this problem would be to use P = 2,500 rather than *A*, and in this case you would obtain an answer greater than \$2,500 for your deposit amount. Clearly the present value should be smaller than the future value in this example.	In exercises 1 & 2, use the simple interest formula $I = Prt$ to find the indicated quantity. 1. P = \$1,000, r = 8%, t = 3 years; find *I*. 2. I = \$700, P = \$3,500, t = 4 years; find *r*. In exercises 3 & 4, use the future value formula $A = P(1+rt)$ to find the indicated quantity. 3. P = \$2,500, r = 8%, t = 3 years; find *A*. 4. A = \$1,770 r = 6%. t = 3 years, find *P*.

Objective 2 – Use the compound interest formula to find a future value.

Key Terms

Compound interest

Summary

In the previous worksheet, we found the future value using the simple interest formula. It seems fair that if money in a bank account has earned interest, the bank should add this interest to your principal and use this as the base when future interest payments are made. This is, in fact, the way most bank accounts work. Interest that is paid on principal plus previously earned interest is called compound interest. In order to determine the future value for this situation, we need to know how often interest is computed and added to your principal. This is called compounding. Assume that an account with principal P is paying rate r and that compounding is being done m times per year If the money remains in the account for n time periods, then the future value, A, of the account is given by the formula $A = P\left(1+\frac{r}{m}\right)^n$.

Notes:

Guided Example

Assume that you invest \$10,000 into an account paying 6% interest for 2 years. Determine the future value, A, if compounding is done as indicated.

1. Monthy

   If interest is compounded monthly, we use $m = 12$ in the formula $A = P\left(1+\frac{r}{m}\right)^n$. Substituting $P = 10{,}000$, $r = 0.06$ and $n = 24$ (because there are 24 monthly periods in 2 years), we find

   $$A = 10,000\left(1+\frac{0.06}{12}\right)^{24}$$

   $$= 10,000(1.005)^{24} = \$11,271.60.$$

2. Quarterly

   Here we use $m = 4$, $n = 8$ and proceed as above:

   $$A = 10,000\left(1+\frac{0.06}{4}\right)^{8}$$

   $$= = 10,000(1.015)^{8} = \$11,264.93.$$

Practice (Exercises 11 & 15 p.406)

In the exercises below, you are given the principal, the annual interest rate, and the compounding period. Use the formula for computing future value using compound interest to determine the value of the account at the end of the specified time period.

1. \$5,000, 5%, yearly; 5 years.

2. \$20,000, 8%, monthly; 2 years.

Objective 3 – Solve the compound interest formula for different unknowns, such as the present value, length, and interest rate for a loan.

Key Terms
Rule of 72

Summary
In the previous worksheet, we introduced the compound interest formula and, given values for P, r, m, and n obtained the future value, A. We will now look at situations where A is given, and the unknown is one of the other variables. We continue to work with the formula $A = P\left(1+\frac{r}{m}\right)^n$.

If P is the unknown, we find our solution by dividing A by $\left(1+\frac{r}{m}\right)^n$.

Example: You wish to have saved \$5,000 at the end 3 years. You will invest at 3%, compounded quarterly. How much money must you invest?

Here we need to find P. We will take the formula $A = P\left(1+\frac{r}{m}\right)^n$ and substitute all known values. Because we have quarterly compounding, $m = 4$. The term is 3 years, which includes 12 quarters, so $n = 12$. We obtain $5,000 = P\left(1+\frac{0.03}{4}\right)^{12}$; performing computations in the parentheses yields $5,000 = P(1.0075)^{12}$. We find P by dividing, $P = \frac{5,000}{(1.0075)^{12}} = 4,571.19$.

If r is the unknown, we find our solution by dividing A by P, then taking the nth root of both sides and solving for r.

Example: You plan to invest \$3,500 in an account with monthly compounding. What rate would be required to have the future value A = \$4,000 in t = 2 years?

Here we need to find r. We will use the formula $A = P\left(1+\frac{r}{m}\right)^n$ and substitute known values. We have $m = 12$ and $n = 24$ since compounding is monthly for 2 years. We obtain $4,000 = 3,500\left(1+\frac{r}{12}\right)^{24}$. Our first step is to divide A by P; the equation that follows is $\frac{4,000}{3,500} = \left(1+\frac{r}{12}\right)^{24}$. We take the 24th root of each side and obtain $1.005579315 \approx \left(1+\frac{r}{12}\right)$ which leads to $r \approx 0.06695$. We conclude that we would need a rate of roughly 6.7%.

If t is the unknown, we will need to use the log function to assist us in solving for the exponent. The exponent property of the log function is $\log y^x = x \log y$. We will look at this case in the guided example.

In cases where we are looking for the time required to double a quantity, we use the Rule of 72, which estimates the doubling time to be approximately $\frac{72}{rate}$ where we express the rate in whole numbers (in other words, for 4% we would use 4).

Notes:

Guided Example	***Practice*** (Exercises 33, 35 & 29 p.406)
You again have \$3,500 to invest and wish to have \$7,000 in the account. The rate is 4% and compounding is done annually. How long will it take? We will use $A = P\left(1+\frac{r}{m}\right)^n$ with $A = 7{,}000$, $P = 3{,}500$, $m = 1$ and $r = 0.04$. We need to find n. $7{,}000 = 3{,}500\left(1+\frac{0.04}{1}\right)^n$. We first divide both sides by 3,500 which results in $\frac{7{,}000}{3{,}500} = \left(1+\frac{0.04}{1}\right)^n$. In order to solve this equation, we use logarithms, yielding $\log\frac{7{,}000}{3{,}500} = \log(1.04)^n$. We use the exponent property of log function to rewrite this as $\log\frac{7{,}000}{3{,}500} = n\log(1.04)$ and finally $n = \frac{\log(2)}{\log(1.04)} = 17.67$. It will take almost 18 years to double the investment. Using the rule of 72, the time to double would be roughly $\frac{72}{4} = 18$, so the rule of 72 gives an answer consistent with ours.	In exercise 1 & 2, use the compound interest formula $A = P\left(1+\frac{r}{m}\right)^n$ and the given information to find either t or r. Use annual compounding. 1. $A = \$2{,}500$, $P = \$2{,}000$, $t = 5$. 2. $A = \$1{,}500$, $P = \$1{,}000$, $r = 4\%$. 3. Imagine that you have \$1,500 to invest. Assume an interest rate of 3.5%. Find the time in years that it will take for your investment to double with annual compounding a) by solving the compound interest formula for n using the log property, and b) using the rule of 72.

Section 8.3 Consumer Loans.

Objective 1 – Determine payments for an add-on loan.
Objective 2 – Compute finance charges on a credit card using the unpaid balance method.
Objective 3 – Use the average daily balance method to compute credit card charges.
Objective 4 – Compare credit card finance charge methods.

Objective 1 – Determine payments for an add-on loan.

Key Terms
Installment loan
Finance charge
Add-on interest method

Summary
In this section we will look at one way that is used to determine the payment amount when you borrow money to make a purchase and pay it back over time. This is called an installment loan, because you make payments, called installments, to pay back your debt. It is important to understand that when you take out a loan, you will pay for the luxury of purchasing something that you don't have the money to afford right now! There will be a finance charge – the interest charged on the loan.

We first look at the add-on interest method for determining payments on a loan. With this method, the payment amount is found by determining the interest owed, adding it to the amount borrowed, and dividing by the number of payments that will be made. The formula is monthly payment = $\frac{P+I}{n}$, where P is the amount of the loan, I is the total interest due on the loan, and n is the number of monthly payments.

Problem solving: We will use the simple interest formula, $I = Prt$ to determine the interest. We use the *Relate a New Problem to an Older One* strategy when finding payments for an add-on loan.

Notes:

Guided Example	***Practice*** (Exercise 5 p.415)
Artis' bank gave her a 3-year add-on interest loan for \$1,300 to assist with college expenses. The annual interest rate is 8.5%. How much interest will she pay? What will her monthly payments be? The interest is found using the simple interest formula $I = Prt$ where $P = 1{,}300$, $r = 0.085$ and $t = 3$. Substituting, we find $I = (1,300)(0.085)(3) = 331.50$. To determine the monthly payments, we use the formula $\frac{P+I}{n}$. The monthly payment will be $\frac{1,300+331.50}{36} = 45.32$.	1. Luis took out an add-on interest loan for \$1,280 to buy a new laptop computer. The loan will be paid back in 2 years and the annual interest rate is 9.5%. How much interest will he pay? What are his monthly payments?

Objective 2 – Compute finance charges on a credit card using the unpaid balance method.

Key Terms

Unpaid balance method

Summary

Determining interest for purchases made on a credit card is more complicated than a loan for a single purchase. We need to consider additional purchases and associated finance charges, as well as giving credit for payments made. One of the ways to compute finance charges on a credit card is called the unpaid balance method. We will be using the formula $I = Prt$, but we compute P using a new formula. To determine P, we take the previous month's balance, add finance charges on that amount and add any additional purchases, then subtract any payments and returns. The formula is P = previous month's balance + finance charges + purchases made – returns – payments. This value for P is substituted into the formula $I = Prt$.

Notes:

Guided Example	***Practice*** (Exercise 27 p.416)
Use the unpaid balance method to determine the finance charge on a credit card where last month's balance was \$600; payment of \$350 was made; interest rate was 18%; bought clothes totaling \$110 and returned pants for \$35. First let's find the finance charge on last month's balance of 600. We use $I = Prt$ with $t = \frac{1}{12}$ and determine the finance charge $I = 600(0.18)\frac{1}{12} = 9$. Using this in the formula for P, above, we find $P = 600 + 9 + 110 - 350 - 35 = 334$. Now we use this value of P in the formula $I = Prt$ and find $I = 334(0.18)\frac{1}{12} = 5.01$.	1. Use the unpaid balance method to find the finance charge on a credit card account where last month's balance was \$640; payment made of \$320; interest rate, 16.5%; bought dog, \$140; bought pet supplies, \$35; paid veterinarian bill \$75.

Objective 3 – Use the average daily balance method to compute credit card charges.

Key Terms
Average daily balance

Summary
A more complicated method for determining the finance charge on a credit card is called the average daily balance method, which is one of the most common methods used by credit card companies. With this method the balance is the average of all daily balances for the previous month. The steps involved in using this method are listed below.

1. Add the outstanding balance for your account for each day of the month.
2. Divide the total in step 1 by the number of days in the month to find the average daily balance.
3. To find the finance charge, use $I = Prt$, where is the average daily balance found in step 2, and t is the number of days in the month divided by 365.

Problem Solving: When computing finance charge using the average daily balance method, use the *Be Systematic Strategy*. You are encouraged to set up a table where you determine the daily balance for each day of the month. Keep information about all transactions in the table.

Notes:

Guided Example

Use the following information to find the finance charge on a credit card for the month of December (31 days) using the average daily balance method. Use an 18% interest rate.

November's unpaid balance was $275.
A payment of $75 was made on Dec. 6.
Charge of $115 for holiday gifts on Dec. 12.
Lunch out with friends; charged $20 on Dec. 19.
Charged $16 for flowers on Dec. 24.

We first make a table listing the daily balances for the month, and which days had these amounts.

Day	Balance	Number of days × Balance
1,2,3,4,5	$275	5 × 275 = 1,375
6,7,8,9,10,11	$200	6 × 200 = 1,200
12,13,14,15,16,17,18	$315	7 × 315 = 2,205
19,20,21,22,23	$335	5 × 335 = 1,675
24,25,26,27,28,29,30,31	$351	8 × 351 = 2,808

To determine the average daily balance, we use the last column in the table to determine the sum and then divide by 31.

Average daily balance

$$=\frac{1,375+1,200+2,205+1,675+2,808}{31}$$

$$=\frac{9,263}{31}\approx 298.81.$$

We use this for P in the formula $I = Prt$ to find the finance charge.

$$I = 298.81(0.18)\left(\frac{31}{365}\right) = 4.57\,.$$

Practice (Exercise 33 p.418)

1. Use the average daily balance method to compute the finance charge on the credit card for the previous month. Assume an annual interest rate of 21%.

   Month of August (31 days) with a previous month's balance of $280.

Date	Transaction
August 5	Made payment of $75
August 15	Charged $135 for hiking boots
August 21	Charged $16 for gasoline
August 24	Charged $26 for restaurant meal

Objective 4 – Compare credit card finance charge methods

Key Terms
None

Summary
When you apply for a credit card, it is important to know what interest rate you will be charged, and how the interest is being computed. We see in the example below that the methods we have studied yield different finance charges. While the differences here may be slight, with larger balances it can be very significant. Buyer beware!

Notes:

Guided Example

Use the scenario below (from the last worksheet) and determine the finance charge using the unpaid balance method. Compare your answer with that using the average daily balance.

Find the finance charge on a credit card for the month of December (31 days) using an 18% interest rate.

November's unpaid balance was $275.
A payment of $75 was made on Dec. 6.
Charge of $115 for holiday gifts on Dec. 12.
Lunch out with friends; charged $20 on Dec. 19.
Charged $16 for flowers on Dec. 24.

Using the unpaid balance method, we will first determine the finance charge for the previous month, $I = 275(0.18)\frac{1}{12} = 4.13$

Now we determine P = previous month's balancc + finance charges + purchases made – returns – payments. We have $P = 275 + 4.13 + 151 - 75 = 355.13$. Now we find the interest $I = 355.13(0.18)\frac{1}{12} = 5.33$.

This is slightly larger than the average daily balance method, where the finance charge was $4.57.

Practice (Exercise 37 p. 417)

1. Use the unpaid balance method to compute the finance charge on the credit card with the given information. Assume an annual interest rate of 21%. Compare your answer to that done in the previous worksheet, where we used the average daily balance method.

   Month of August (31 days) with a previous month's balance of $280.

Date	Transaction
August 5	Made payment of $75
August 15	Charged $135 for hiking boots
August 21	Charged $16 for gasoline
August 24	Charged $26 for restaurant meal

Section 8.4 Annuities

Objective 1 – Calculate the future value of an annuity.
Objective 2 – Perform calculations regarding sinking funds.

Objective 1 – Calculate the future value of an annuity.

Key Terms

Annuity
Ordinary annuity
Future value

Summary

For most large expenses, we save over a period of time rather than making a one-time lump sum deposit. An annuity is an interest-bearing account into which we make a series of payments of the same size. If one payment is made at the end of every compounding period, the annuity is called an ordinary annuity. The future value of an annuity is the amount in the account, including interest, after all payments are made. To find the future value, we use the formula $A = R\left(\frac{\left(\left(1+\frac{r}{m}\right)^n - 1\right)}{\left(\frac{r}{m}\right)}\right)$ where R is the regular payment, r is the annual interest rate, m is the number of times interest is compounded per year, and n is the total number of compounding periods. There are some redundant parentheses in the formula above, but if you are typing the formula into your calculator directly they will ensure that proper order of operations is followed.

Notes:

Guided Example	***Practice*** (Exercise 35 p.424)
You have decided that, rather than stopping on the way to class for coffee and a bagel, you will have breakfast at home. You determine that this will give you an extra \$50 per month and decide to deposit this amount (monthly) into an ordinary annuity paying 3% annual interest. How much money will you have saved in 4 years? We will use the formula above with $R = 50$, $r = 0.03$, $m = 12$ and $n = 48$. $A = 50\left(\frac{\left(\left(1+\frac{0.03}{12}\right)^{48}-1\right)}{\left(\frac{0.03}{12}\right)}\right) = \$2{,}546.56.$ If you prefer to do the problems in steps, do the following: 1. Find $\frac{r}{m}$ and add 1. 2. Raise $1+\frac{r}{m}$ to the power n and subtract 1. 3. Divide the amount that you found in step 2 by $\frac{r}{m}$. 4. Multiply the quantity that you found in step 3 by R. Note that you deposited 48 payments of 50, so the sum of the deposits was \$2,400. The difference, \$146.56 is earned interest.	1. Matt is saving to buy a new Vespa. If he deposits \$75 at the end of each month in an ordinary annuity that pays an annual interest rate of 6.5%, how much will he have saved in 30 months?

Objective 2 – Perform calculations regarding sinking funds

Key Terms
Sinking fund

Summary
Hopefully we approach annuities with goals in mind! You may want to save regularly to have a fixed amount available for a future purchase, trip, or other splurge! The question is, how much should you save each month to accomplish this? The account that you establish for your deposits is called a sinking fund. Because a sinking fund is a type of annuity we do not need to determine a new formula, but rather need to take the one we had for ordinary annuities, and in this case, solve for the payment amount, R. We recall that the formula is

$$A = R\left(\frac{\left(\left(1+\frac{r}{m}\right)^n - 1\right)}{\left(\frac{r}{m}\right)}\right).$$

Problem Solving: When determining the deposits for a sinking fund, we use the *Relate a New Problem to an Older One Strategy*. The problem solving is very similar, but in this case, we are solving for R with a known value for A.

Notes:

Guided Example	***Practice*** (Exercise 47 p.425)
You and some friends are planning a trip to Europe after graduation. You anticipate needing \$4,000 and the trip would be in 3 years. How much must you deposit monthly into a sinking fund in order to have the desired amount? Assume that the annual interest rate is 3%. We substitute $A = 4{,}000$, $r = 0.03$. $m = 12$ and $n = 36$ into the annuity formula, and solve for R. We obtain $4{,}000 = R\left(\frac{\left(\left(1+\frac{0.03}{12}\right)^{36} - 1\right)}{\left(\frac{0.03}{12}\right)}\right)$, or $4{,}000 = R(37.62056)$ Dividing both sides of the equation by 37.62056 gives us $R =$ \$106.32. Note that if we make 36 payments of \$106.32, we will have deposited \$3,827.52. Thanks to the power of compound interest, we will have \$4,000 in the account at the end of the 3 year period!	1. Sandra Lee is making monthly payments into an ordinary annuity. She wants to have \$600 in the fund to buy a new convection range in 6 months, and the account pays 8.2% annual interest. What are her monthly payments to the account?

Section 8.5 Amortized Loans

Objective 1 – Calculate the payment to pay off an amortized loan.

Objective 2 – Construct an amortization schedule.

Objective 3 – Find the present value of an annuity.

Objective 4 – Calculate the benefit of refinancing a loan.

Objective 1 – Calculate the payment to pay off an amortized loan.

Key Terms
Amortization
Amortized loan

Summary
The process of paying off a loan (plus interest) by making a series of regular, equal payments is called an amortization, and such a loan is called an amortized loan. In order to find the payments on an amortized loan, we use the formula $P\left(1+\frac{r}{m}\right)^n = R\left(\frac{\left(1+\frac{r}{m}\right)^n - 1}{\frac{r}{m}}\right)$ where R is the amount of your payment, P is the amount you borrowed, r is the annual interest rate, m is equal to the number of compoundings per year and n is the number of total compounding periods. We could solve this formula for R or P, but we can shorten the list of formulas by giving this one, and using it to find P or R, given the circumstances of the problem.

Problem Solving: When determining loan amounts, we are using the *Relate a New Problem to an Older One* strategy. The formula on the left hand side of the equation is the compound interest formula from section 8.2; that on the right is the annuity formula from section 8.4. We are putting them together to establish the payments on an amortized loan.

Notes:

Guided Example	***Practice*** (Exercises 3 & 5 p.433)
Congratulations! You are buying a home. In order to make this a reality, you need to borrow \$159,000. The bank has offered a 30 year loan charging 5% annual interest. Calculate the amount of the loan payments and the total of the payments if you were to repay the loan over 30 years. Using the formula $P\left(1+\frac{r}{m}\right)^n = R\left(\frac{\left(1+\frac{r}{m}\right)^n - 1}{\frac{r}{m}}\right)$ we substitute $P = 159{,}000$, $r = 0.05$, $m = 12$ and $n = 360$. $159{,}000\left(1+\frac{0.05}{12}\right)^{360} = R\left(\frac{\left(1+\frac{.05}{12}\right)^{360} - 1}{\frac{0.05}{12}}\right)$ We update the left and right hand side as follows: $710{,}371 = R(832.2586354)$ so $R = \frac{710{,}371}{832.2586354} = \$853.55.$ If you repay this loan over 30 years, your payments will total $(360)(853.55) = 307{,}278$. The finance charge is the difference between this and the amount you borrowed, \$159,000, which is \$148,278.	In the examples below, you are given a loan amount, P, an annual interest rate, r, and the length of the loan in years. Find the monthly payment R necessary to pay off the loan. 1. $P = \$8{,}000$ $r = 8\%$, $t = 10$ years. 2. $P = \$120{,}000$ $r = 6\%$, $t = 30$ years.

Objective 2 – Construct an amortization schedule

Key Terms
Amortization schedule

Summary
When you have an amortized loan, your payments are always the same, but the amount being paid toward the principal and interest changes every month. At the beginning, you pay much more interest than you do towards reducing the principal. At the end of your loan, nearly all of the payment goes to toward principal. In order to see how much of the payment is going to each of these, we often use an amortization schedule. In an amortization schedule (or table), we compute the amount of the payment that is being paid towards the principal using $I = Prt$ with P = unpaid balance on the loan, r = annual interest rate, and $t = \frac{1}{12}$ for one month.

Notes:

Guided Example

You have taken out a loan for a new car. The loan amount was $12,250 and the interest rate is 3%. Your payment amount is $220.12.
Complete the first row of the amortization table below.

Annual Interest Rate	Payment	Interest Paid	Paid on Principal	Balance
3.0%	$220.12			

To determine the amount of interest, we use $I = Prt$ with $P = 12{,}250$, $r = 0.03$ and $t = \frac{1}{12}$.
The amount paid toward interest is
$I = 12,250\left(0.03\right)\frac{1}{12} = 30.63$. The amount paid on principal is $220.12 – $30.63 = $189.49. The new balance is $12,250 – 189.49 or $12,060.51.

Practice (Exercise 15 p.433)

1. We give you the annual interest rate and a line in the amortization table. Complete the next line of the table.

Annual Interest Rate	Payment	Interest Paid	Paid on Principal	Balance
8.4%	$246.01	$32.04	$213.97	$4,362.49
8.4%	$246.01			

Objective 3 – Find the present value of an annuity.

Key Terms
Present value of an annuity

Summary
In a loan situation, the amount that we borrow, *P*, is called the present value of the annuity. If we know the monthly payment, the interest rate, and the number of payments, then we can solve for this unknown value. As a consumer this is an important calculation for you to make! You know best how much you can afford per month; do a little research and see what interest rates are being offered and decide how long you want the loan term to be. Then YOU can figure out how much you can afford to borrow!

Problem Solving: We don't need a new formula! Use the *Relate a New Problem to an Older One* strategy.

Notes:

Guided Example	***Practice*** (Exercise 29 p.434)
After careful consideration, you have decided that you can afford \$200 per month for a car payment. Your credit union is offering a rate of 8% on a 4 year loan. How much can you afford to borrow for your vehicle? We use the formula $P\left(1+\frac{r}{m}\right)^n = R\left(\frac{\left(1+\frac{r}{m}\right)^n - 1}{\frac{r}{m}}\right)$ with $R = 200$, $r = 0.08$, $m = 12$ and $n = 48$. $P\left(1+\frac{0.08}{12}\right)^{48} = 200\left(\frac{\left(1+\frac{0.08}{12}\right)^{48} - 1}{\frac{0.08}{12}}\right)$, and find $P(1.3756661) = 200(56.34991507)$; $(1.3756661)P = 11269.98301$; $P = 8{,}192.38$. Based on your finances, you can borrow \$8,192.38 for your vehicle.	1. If Addison can afford car payments of \$350 per month for 4 years, what is the price of a car she can afford now? Assume an interest rate of 10.8%.

Objective 4 – Calculate the benefit of refinancing a loan.

Key Terms
Refinancing

Summary
At some point, you may receive an offer to refinance your loans; if you take it, you are taking out a second loan to pay off the obligation on the first one. This is particularly attractive if the rate is lower and the term is something you can work with. We will look at some examples to look at the amount you can save. In the examples and exercises on this topic, we will use the table below. It gives you the amount of the loan payment per thousand borrowed for various terms and interest rates.

	Number of Years for the Loan				
Annual Interest Rate	3	4	10	20	30
4%	\$29.53	\$22.58	\$10.12	\$6.06	\$4.77
5%	29.97	23.03	10.61	6.60	5.37
6%	30.42	23.49	11.10	7.16	6.00
8%	31.34	24.41	12.13	8.36	7.34
10%	32.27	25.36	13.22	9.65	8.78
12%	33.21	26.33	14.35	11.01	10.29

Notes:

Guided Example	***Practice*** (Exercise 39 p.434)
1. Compute the monthly payment on a loan for 20 years at 10% if the loan amount is \$10,500. From the table, we see that the payment per thousand is 9.65; with a loan amount of \$10,500 we multiply 9.65 by 10.5 to find \$101.33. 2. After making payments for 10 years, you have an offer to refinance the balance of \$7,667.76 at a rate of 8% for 10 years. What will the payment be if you take this option? Using the table with number of years = 10 and interest rate of 8% we find \$12.13 per thousand so $(12.13)(7.66776) = 93.01$. Taking this option for the next 10 years, you will save $120(101.33 - 93.01) = \$998.40$.	1. In order to buy a vacation home, Neal and Lilly took out a 30-year mortgage for \$120,000 at an annual interest rate of 8%. After 10 years, they refinanced the unpaid balance of \$105,218 at an annual rate of 6%. Determine the monthly payments on the original loan, those on the new loan, and then determine how much will be saved by refinancing.

Section 8.6 Annual Percentage Rate.

Objective 1 – Calculate the annual percentage rate from a table.

Objective 2 – Estimate the annual percentage rate.

Objective 1 – Calculate the annual percentage rate from a table.

Key Terms
Finance charge
Annual Percentage Rate (APR)

Summary
The Truth in Lending Act requires that, prior to consummating a loan transaction, you be given the finance charge and the annual percentage rate (APR). The finance charge is the total amount the borrower pays to use the money. The APR is the "true" interest rate; it is the annual rate, including consideration for any fees paid.

You can use the table below to determine the APR.

	Finance Charge per $100						
	APR						
Number of Payments	10%	11%	12%	13%	14%	15%	16%
6	$2.94	$3.23	$3.53	$3.83	$4.12	$4.42	$4.72
12	$5.50	$6.06	$6.62	$7.18	$7.74	$8.31	$8.88
24	$10.75	$11.86	$12.98	$14.10	$15.23	$16.37	$17.51
36	$16.16	$17.86	$19.57	$21.30	$23.04	$24.80	$26.57
48	$21.74	$24.06	$26.40	$28.77	$31.17	$33.59	$36.03

1. Find the finance charge on the loan.
2. Determine the finance charge per $100 on the loan.
3. Use the line of the table that corresponds to the number of payments to find the number closest to the amount in step 2.
4. The column heading for the column containing the number found in step 3 is the APR.

Notes:

Guided Example	***Practice*** (Exercises 7 & 17 pp. 440 - 441)
You have borrowed \$6,000 to pay for unexpected expenses with an add-on loan charging 8% for 2 years. Calculate the APR. With an add-on loan, your interest over the two year period would be $I = Prt = (6{,}000)\ (0.08)\ (2) = 960$. In order to use the table above, we need the finance charge per \$100 borrowed. To find this, we take the finance charge (interest) found, divide by the amount borrowed (6,000) and then multiply by 100. We find $\frac{960}{6000} \times 100 = 16$. We then look at the table in the row for 24 payments and try to find \$16. It is not there; the closest value is \$16.37 which is in the column for an APR of 15%. You might ask, "how can this be"? The add-on loan has determined loan payments based on owing \$6,000 over 2 years; in truth, as you make payments, you erode the principal and thus those payments (in this case \$290 per month) are being paid on a balance due that is decreasing!	For these exercises, use the table to find the APR to the nearest whole percent. Assume all rates are annual rates. 1. Michael has agreed to pay off a \$3,000 loan by making 24 monthly payments. The total finance charge on his loan is \$420 2. Pete took out a \$4,500 add-on loan to buy a sound system for his band and will repay it by making 48 payments of \$116.50.

Objective 2 – Estimate the annual percentage rate.

Key Terms
None

Summary
In the case of an add-on loan, we can estimate the APR using the formula $\text{APR} \approx \dfrac{2nr}{n+1}$ where r is the annual interest rate and n is the number of payments.

Notes:

Guided Example	***Practice*** (Exercises 11 & 13 p.441)
You have borrowed \$6,000 to pay for unexpected expenses with an add-on loan charging 8% for 2 years. Estimate the APR. Using the formula $\text{APR} \approx \dfrac{2nr}{n+1}$ we find $\text{APR} \approx \dfrac{2(24)(0.08)}{24+1} = 0.1536$. This is consistent with the answer we found in the previous worksheet using the table; we found the interest rate, to the nearest whole percent, to be 15.	Estimate the APR for the add-on loan using the given number of payments and annual interest rate. 1. $n = 36$; $r = 6.4\%$ 2. $n = 42$; $r = 7\%$

Chapter 8 Answers

Section 8.1

Objective 1	1) 0.2735 2) 0.2% 3) 31.25%
Objective 2	1) 6% 2) She divided by 21,065 rather than 19,875.
Objective 3	1) 9.4% 2) 1,269
Objective 4	1) $34,476.75 2) $7,721.25
Objective 5	1) 23.8% 2) 3.242

Section 8.2

Objective 1	1) $240 2) 5% 3) $3,100 4) $1,500
Objective 2	1) $6,381.41 2) $23,457.76
Objective 3	1) 4.56% 2) 10.34 years 3) a. 20.15 b. 20.6

Section 8.3

Objective 1	1) $243.20; $63.47
Objective 2	1) $7.96
Objective 3	1) $5.37
Objective 4	1) $6.77; larger than the charge with average daily balance ($5.37).

Section 8.4

Objective 1	1) $2,435.99
Objective 2	1) $98.31

Section 8.5

Objective 1	1) $97.06 2) $719.46
Objective 2	1)

Payment	Interest Paid	Paid on Principal	Balance
$246.01	$30.54	$215.47	$4,147.02

Objective 3	1) $13,593.02
Objective 4	1) $880.80; $753.36; $30,585.60

Section 8.6

Objective 1	1) 13% 2) 11%
Objective 2	1) 12.45% 2) 13.67%

Chapter 9: Geometry

Section 9.1 Lines, Angles, and Circles

Objective 1 – Solve problems using the basic properties of geometric objects such as points, lines, and planes.
Objective 2 – Work with the fundamental properties of angles.
Objective 3 – Solve problems involving the relationships among angles, arcs, and circles.

Objective 1 – Solve problems using the basic properties of geometric objects such as points, lines, and planes.

Key Terms
Point
Line
Plane
Ray
Line segment
Parallel lines
Intersecting lines

Summary
The foundation for the work we will do in this chapter was set around 300 BC in Euclid's *Elements*. Euclid defined a point as "that which has no part," a line has "length but no breadth," and a plane has "length and breadth only."

A point on a line defines two half lines, as seen below. If the half line has the point included, we call it a ray. A piece of line joining two points and including the points is called a line segment. We don't formally define a plane, but you can think of it as being an infinite two-dimensional surface (like an infinitely large flat sheet of paper). Parallel lines are lines which lie on the same plane but have no points in common; two different lines that lie on the same plane but are not parallel are called intersecting lines.

(a) line

(b) half line

(c) ray

(d) line segment *AB*

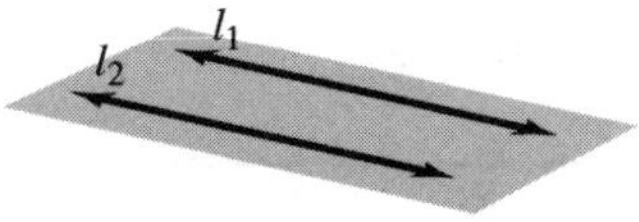

parallel lines

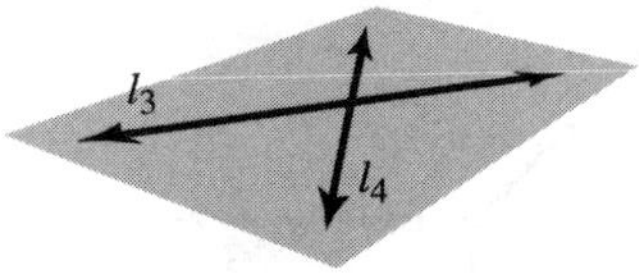

intersecting lines

Notes:

Objective 2 – Work with the fundamental properties of angles.

Key Terms
Angle
Vertex
Initial side
Terminal side
Acute angle
Right angle
Obtuse angle
Straight angle
Vertical angles
Complementary angles
Supplementary angles
Perpendicular lines
Transversal
Alternate interior angles
Alternate exterior angles

Summary
Two rays having a common endpoint form an angle. The common point is called the vertex. As we see in the diagram below, an angle has an initial side and a terminal side. We measure the size of the angle in degrees from the initial side to the terminal side.

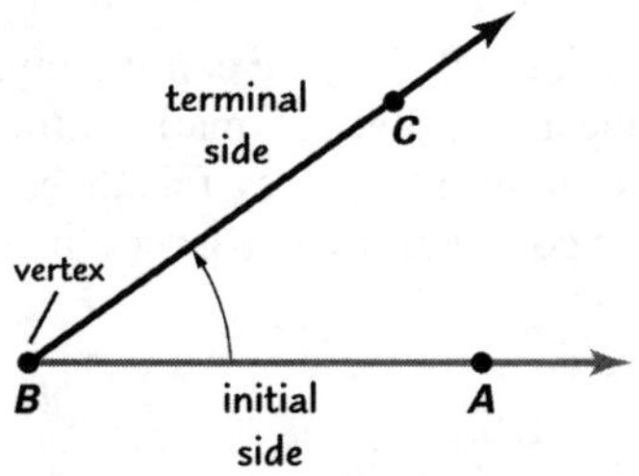

We classify angle types based on the measure of the angle. Angles with degree measure less than 90 degrees are called acute. If the angle measure is 90°, we call the angle a right angle. An obtuse angle is one with angle measure between 90° and 180°, and an angle measuring 180° is called a straight angle. We use $m\angle ABC$ to represent the measure of angle ABC.

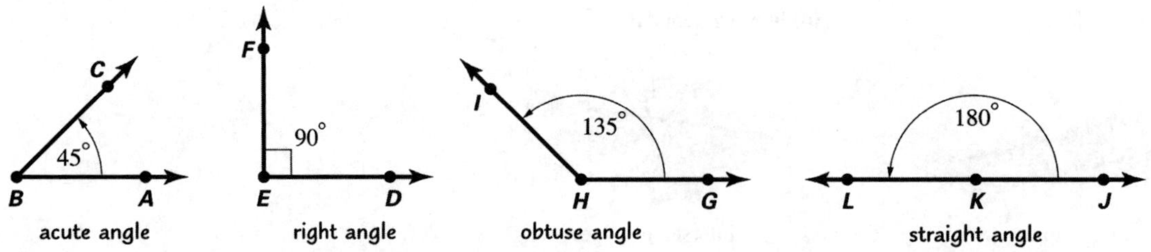

Two angles are called complementary angles if their measures sum to 90° and supplementary if their measures sum to 180°.

If two lines intersect, four angles are formed. The angles which are opposite one another are called vertical angles and have the same measure. If the angles formed when the lines intersect are 90° angles, the lines are defined to be perpendicular lines.

If we intersect a pair of parallel lines with another line not parallel to them, called a transversal, eight angles are formed. There are four pairs of vertical angles (in the diagram on the next page 1 & 4, 2 & 3, 5 & 8, 6 & 7). Angles 1 and 5 are examples of corresponding angles, which have the same degree measure. Angles 2 and 6 are also corresponding angles. Angles which are on the exterior (above the top line or below the bottom line) and on opposite sides are called alternate exterior angles, and they have the same measure. Angles 1 and 8 are alternate exterior angles. If they are on the interior sides of the lines and opposite sides, they are called alternate interior angles, and again have the same measure. Angles 3 and 6 are alternate interior angles. Interior angles on the same side of the transversal, such as 4 and 6 are supplementary angles.

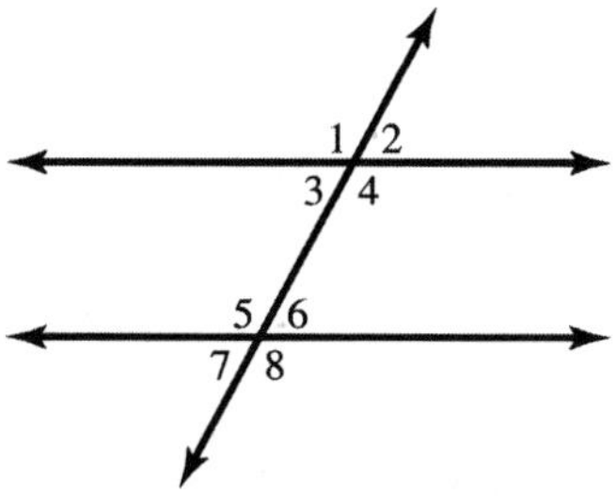

Problem Solving: Rather than trying to memorize all of the angle types and names, use the *Analogies Principle* to help you remember the concepts interior angles, exterior angles, alternate, and opposite.

Notes:

<table>
<tr><th><u>Guided Example</u></th><th><u>Practice</u> (Exercises 1, 3, 5, & 7 p.455)</th></tr>
<tr><td>

Using the diagram below, assume that line l and m are parallel.

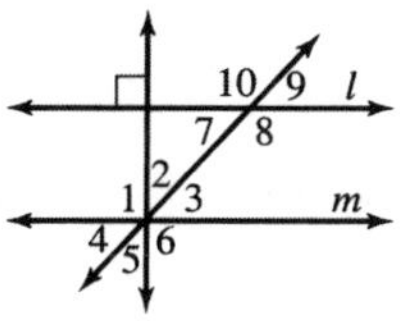

1. Find a pair of alternate exterior angles.

   Alternate exterior angles must be above line l and below m and on opposite sides. Angles 4 and 9 would be one example.

2. What angle, when paired with angle 8, would be vertical angles?

   To be vertical angles, the angles must share the same point as a vertex but be on opposite side of the intersecting lines. Angles 8 and 10 are vertical angles.

3. What is the sum of the measures for angles 8 and 3?

   Angles 3 and 8 are interior angles on the same side of the transversal, so their angle measure sums to 180°. so we could write $m\angle 3 + m\angle 8 = 180°$.

</td><td>

Using the diagram at left, where lines l and m are parallel, find the following:

1. Pair of vertical angles.
2. Alternate interior angles.
3. An Obtuse angle
4. Supplementary angles.

</td></tr>
</table>

Objective 3 – Solve problems involving the relationships among angles, arcs, and circles.

Key Terms

Circle
Radius
Diameter
Circumference
Central angle

Summary

A circle is the set of all points lying on a plane that are located at a fixed distance, called the radius, from a given point, called the center. A diameter of a circle is a line segment passing through the center with both endpoints lying on the circle. The circumference is the distance around the circle.

An angle that has its vertex at the center of a circle is called a central angle. The arc on the circle that the angle creates is proportional to the measure of the angle. We will use this relationship in the guided example below.

Notes:

Guided Example	***Practice*** (Exercises 31 & 33 p.455)
In the circle below, C is at the center. EB and DA are diameters of the circle, and CA is a radius. Answer the following questions assuming that $m\angle ACB = 45°$ and the length of arc AB is 3 inches, and the length of EB is approximately 7.6 inches. 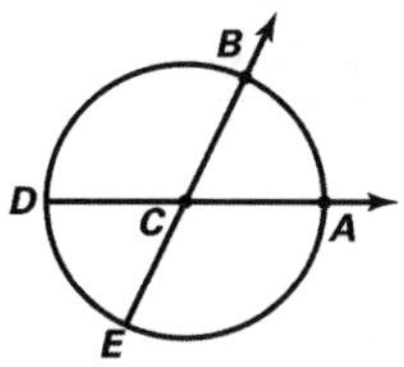 1. What is the length of the radius CA? Since we know that the length of EB is 7.6 inches, the length of any diameter is the same. Thus the length of DA is 7.6 and the radius CA would be half of that, or 3.8 inches. 2. What is the circumference of the circle? Because arc AB is proportional to $m\angle ACB$, we set up the ratio $\frac{m\angle ACB}{360°} = \frac{\text{length of arc } AB}{\text{circumference of circle}}$. Substituting the known information, and c for the unknown circumference, we have $\frac{45°}{360°} = \frac{3 \text{ inches}}{c}$. We cross multiply to solve this ratio, $45c = (360)(3)$, and then divide both sides by 45 to find $c = \frac{(360)(3)}{45} = 24$. The circumference of the circle is 24 inches.	Use the circle at left, with center at C, to answer the questions. 1. Given a circumference of 24 feet and $m\angle ACB = 90°$ find the length of arc AB. 2. Given the circumference = 12 meters and the length of arc $AB = 4$ meters, find $m\angle ACB$.

Section 9.2 Polygons

Objective 1 – Be able to use the basic terminology and properties of polygons to solve problems.

Objective 2 – Solve problems involving angle relationships of polygons.

Objective 3 – Use similar polygons to solve problems.

Objective 4 – Be aware of some differences between Euclidean and non-Euclidean geometries.

Objective 1 – Be able to use the basic terminology and properties of polygons to solve problems.

Key Terms
Closed
Simple
Polygon
Edges
Vertex
Regular
Convex
Triangle (scalene, isosceles, equilateral)
Quadrilateral (trapezoid, parallelogram, rectangle, rhombus, square)

Summary
A polygon is a simple, closed figure in the plane consisting only of line segments, called edges, such that no two consecutive edges lie on the same line. A closed figure is one that we can draw without lifting our pencil, and begin and end at the same vertex. In addition, a simple figure is one where we can draw it without lifting our pencil, and in drawing it we never pass through the same point twice, with the possible exception of the starting and end points. The polygon is regular if all of its edges are the same length and all angles have the same measure. Edges of any polygon meet at a vertex. A polygon is convex if, for any two points *X* and *Y* inside the polygon, the entire line segment *XY* also lies inside the polygon.

Polygons with three sides are called triangles. We classify triangles according to the length of the edges. If all three edges have the same length, the triangle is equilateral. If at least two sides are equal in length, the triangle is isosceles. Triangles with all three sides of different length are called scalene.

Quadrilaterals are polygons with four sides. There are several specific types of quadrilaterals. A trapezoid has one pair of parallel sides; a parallelogram has two pairs of parallel sides. A parallelogram where all sides are equal in length is called a rhombus; in addition if all four angles are right angles we call the quadrilateral a square. A rectangle is a parallelogram with four right angles, but we don't have the requirement that all sides are the same length.

Notes:

Guided Example

1. Is the figure below a polygon? If not, explain what part of the definition fails. If it is a polygon, determine if it is convex.

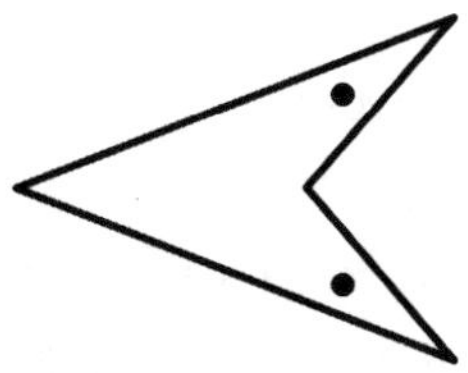

The figure is a polygon but it is not convex. If you draw a segment from the two points drawn in the interior of the polygon, the line segment would not be entirely inside of the polygon.

2. Give an example of a triangle that is neither scalene nor equilateral.

The triangle below has exactly two sides of the same length, so it is not scalene (where all sides would have different lengths) and it is not equilateral (where all sides would have the same length).

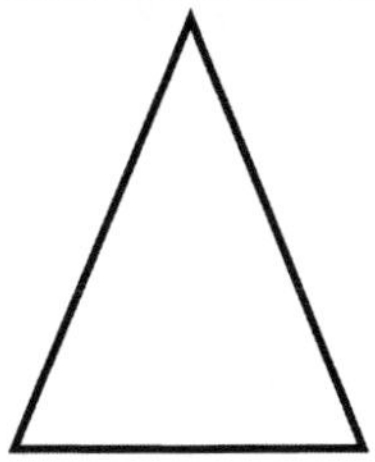

Practice (Exercises 3, 5 & 7 p.465)

1. Is the figure below a polygon? If it is not, explain what part of the definition fails.

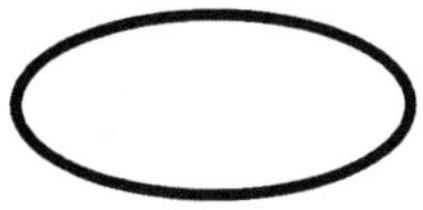

For exercises 2 and 3, provide a counterexample to each statement

2. If a polygon has four equal sides, then it is a square.

3. If a pentagon is convex, then it must be regular.

Objective 2 – Solve problems involving angle relationships of polygons.

Key Terms

Angle sum of a polygon

Summary

The sum of the measures of the interior angles of a convex polygon having n sides is $(n-2)\times 180°$. As an example, in a triangle, which has $n = 3$ sides, the interior angles sum to $(3-2)\times 180° = 180°$. A five sided convex polygon, called a pentagon, would have interior angles that summed to $(5-2)\times 180° = 3\times 180° = 540°$. If the polygon is regular, which means all sides are the same length, then all of the angles will have the same measure. In this case, each interior angle will have measure $\frac{(n-2)\times 180°}{n}$.

Notes:

Guided Example

1. Consider the quadrilateral below. Find the measure of angle D

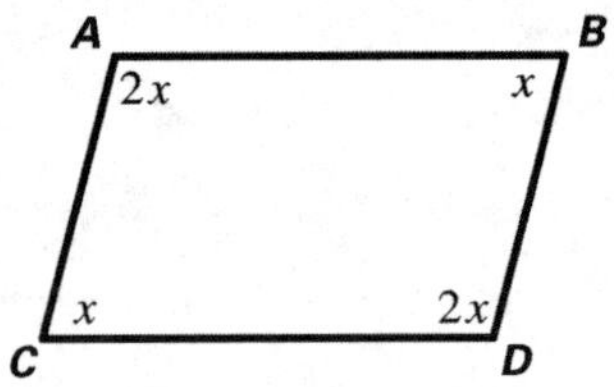

Since the figure is a quadrilateral we know the sum of the interior angles $(4-2)\times 180° = 360°$. The angle measures are given in terms of x; so we have the equation $x + 2x + x + 2x = 360$. Combining like terms yields $6x = 360$ so $x = 60$. Angle D measures $2x = 120°$

2. A 12 sided polygon is called a dodecagon. Find the sum of the interior angles of a dodecagon and the measure of an interior angle of a regular dodecagon.

The sum of the interior angles would be $(12-2)\times 180° = 10\times 180° = 1800°$. If the dodecagon is regular, each angle would be the same, $\frac{(12-2)\times 180°}{12} = \frac{1800°}{12} = 150°$.

Practice (Exercises 11 & 17 p.465)

1. We have indicated the measures of the angles of the triangle below in terms of x. Find the measure of angle A.

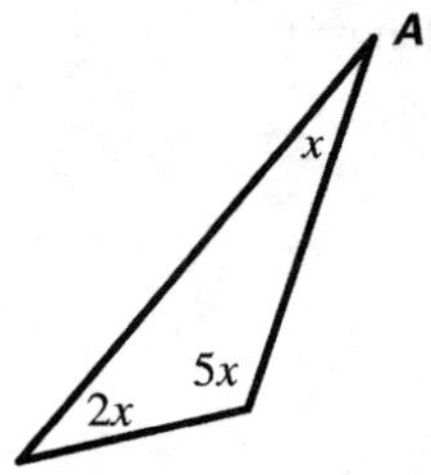

2. What is the measure of an interior angle of a regular 20-sided polygon?

Objective 3 – Use similar polygons to solve problems.

Key Terms
Similar

Summary
Two polygons are similar if their corresponding sides are proportional and their corresponding angles have equal measures. Identifying similar polygons is particularly helpful in solving applied problems. Many of these, as we will see in the guided example, involve modeling.

Notes:

Guided Example

Bob and his friends are lining a field. His friends Carlos and Faiza are standing 9 feet apart; Carlos is standing 5 feet from Bob, and Devin is 8 feet from Carlos, on the same line as Bob and Carlos. Eben is standing on the same line as Bob and Faiza. How far apart are Devin and Eben? Use the diagram below and properties of similar triangles.

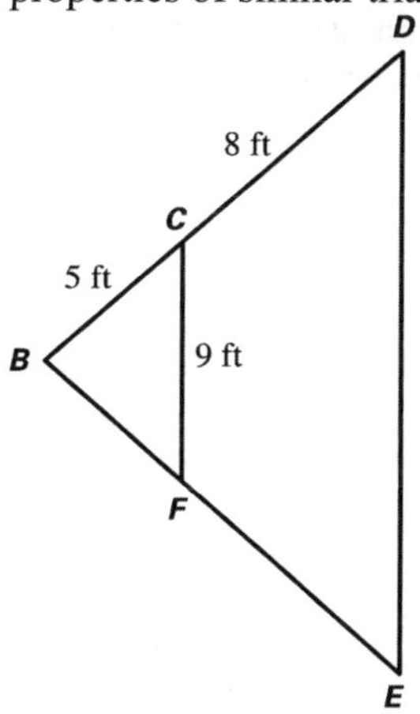

Note that triangles *BCF* and *BDE* are similar, so we can set up the ratio $\frac{\text{length of } BC}{\text{length of } BD} = \frac{\text{length of } CF}{\text{length of } DE}$. Substituting the known values, and assigning x to the length of *DE,* we have $\frac{5}{13} = \frac{9}{x}$. To solve this we cross multiply and get $5x = (13)(9)$. Dividing both sides of the equation by 5 yields $x = \frac{(13)(9)}{5} = 23.4$ ft.

Practice (Exercise 31 p.486)

1. Use similar triangles to determine the width of the river at point *A*.

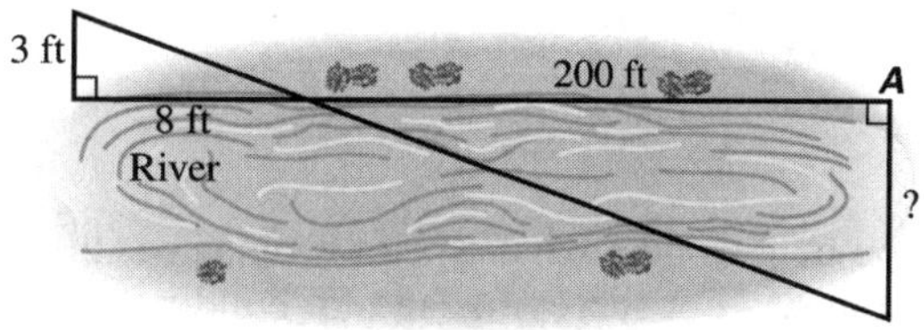

Objective 4 – Be aware of some differences between Euclidean and non-Euclidean geometries.

Key Terms
Euclidean
Non-Euclidean
Great circle

Summary
When you hear the term "geometry" you most likely think of the Euclidean, or planar, geometry we have been working on. In Euclidean geometry, lines can be parallel and the sum of the measures of the interior angles of a triangle is 180°. But what if we aren't looking at a plane?

Imagine taking an un-inflated balloon and drawing a triangle with a marker. What happens when you inflate the balloon? The shape of the triangle has changed, and the sum of the interior angles is more than 180°. What has happened? We are no longer working on a plane! This leads to spherical geometry, where the shortest distance between two points is not a line, but an arc on a great circle (see the diagram below). This is one example of a geometry that is non-Euclidean.

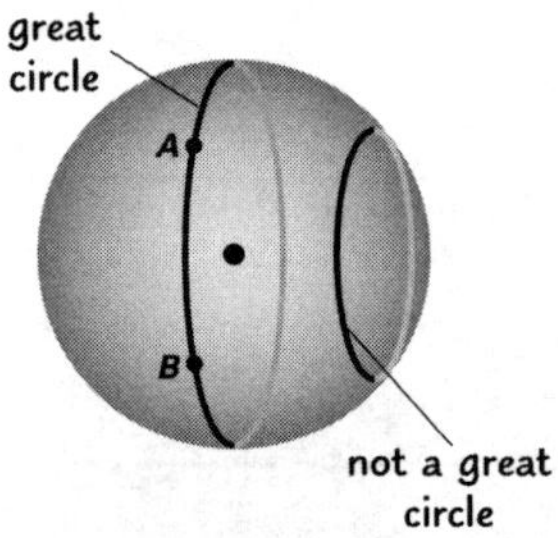

You may wonder what type of geometry is the "right" one. The answer is that there is no one "correct" geometry. For surveying land for building structures, Euclidean geometry is appropriate. To understand why your flight from New York City to Paris is not a straight line, non-Euclidean geometry holds the answer.

Notes:

Section 9.3 Perimeter and Area

Objective 1 – Calculate the perimeter and area of geometric objects.
Objective 2 – Be able to explain how area formulas for geometric figures are related.
Objective 3 – Use the Pythagorean theorem to solve problems involving right triangles.
Objective 4 – Calculate the circumference and area of circles.

Objective 1 – Calculate the perimeter and area of geometric objects.

Key Terms
Perimeter
Area

Summary
We will review the formulas for the perimeter and area of some basic polygons. The perimeter of a polygon is the sum of the lengths of the sides of the polygon. The area is a measure of the amount of surface that the polygon covers.

We will start with some common quadrilaterals. For a rectangle with length l and width w, the perimeter is $P = 2l + 2w$ and the area is $A = l \cdot w$. The area of a parallelogram with height h and base b is $A = h \cdot b$. The area of a trapezoid with lower base b_1, upper base b_2 and height h has area $A = \frac{1}{2}(b_1 + b_2) \cdot h$

A triangle with height h and base b, has area $A = \frac{1}{2} h \cdot b$. The height is perpendicular to the base; if we don't know the height, we can use the lengths of the three sides, a, b, and c to find the area of the triangle. We first compute $s = \frac{1}{2}(a + b + c)$ and then find the area using Heron's formula, $A = \sqrt{s(s-a)(s-b)(s-c)}$.

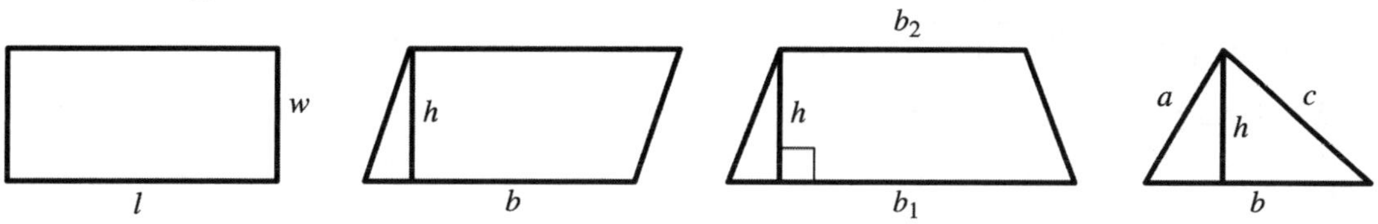

Notes:

Guided Example	***Practice*** (Exercises 3, 5, & 7 p.478)
In the figure below 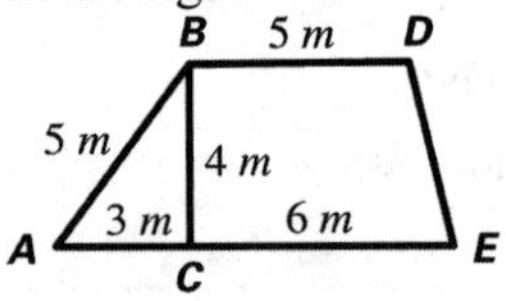 1. Find the area of ΔABC using $A=\frac{1}{2}h\cdot b$ and verify using Heron's formula. ΔABC has height h = 4m and base b = 3m. Substituting into the formula, we find $A=\frac{1}{2}h\cdot b=\frac{1}{2}(4)(3)=6$, so the area is 6 m^2. To use Heron's formula, we must first find s. We will assign b = 3, a = 5, c = 4 in the formula for s to obtain $s=\frac{1}{2}(a+b+c)=\frac{1}{2}(5+3+4)=6$. This is not the area; for that we need to substitute s into $A=\sqrt{s(s-a)(s-b)(s-c)}$. We find $A=\sqrt{6(6-5)(6-3)(6-4)}$ Which simplifies to $A=\sqrt{36}=6$ m^2. This verifies our answer for the area. 2. Find the area of trapezoid $BDEA$. The trapezoid has h = 4m, b_1 = 9m and b_2 = 5m. Substituting into the formula $A=\frac{1}{2}(b_1+b_2)\cdot h$ we obtain $A=\frac{1}{2}(9+5)\cdot 4=\frac{1}{2}(56)=28$ m^2.	Find the area of each figure. 1. 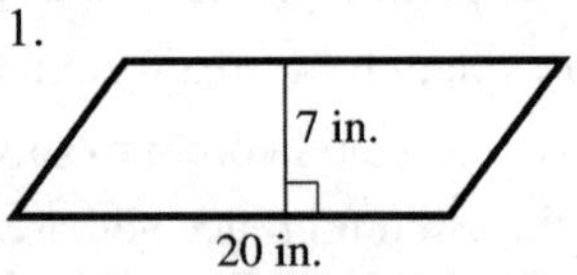 2. 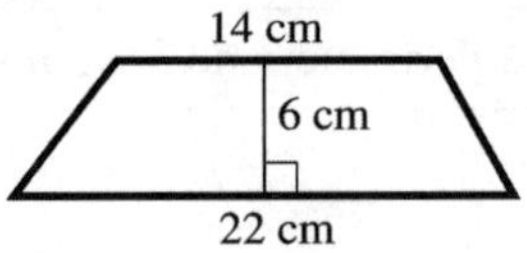 3. 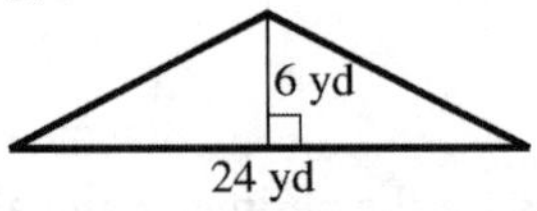

Objective 2 – Be able to explain how area formulas for geometric figures are related

Key Terms
None

Summary
In the previous worksheet, we used formulas for the area of a rectangle, parallelogram, and triangle.

Looking at the figure below of the parallelogram with base b and height h, we can see why the area of the parallelogram is the same as the area of a rectangle. We can think of cutting off the triangle on the left, sliding it over, and attaching it to the right hand side of the parallelogram, yielding a rectangle. We know the area of this rectangle is $h \cdot b$ so we have the area of the parallelogram!

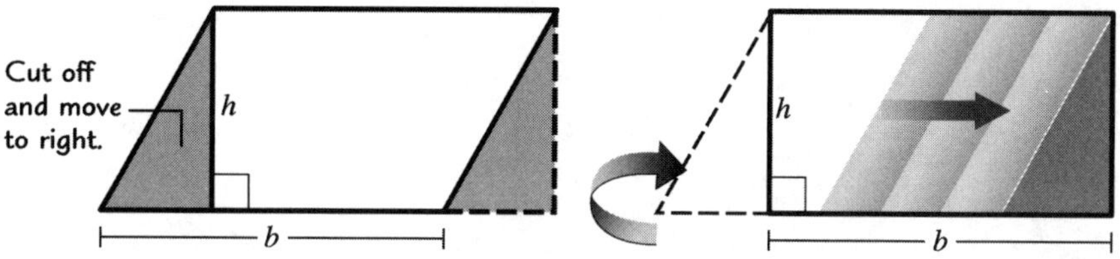

We can derive the area of a triangle by taking a parallelogram and dividing it in half, as seen in the diagram. Triangle ABC is half of the parallelogram with area $h \cdot b$ so its area is $\frac{1}{2}h \cdot b$.

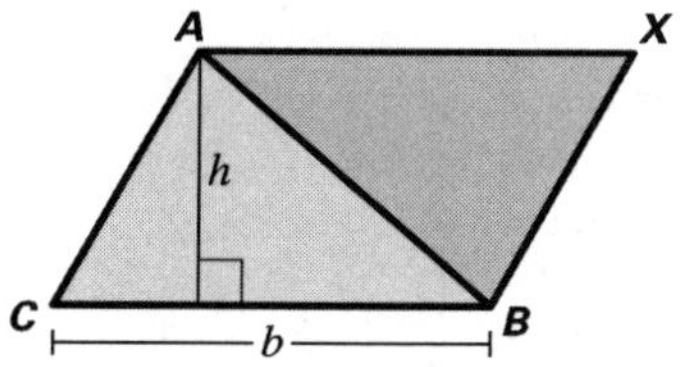

Notes:

Objective 3 – Use the Pythagorean theorem to solve problems involving right triangles.

Key Terms
Pythagorean Theorem

Summary
One of the most famous theorems in mathematics is the Pythagorean theorem. The Pythagorean theorem states that the sum of the squares of the lengths of the legs of a right triangle equals the square of the length of the hypotenuse, which is the side opposite the 90° angle in the triangle.

Notes:

Guided Example

Use the Pythagorean theorem to find the length of the side, x.

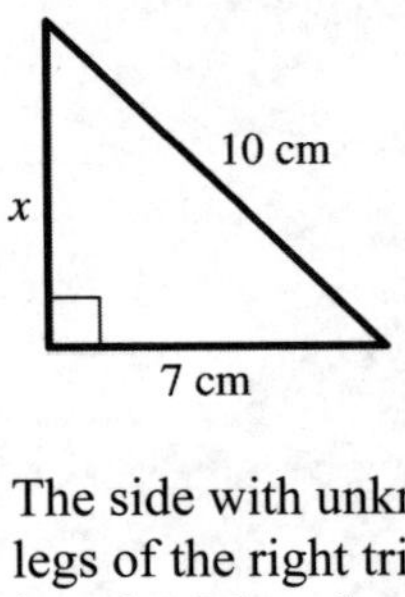

The side with unknown length, x, is one of the legs of the right triangle. We substitute the length of the other leg and the hypotenuse into the Pythagorean formula,

$x^2 + 7^2 = 10^2$; $x^2 + 49 = 100$

$x^2 = 51$, so $x = \sqrt{51} \approx 7.14$m.

Practice (Exercise 27 p.479)

1. Find the length of side x in the triangle below.

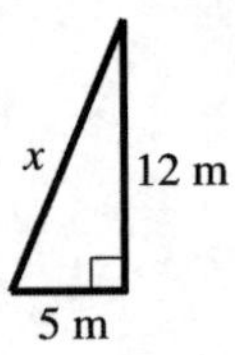

Objective 4 – Calculate the circumference and area of circles.

Key Terms

Circumference

Area

Summary

The formulas for the circumference and area of a circle involve the number pi, which we represent using the Greek letter π. Pi is equal to the ratio of the circumference to the diameter of any circle. We approximate pi by 3.14 when doing our calculations.

A circle with radius r has circumference $C = 2\pi r$ and area $A = \pi r^2$.

Notes:

Guided Example

Consider the following figure, where the circle is inside a square with side length 6 meters. Find the circumference and area of the circle. Then find the area of the shaded region.

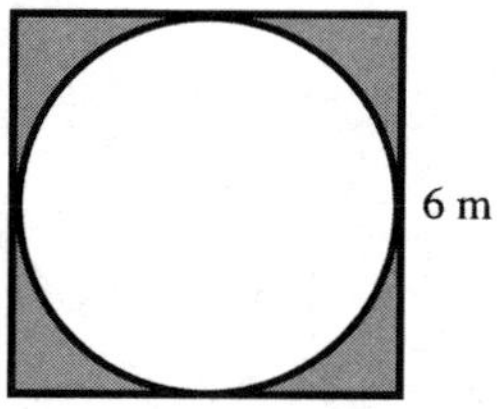

The diameter of the circle would be 6 meters so the radius $r = 3$. The circumference of the circle is $C = 2\pi r \approx 2(3.14)(3)$ which is 18.84 meters.

The area of the circle would be $A = \pi r^2 \approx (3.14)(3^2) = 28.26$ square meters.

The shaded region would have area $36 - 28.26 = 7.74$ square meters. We first find the area of the square and then subtract the area of the circle that we just found.

Practice (Exercises 9 & 67 p.478 & p. 482)

1. Find the area of the circle below.

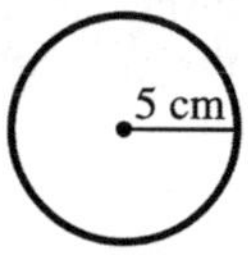

2. A design for a rectangular Japanese stone garden has a circular pond it its center as shown. How many square yards will be covered with stones?

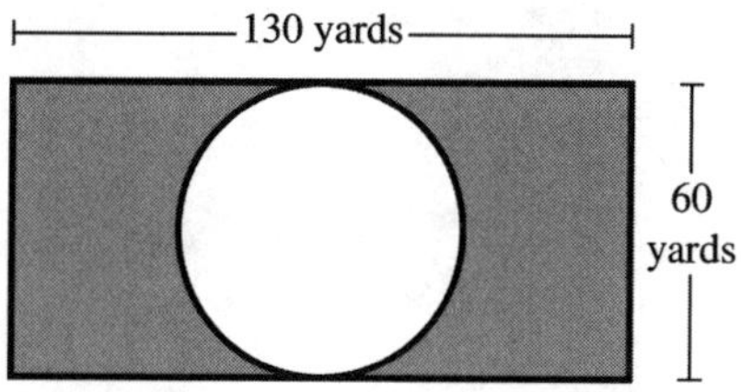

Section 9.4 Volume and Surface Area

Objective 1 – Use the formulas for the volume of three-dimensional objects to solve problems.

Objective 2 – Be able to apply the volume and surface area formulas for cylinders.

Objective 3 – Apply the relationship between the volume and surface area formulas for cylinders, spheres, and cones to solve real-world problems.

Objective 1 – Use the formulas for the volume of three-dimensional objects to solve problems.

Key Terms
Volume
Surface area

Summary
We now consider three dimensional figures and give formulas for the volume and surface area. The volume of a three-dimensional figure represents the amount of space that it occupies (you can also think about the amount of material that could fill the figure). Looking first at a rectangular solid with length l, width w, and height h, the volume is $V = lwh$. Generally speaking, if an object has a flat top and base and sides perpendicular to the base, then the volume is $V = A \cdot h$, where A is the area of the base and h is the height. Note that in a rectangular solid, the area of the base is given by $A = lw$.

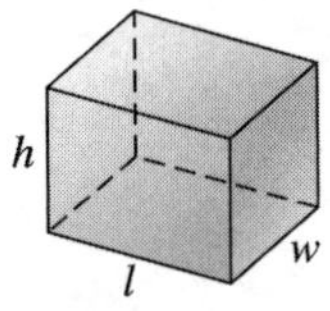

The surface area of a three-dimensional figure is the measure of the outermost part of the figure. In the case of a rectangular solid, the surface area is $S = 2lw + 2lh + 2wh$.

Problem solving: when finding the surface area of a rectangular solid, we *Relate a New Problem to an Older One*. We go back to our formula for the area of a rectangle and apply it to all six sides of the solid.

Notes:

Guided Example

1. What is the volume of a rectangular cake with a length of 9 inches, width 5 inches, and height 4 inches?

   The volume would be
   $V = lwh = 9 \cdot 5 \cdot 4 = 180$

   We measure volume in cubic units, so the volume is 180 in^3.

2. If you going to dip the cake in melted chocolate, covering all sides, what is the surface area covered in chocolate?

   We use the surface area formula,
   $S = 2(9)(5) + 2(9)(4) + 2(5)(4)$,
   and find the surface area covered in chocolate is 202 in^2.

Practice (Exercises 1 & 11 p.490)

1. Find the surface area and volume of the figure.

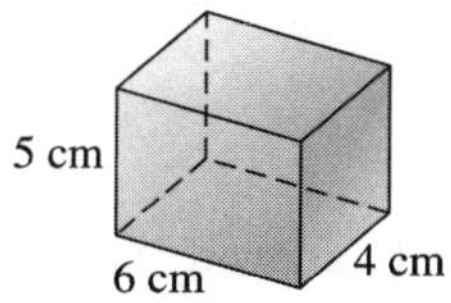

2. Find the volume of the figure.

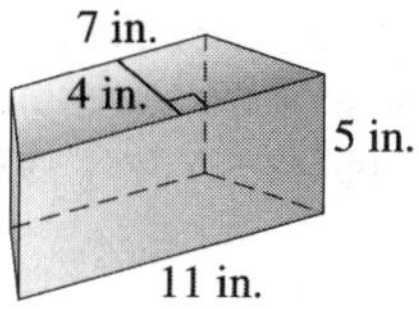

Objective 2 – Be able to apply the volume and surface area formulas for cylinders.

Key Terms
Cylinder

Summary
A right circular cylinder is a figure with a circular base and top, with sides perpendicular to the base. A soup can is a good example of such a figure. A right circular cylinder with radius r and height h has volume $V = \pi r^2 h$ and surface area $S = 2\pi rh + 2\pi r^2$.

Problem Solving: Since a right circular cylinder has a top and bottom in the shape of a circle, the volume formula is an application of area of the base times the height as we saw with rectangular solids. We are using the *Relate a New Problem to and Older One* strategy.

Notes:

Guided Example	***Practice*** (Exercise 23 p.490)
Many Bluetooth speakers are in the shape of a right circular cylinder. Find the volume and surface area of such a speaker with radius 2 inches and height 5 inches. Compare the volume to that of a speaker with a triangular base with area 11 in^2 and sides perpendicular to the base with height 7 in. The speaker in the shape of a right circular cylinder has volume $V = \pi r^2 h \approx (3.14)(2)^2(5)$ which is 62.8in^3. The surface area of this speaker is $S = 2\pi rh + 2\pi r^2 \approx 2(3.14)\cdot 2\cdot 5 + 2(3.14)\cdot 2^2$ $= 87.92\ \text{in}^2$. The speaker with a triangular base and height 7in has volume $V = A\cdot h = (11)(7) = 77\ \text{in}^3$. The speaker with the triangular base has a larger volume than that with a circular base.	1. One cake is rectangular with a length of 14 inches, a width of 9 inches, and a height of 3 inches. The other cake is a round cake with a radius of 5 inches and a height of 4 inches. Which cake has more volume?

Objective 3 – Apply the relationship between the volume and surface area formulas for cylinders, spheres, and cones to solve real-world problems.

Key Terms

Right circular cone
Sphere

Summary

A right circular cone is a figure whose base is a circle, the figure is in the shape of a cone, and the line segment from the tip to the center of the base is a right angle. See the figure below for an example of a right circular cone.

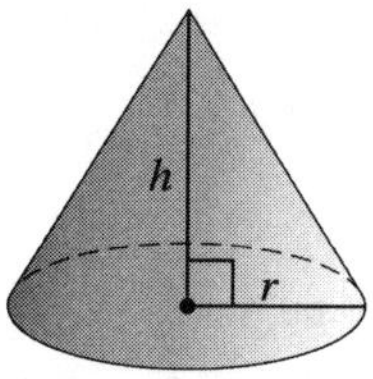

The volume of the cone is $V = \frac{1}{3}\pi r^2 h$ and the surface area is $S = \pi r\sqrt{r^2 + h^2}$. If we wish to include the area of the base in the surface area, then we add πr^2 to the formula for *S*, obtaining $S = \pi r\sqrt{r^2 + h^2} + \pi r^2$.

The final three-dimensional figure we consider is a sphere of radius *r*. The volume of the sphere is $V = \frac{4}{3}\pi r^3$ and the surface area is $S = 4\pi r^2$.

Notes:

Guided Example

The right circular cone below has radius 2m and height 5m. Find the surface area (excluding the base) and volume of the cone.

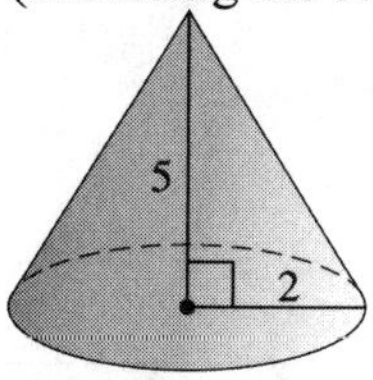

The surface area is $S = \pi r\sqrt{r^2 + h^2}$.
Substituting $r = 2$ and $h = 5$ we find
$S = \pi \cdot 2\sqrt{2^2 + 5^2} \approx (3.14)(2)\sqrt{29} \approx 33.8 \text{ m}^2$.

The volume of the cone is $V = \frac{1}{3}\pi r^2 h$.

Substituting as above, we find
$V = \frac{1}{3}\pi 2^2 \cdot 5 \approx \frac{1}{3}(3.14) \cdot 4 \cdot 5$ or approximately 20.9 cubic meters.

Practice (Exercises 7 & 19 p.490)

1. Find a) the surface area and b) the volume of the figure below.

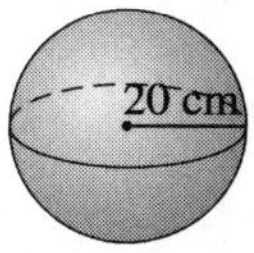

2. A cylindrical pitcher with radius 2.5 inches and 8 inches high is filled with a tropical drink. How many glasses shaped like an inverted cone with a height of 1.5 inches and a radius of 2 inches can be filled from this pitcher?

Section 9.5 The Metric System and Dimensional Analysis

Objective 1 – Be able to use basic units of measurement in the metric system.

Objective 2 – Make conversions between metric measurements.

Objective 3 – Use dimensional analysis to make conversions between different measurement systems to solve applied problems.

Objective 1 – Be able to use basic units of measurement in the metric system.

Key Terms

Meter
Liter
Gram
Kilo-
Hecto-
Deka-
Deci-
Centi-
Milli-

Summary

Outside of the United States, most of the world uses the metric system of measurement. In the metric system, the base unit of length is the meter, which is slightly more than a yard. The basic metric for weight is the gram, which is about the weight of a paper clip. The unit of measurement for volume is the liter, which is slightly more than a quart.

Although measurement in the metric system is based on meters, liters, and grams, we often use variations of these basic units to measure different quantities. The table below shows how the prefixes kilo-, hecto-, deka- deci-, centi-, and milli- impact the base unit.

kilo-(k)	hecto-(h)	deka-(da)	base unit	deci-(d)	centi-(c)	milli-(m)
× 1,000	× 100	× 10		× 1/10 or × 0.1	× 1/100 or × 0.01	× 1/1,000 or × 0.001

Notes:

Guided Example	***Practice*** (Exercises 17 & 19 p.501)
One thousand grams is about 2.2 pounds. If you were at the candy shop outside Buckingham Palace in London buying fudge, would you be more likely to order 5 grams or 500 grams? Since 1,000 grams is about 2.2 pounds, if you were purchasing 500 grams you would have just over 1 pound of fudge. If you ordered 5 grams, you would just be getting the crumbs!	Pick the most appropriate measurement for each of the following items. 1. The weight of books in your backpack. a. 500 g b. 80 hg c. 6 kg 2. The volume of a bottle of wine. a. 0.25L b. 0.2kL c. 750mL

Objective 2 – Make conversions between metric measurements.

Key Terms
None

Summary
When converting from one unit of measurement in the metric system to another, it is important to think about your answer. Are you moving from a larger unit to a smaller unit, in which case the answer will be larger? If you are moving from a smaller unit to a larger unit, the converted number of units will be smaller.

It may help to think about your conversion in terms of the way you are moving in the table below. Each time you move from one unit to another to the right, you require 10 times the number of objects, so you multiply by 10. Moving in the opposite direction, you divide the number of objects by 10.

kilo-(k)	hecto-(h)	deka-(da)	base unit	deci-(d)	centi-(c)	milli-(m)

Notes:

Guided Example	***Practice*** (Exercises 1, 3 & 5 p.501)
1. Convert 1.24 kilograms to decigrams. Let's first convert from kilograms to grams. Since 1000 grams are in a kilogram, there are 1240 grams in 1.24 kilograms. Now, to convert from grams to decigrams, we multiply by 10 yielding 12,400 decigrams in 1.24 kilograms. Note that we could have started with 1.24 and multiplied by 10^4since decigrams are 4 steps to the right of kilograms. 2. Convert 32,485 milligrams to hectograms. Moving from milligrams to hectograms in the table above requires 5 steps to the left. Since our answer must be smaller, we divide 32,485 by 10^5= 100,000 to obtain 0.32485 hectograms.	Make the following conversions: 1. 2.4 kiloliters to deciliters. 2. 28 decimeters to millimeters. 3. 3.5 dekaliters to deciliters.

Objective 3 – Use dimensional analysis to make conversions between different measurement systems to solve applied problems.

Key Terms
Unit fraction

Summary
In order to make conversions between different measurement systems, we need to create unit fractions, where the numerator and denominator have different units of measurement but the value is equal to 1. An example of a unit fraction is $\frac{5,280 \text{ feet}}{1 \text{ mile}}$. We will focus on conversions concerning length, but similar calculations could be done for mass and volume.

Example: Use the unit fraction above to determine how many yards you walked if you walked one-half mile.

Using the unit fractions, we need one converting feet to miles and another from yards to feet. We then multiply to find our answer; $0.5 \cancel{\text{mile}} \cdot \frac{5,280 \cancel{\text{feet}}}{1 \cancel{\text{mile}}} \cdot \frac{1 \text{ yard}}{3 \cancel{\text{feet}}} = 880 \text{ yards}$. Note that the units cancel when we set up our unit fractions correctly.

In order to do the examples and exercises, you will need to know some key relationships which are summarized below.

1 inch = 2.54 centimeters
1 foot = 30.48 centimeters
1 yard = 0.9144 meter
1 mile = 1.6 kilometers

Notes:

Guided Example

You have purchased an insecticide to spray on your lawn to kill ticks. The package says the product will cover an area of 300 square meters. Your yard is a rectangle with length 75feet and width 40 feet. Will one package be sufficient to treat your lawn?

We will use the information above, 1 yard = 0.9144 meter to find the number of square meters in one square foot. Using unit fractions we find

$1\ \cancel{\text{foot}} \cdot \frac{1\ \cancel{\text{yard}}}{3\ \cancel{\text{feet}}} \cdot \frac{0.9144\ \text{meter}}{1\ \cancel{\text{yard}}} = 0.3048\ \text{meter}$. We can then establish that 1 square foot = $(0.3048)^2$ or 0.09290304 square meters.

Your yard is $(70)(45) = 3{,}150$ square feet, so converting to square meters we have

$3{,}150\ \cancel{\text{ft}^2} \cdot \frac{0.09290304\ \text{m}^2}{1\ \cancel{\text{ft}^2}} \approx 292.64\ \text{m}^2$. It looks like you will have just enough for your lawn.

Practice (Exercises 43 & 49 p.502)

1. The highest mountain in the world is Mount Everest, which is 8,850 meters high. How high is it in feet?

2. In *The Amazing Race*, Rachel is driving a car that shows speed in both miles per hour and kilometers per hour. If she is traveling in Italy at 80 kilometers per hour, what is her speed in miles per hour?

Section 9.6 Geometric Symmetry and Tessellations

Objective 1 – Be able to solve problems using the relationship between symmetry and rigid motions.

Objective 2 – Recognize the symmetries of an object.

Objective 3 – Determine when it is possible to tessellate a plane with polygons.

Objective 1 – Be able to solve problems using the relationship between symmetry and rigid motions.

Key Terms

Rigid motion
Reflection
Translation
Translation vector
Glide reflection
Rotation

Summary

A rigid motion is the action of taking a geometric object in the plane and moving it in some fashion to another position in the plane without changing its shape or size. Every rigid motion is essentially a reflection, a translation, a glide reflection, or a rotation.

A reflection is a rigid motion in which we move an object so that the ending position is a mirror image of the object in its starting position. In a reflection, there is a line, called the axis of reflection, that acts as a mirror to transform the figure from its original position to its final position.

A translation is a rigid motion in which we move a geometric object by sliding it along a line segment in the plane. The direction and length of the line segment completely determine the translation. We represent the distance and the direction of a translation by a line segment with an arrow on it, called the translation vector.

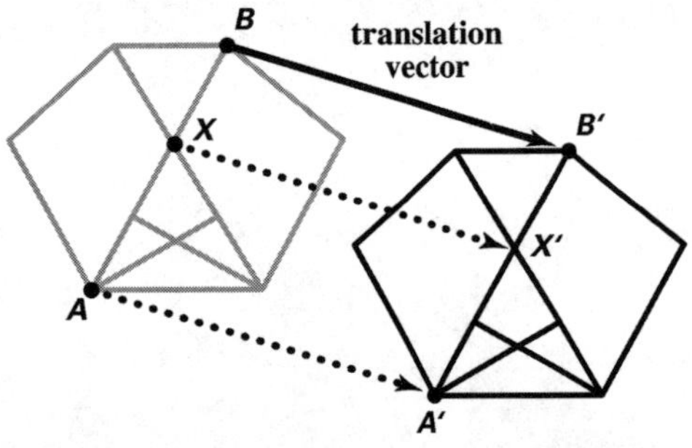

A glide reflection is a rigid motion formed by performing a translation (the glide) followed by a reflection.

Problem Solving: Use *The Order Principle* – performing a translation and then a reflection is not the same as if we reflect first and then translate.

A rotation is done when we first select a point, called the center of the rotation, and then while holding this point fixed, rotate the plane about this point through an angle called the angle of rotation.

Notes:

Guided Example

Perform the indicated glide reflection on figure A.

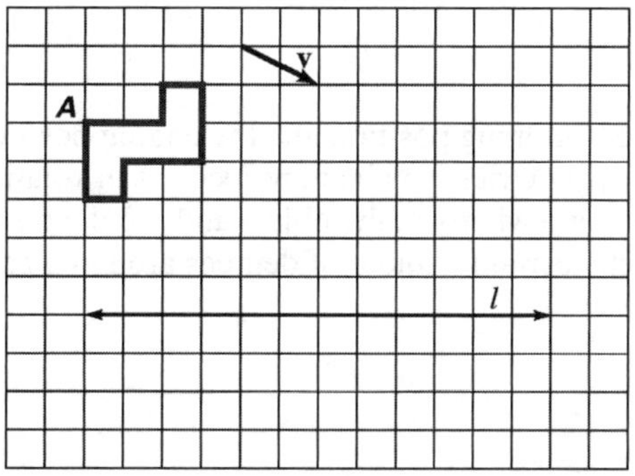

Using the Order Principle, we first glide the figure as shown below. Note that each corner of the figure moves down one unit, and to the right 2 units. We then reflect the figure in line *l*.

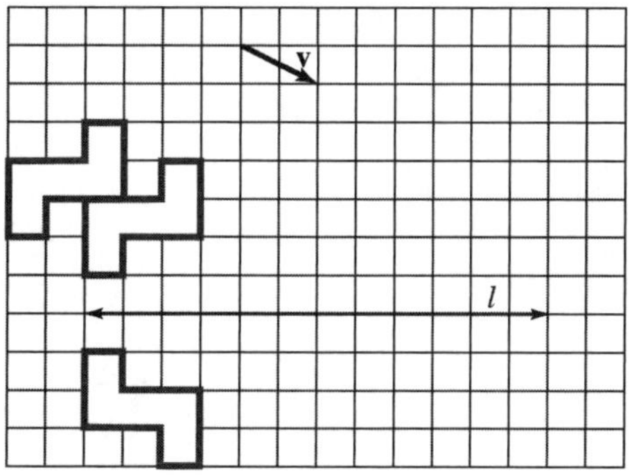

Practice (Exercise 7, 11, 17 p.515)

1. Reflect figure *A* about line *l*. Call the resulting figure *A'*. Then reflect figure *A'* about line *m*. Call the resulting figure *A''*.

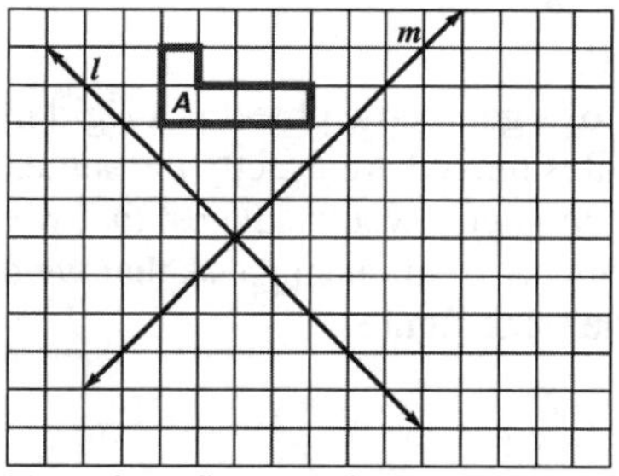

2. Translate figure *A* by vector **w**.

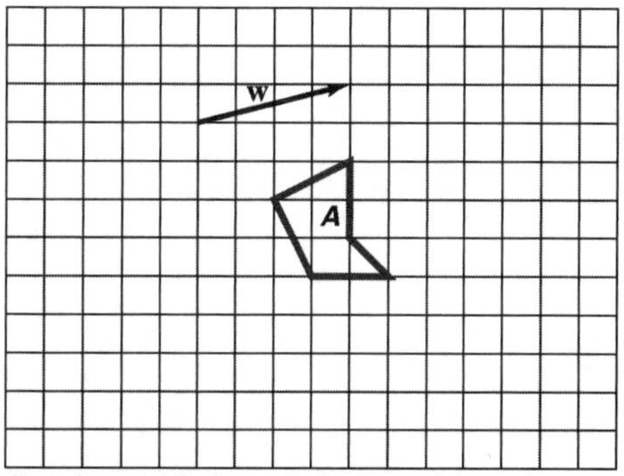

3. Starting with the original position shown, rotate each shape about point P for each of the following rotations: 45º; 90º; 180º.

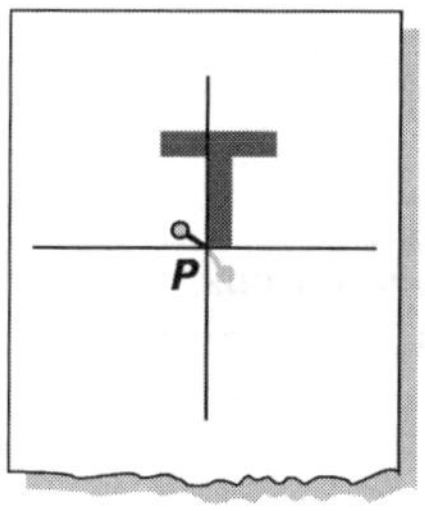

Objective 2 – Recognize the symmetries of an object.

Key Terms
Symmetry

Summary
A symmetry of a geometric object is a rigid motion such that the beginning position and the ending position of the object that is moved are exactly the same. We will consider symmetries which are reflections or rotations. A reflection symmetry would suggest that we could put a mirror somewhere on the object and see the other half in the reflection. A rotation suggests that we can rotate the object a certain number of degrees around a point and obtain the same figure.

Notes:

Guided Example	***Practice*** (Exercise 39 page 516)
Consider the object below. Discuss any reflectional or rotational symmetries. 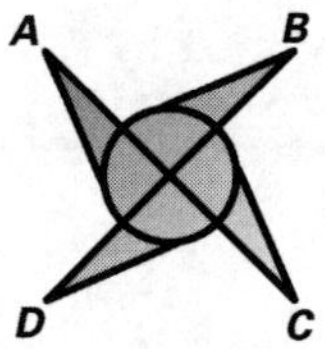 We cannot reflect the figure in any axis and have the same figure on the other side. This is like saying that we can't place a mirror anywhere and have the hidden portion of the figure come up in our mirror. There are no reflectional symmetries. For rotations, we can fix the center point, and rotate the figure 1/4 turn, 1/2 turn, or 3/4 turn and the same object is shown. This means that we have 90º, 180º, and 270º rotational symmetries.	1. List all of the reflectional symmetries for the object below. Also find all rotational symmetries of the object using angles between 1º and 359º.  (Olympic symbol)

Objective 3 – Determine when it is possible to tessellate a plane with polygons

Key Terms
Tessellations
Regular tessellation

Summary
A tessellation (also called a tiling) of the plane is a pattern made up entirely of polygons that completely cover the plane. The pattern must have no holes or gaps, and polygons cannot overlap except at their edges. A regular tessellation consists of regular polygons of the same size and shape such that all vertices of the polygons touch other polygons only at their vertices. The only regular tessellations that can be done are with equilateral triangles, squares, and regular hexagons. Consider the tessellations in the figure below; the measure of an interior angle of the regular polygon is shown.

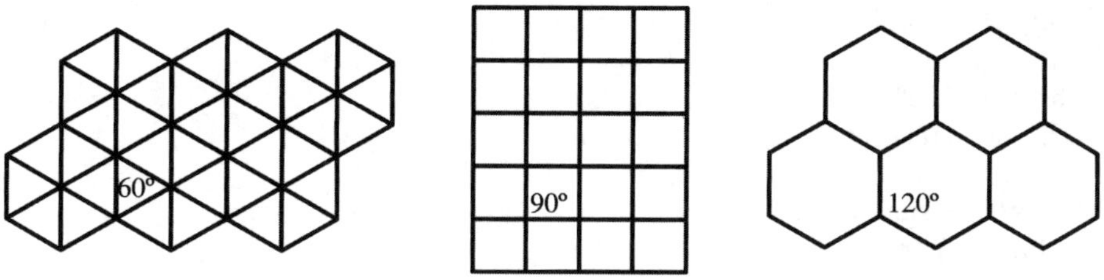

If we look at the tessellation with the regular hexagon (six-sided polygon), each interior angle measures 120°. At the point where three hexagons meet, the angle sum around the point measures 360°.

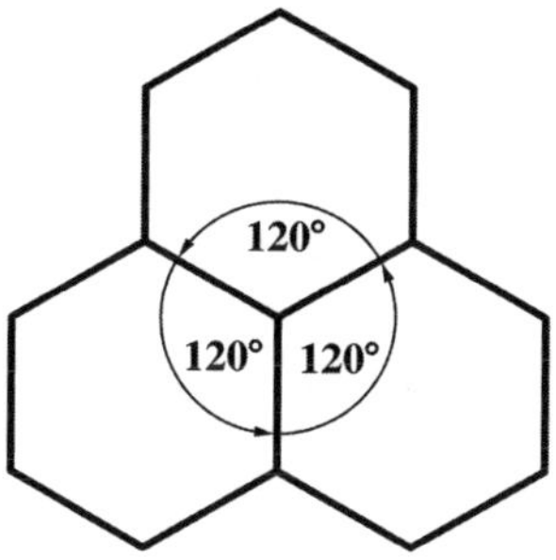

What if we were to combine different polygons? We could also consider polygons which aren't regular, but have the same side length (recall that to be regular, all angles of the polygon must have the same measure). In order to address this question, we need to think about ways we can arrange the polygons so that the sum of the interior angles around a point measures 360°.

Problem Solving: In order to determine if the polygons can be arranged around a point with an interior angle sum of 360° we will *Relate a New Problem to and Older One* and use the fact that the sum of the interior angles of a convex polygon with *n* sides is $(n-2)\times 180°$ and the interior angle of a regular polygon measures $\frac{(n-2)\times 180°}{n}$

Notes:

Guided Example	***Practice*** (Exercise 27 p.516)
Is it possible to tessellate the plane using just regular octagons and squares? If so, how many of each would meet at each vertex? To answer this question, we need to think about ways we can arrange the polygons so that at any point where they meet, the angles around the point sum to 360º. The interior angle of a regular octagon measures $\frac{(8-2)\times 180°}{8} = 135°$ If we had one octagon at each vertex, we would need to make up the remaining $360° - 135° = 225°$ entirely with angles from the square. We cannot do this, however, because 225º is not divisible by 90, which is the angle measure for a square. What if we had two octagons? The sum of the interior angle measure for the octagons would be $2(135°) = 270°$, and the remaining angle measure would be $360° - 270° = 90°$, precisely the angle measure for one square. Thus we can tessellate the plane with two regular octagons and a square with side measure equal to the side measure of the octagon, as shown below. 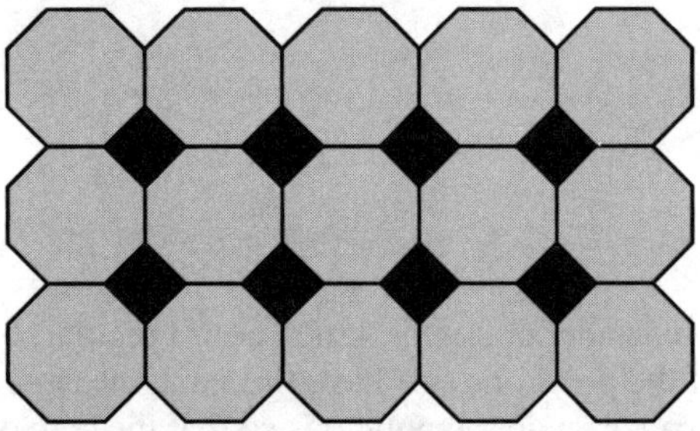	1. Explain why we can tessellate the plane with a regular hexagon, but not with a regular pentagon.

Section 9.7 Fractals

Objective 1 – Be able to describe the self-similar nature of fractals.
Objective 2 – Calculate the length and area of fractal objects.
Objective 3 – Compute fractal dimension.

Objective 1 – Be able to describe the self-similar nature of fractals.

Key Terms
Fractal geometry
Self-similar

Summary
Imagine looking at the coastline from an airplane as you are descending. As you get closer and closer to the ground, the shape of the coastline and inlets may repeats itself, but on a different scale. A newer type of geometry, called fractal geometry enables us to describe this phenomenon more accurately than can be done using Euclidean geometry. With a fractal object, as we magnify the object's picture, we see patterns that are similar to the patterns present in the overall object. An example of a fractal object is the Koch curve, pictured below. Note that smaller portions of the Koch curve repeat the same pattern that was present in the original curve.

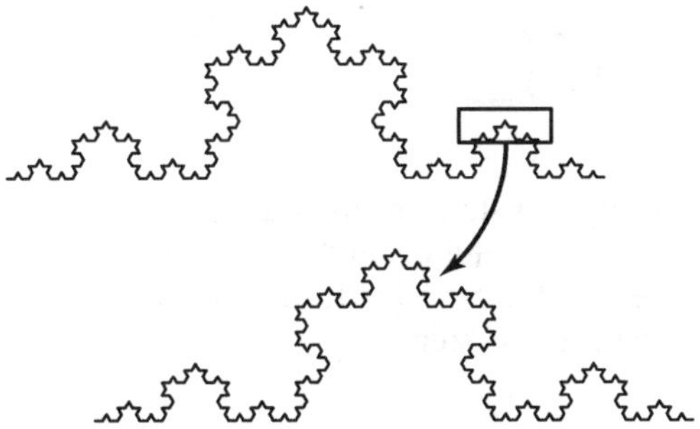

You can also see the self-similar property in the picture of the rivers in Norway.

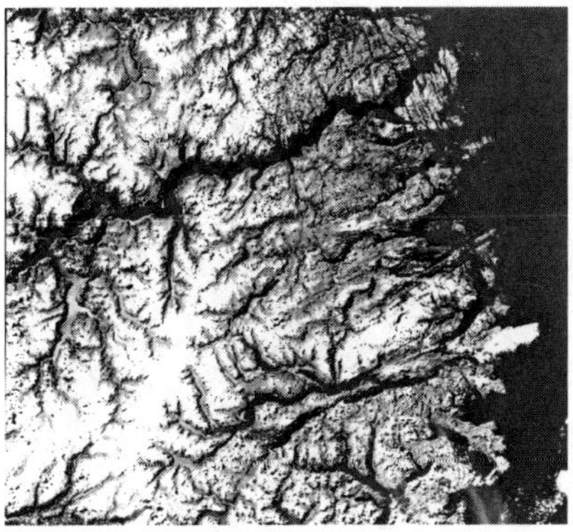

Notes:

Objective 2 – Calculate the length and area of fractal objects.

Key Terms

None

Summary

Just as in Euclidean geometry, length and area of fractal objects is of interest. Unlike Euclidean geometry, however, in fractal geometry we measure the length or area as steps in the construction of the object. We can then make a conjecture about the length or area of the entire object.

Notes:

Guided Example

Below are steps 0 and 1 for the construction of a fractal. We will create step 2, and then find the length for each step.

Assume the length at step 0 is 1.

To obtain step 2, we take the replace the middle 2 sections (so $\frac{2}{4}$ of the original length) with 4 sections. The length at this step is $\frac{6}{4}$.

In step 2, we will take each section from step 1 and repeat the pattern, yielding

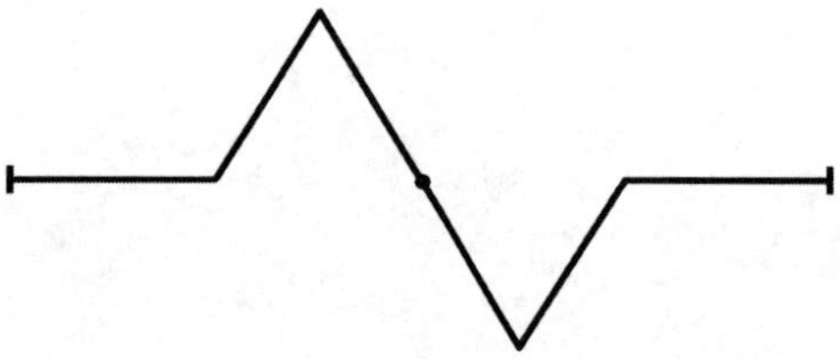

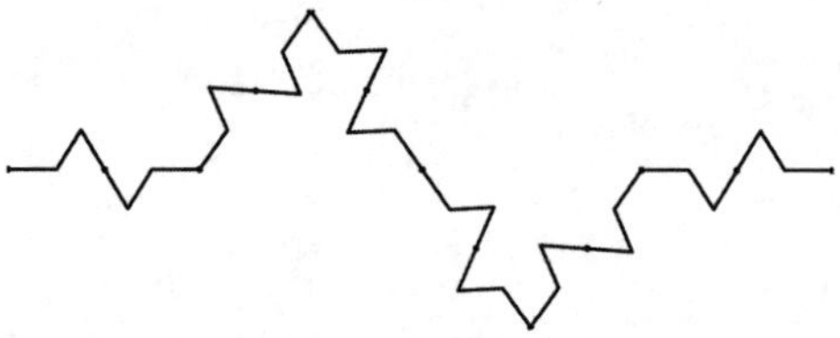

The length at this step is $\frac{36}{16}$. If we were to continue, at each stage the length would be $\frac{6}{4}$ times the length at the previous stage. We could continue indefinitely; the length of this curve is infinite.

Practice (Exercise 5 p.526)

1. You are given steps 0 and 1 for constructing a fractal.

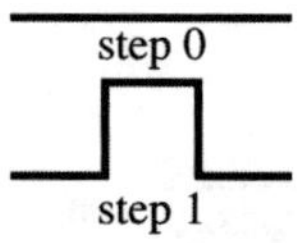

a. Construct step 2 of the fractal.
b. Assume that the length of the line segment in step 0 is 1. Find the length of the curve in step 5.

Objective 3 – Compute fractal dimension.

Key Terms

Fractal dimension

Summary

Due to the intricacies of a fractal curve, its dimension is considered larger than the one-dimensional line segment that generates it, but less than the two dimensional region in the plane that contains it. We define the fractional dimension of an object as a number *D* that satisfies the equation $n = s^D$, where *s* is the scaling factor of the object and *n* is the amount by which the quantity we are measuring (length, area, volume) of the object changes when we apply the scaling factor to the object. In order to solve for *D*, we will need to use the property of the log function $\log a^x = x \log a$.

Notes:

Guided Example

Find the dimension of the fractal curve we discussed in the previous worksheet (stage 0 and stage 1 are reprinted here).

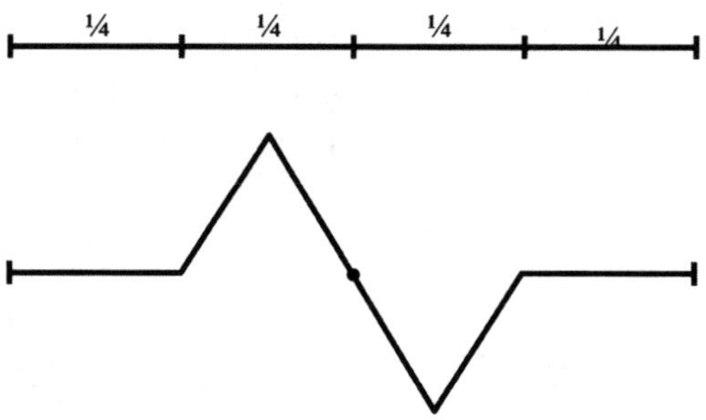

Recall that at each stage, the length of the curve was increased by a factor of $\left(\frac{6}{4}\right)$. A scale factor of 4 increases the length by 6. Thus we need to solve $n = s^D$ where $6 = 4^D$. Using the exponent property of the log function, we have $\log 6 = D \log 4$, so

$$D = \frac{\log 6}{\log 4} \approx 1.292.$$

Practice (Exercise 9 p. 526)

Find the fractal dimension of the curve considered in the previous worksheet, reproduced here.

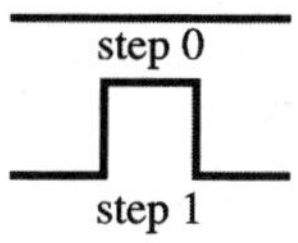

Chapter 9 Answers

<u>Section 9.1</u>

Objective 2 — 1) 7 & 9 2) 3 & 7 3) 10 4) 9 & 10 (there are other answers for these; an example is given)

Objective 3 — 1) Arc AB has length 6 feet 2) $m\angle ACB = 120°$

<u>Section 9.2</u>

Objective 1 — 1) It is not a polygon; not made of line segments 2) the figure could be a rhombus, which is not a square. 3) A pentagon could be convex and not have all sides the same length.

Objective 2 — 1) 22.5° 2) 162°

Objective 3 — 1) 75 ft.

<u>Section 9.3</u>

Objective 1 — 1) 140 in^2 2) 108 cm^2 3) 72 yd^2

Objective 3 — 1) 13 m

Objective 4 — 1) 78.5 cm^2 2) 4,974 yd^2

<u>Section 9.4</u>

Objective 1 — 1) a. 148 cm^2 b. 120 cm^3 2) 180 in^3

Objective 2 — 1) the rectangular cake

Objective 3 — 1) a. 5,024 cm^2 b. 33,493.33cm^3 2) 25

<u>Section 9.5</u>

Objective 1 — 1) c 2) c

Objective 2 — 1) 24,000 dL 2) 2,800 mm 3) 350 dL

Objective 3 — 1) 29,035 ft 2) 50 mph.

<u>Section 9.6</u>

Objective 1

1)

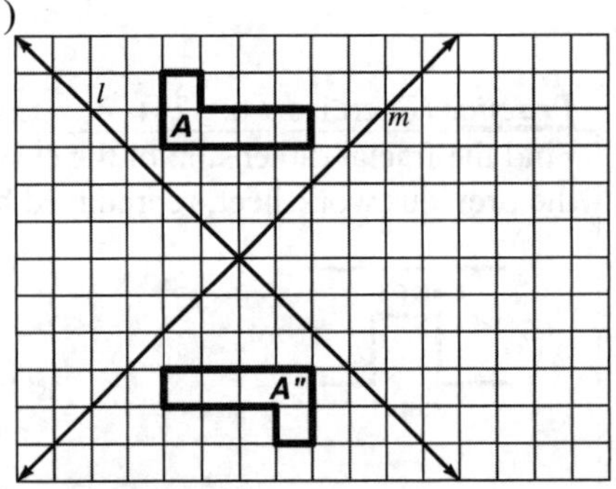

2)

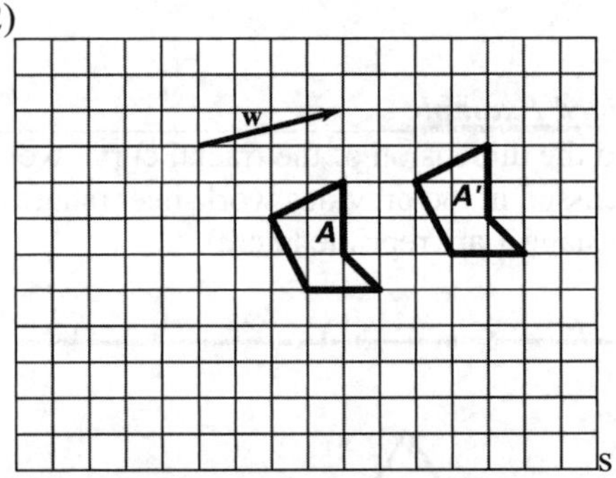

3)

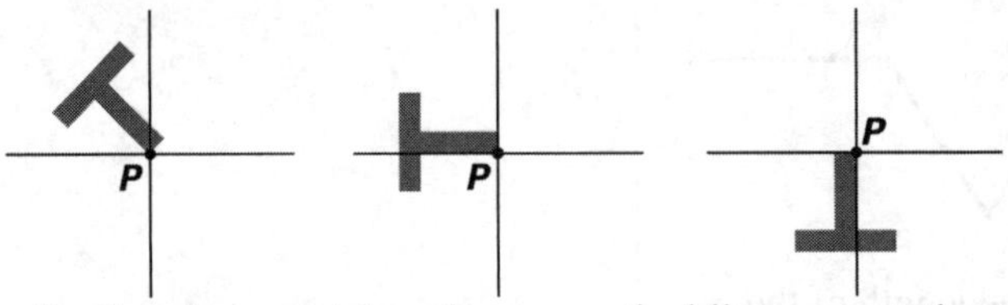

Objective 2 — 1) reflectional symmetries: about a vertical line; no rotational symmetries.

Objective 3 — 1) Using the interior angles of a regular pentagon, we cannot obtain an angle sum of 360° around a point.

<u>Section 9.7</u>

Objective 2 — 1) a.

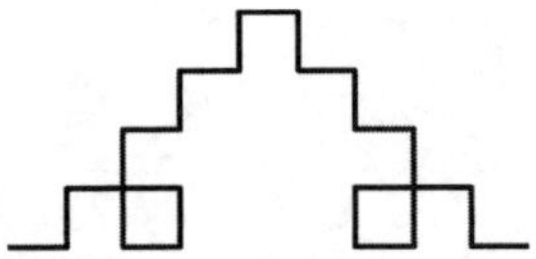

b. $\left(\frac{5}{3}\right)^5$

Objective 3 — 1) 1.46

Chapter 10: Apportionment

Section 10.1 Understanding Apportionment

Objective 1 – Show how the Hamilton apportionment method can lead to the Alabama Paradox.
Objective 2 – Calculate the absolute and relative unfairness of an apportionment.

Objective 1 – Show how the Hamilton apportionment method can lead to the Alabama Paradox.

Key Terms
Alabama Paradox
Hamilton apportionment method

Summary
In this chapter we look at how we allocate items proportionately. A great example of apportionment is the United States House of Representatives, where, generally speaking, the number of representatives is proportional to state population. Unusual things can happen, however, when we take the apportionment and make changes to the system, such as changing the total number of representatives. The Alabama paradox refers to the fact that when the size of the House of Representatives changed from 299 to 300 in 1881, Alabama actually lost representatives; going from 8 to 7 even when no state had a change in population. This resulted from the apportionment method used; the Hamilton method.

To determine representation of a group in a larger organization using the Hamilton method:

1. Determine the number of representatives that are due to each group by computing
   percent in the group × size of overall organization.
2. Assign the integer part of the exact number of members to each group.
   If there are more representatives in the overall organization to be assigned, go to step 3
3. The first additional member goes to the group having the largest fractional part; the second additional member goes to the group with the second largest fractional part. Continue in this manner until all representatives have been assigned.

Notes:

Guided Example	***Practice*** (Exercise15 p.540)
The student judiciary board at your institution has 9 student representatives. The representatives are assigned by college in proportion to the number of students in the institution. There are 11,000 students in the institution; 5,120 in the College of Arts, 4,120 in the College of Science, and 1,760 in the College of Business and Pre-Professional Programs. Determine the number of representatives from each college. First we determine the number of representatives for each group using the formula	1. Suppose an electronics company has three divisions: (D)igital, (C)omputers, and (B)usiness products. Division D has 140 employees, C has 85, and B has 30. Assume that a 12-member quality improvement council has membership on the council proportional to the number of employees in the three divisions. a. Apportion this council using the Hamilton method. b. Now increase the council's size to 13, and then to 14, and so on, redoing the apportionment each time until an Alabama paradox occurs.

percent in the group × size of overall organization .
In this case the size of the organization is 9, the number of student representatitives.

College of Arts: $\frac{5,120}{11,000}\times 9 = 4.19$

College of Science & Technology

$\frac{4,120}{11,000}\times 9 = 3.37$

College of Business & Pre-Professional Programs $\frac{1,760}{11,000}\times 9 = 1.44$.

We assign as follows:
College of Arts - 4;
College of Science - 3;
College of Business & Pre-Professional Programs - 1.

There is still 1 representative to assign; this would go to the College of Business and Pre-Professional Programs because they have the largest fractional allocation. They would thus have 2.
Now imagine that the judiciary board will have an additional member, thus a total of 10. With the same breakdown in the colleges, how many representatives does each have?

College of Arts: $\frac{5,120}{11,000}\times 10 = 4.65$

College of Science & Technology

$\frac{4,120}{11,000}\times 9 = 3.75$

College of Business & Pre-Professional Programs $\frac{1,760}{11,000}\times 6 = 1.60$ .We would have 4 from College of Arts, 3 from the College of Science, and 1 from the College of Business and Pre-Professional programs. Assigning the last two, one would go to the College of Science and 1 to College of Arts. The College of Business and Pre-Professional Programs would go from 2 representatives out of 9 to 1 out of 10. One might ask how such a thing is possible and the answer is quite simple. When the judiciary board increased in size, they reallocated the original 9 members, rather than looking solely at where to assign the new member.

c. What is the first size above 12 when the paradox occurs, and what division loses a seat when the size of the council increases?

Objective 2 – Calculate the absolute and relative unfairness of an apportionment.

Key Terms
Average constituency
Absolute unfairness

Summary
As we move to other methods of apportionment (so as to avoid the Alabama paradox) we need to develop some definitions around apportionment. For this discussion, we will consider apportionment for state representation. The average constituency of a state is the quotient $\frac{\text{population in the state}}{\text{number of represenatives from the state}}$. In comparing the representation between two states A and B, we say that A is more poorly represented than state B if the average constituency of A is greater than the average constituency of state B.

Example: According to 2010 census data, the population of Massachusetts was 6,547,629, while Missouri had a population of 5,988,927. The allocation of Representatives for each state was 9 and 4 respectively. Determine the average constituency for each state.

We compute the average constituency for each state below:
for Massachusetts it is $\frac{6,547,629}{9} \approx 727,514.33$; for Missouri it is $\frac{5,988,927}{4} = 1,497,231.75$. It appears that Missouri is more poorly represented than Massachusetts.

The absolute unfairness of an apportionment is found by subtracting the smaller average constituency from the larger. Although absolute unfairness measures the imbalance of an apportionment between two states, it is inadequate to compare the unfairness of two apportionments. It is important to take into consideration the size of the states involved. For that reason, we define the relative unfairness, which is the quotient
$\frac{\text{the absolute unfairness of the apportionment}}{\text{the smaller average constituency of the two states}}$.

Notes:

Guided Example	***Practice*** (Exercise 9 p.540)
Compute the absolute unfairness and the relative unfairness of the apportionment of Massachusetts and Missouri. The absolute unfairness is $1,497,231.75 - 727,514.33 = 769,717.42$. The relative unfairness of the apportionment is $\frac{769,717.42}{727,514.33} \approx 1.06$	1. Which state is more poorly represented: State A with a population of 27,600 and 16 representatives, or state B with a population of 23,100 and 14 representatives? What is the absolute unfairness of this apportionment? What is the relative unfairness of this apportionment?

Section 10.2 The Huntington-Hill Apportionment Principle

Objective 1 – Use the apportionment criterion in allocating representatives.

Objective 2 – Compute Huntington-Hill numbers.

Objective 3 – Use the Huntington-Hill method to do apportionment.

Objective 1 – Use the apportionment criterion in allocating representatives.

Key Terms

Apportionment criterion

Summary

The Alabama paradox occurred because we reapportioned seats that had already been assigned. We now develop a method that avoids this by assigning only new seats. To do this we use the apportionment criterion, which says that when assigning a representative among several parties, make the assignment so as to give the smallest relative unfairness. Recall that we measure the relative unfairness with the formula

$$\text{relative unfairness} = \frac{\text{the absolute unfairness of the apportionment}}{\text{the smaller average constituency of the groups}}.$$

Notes:

Guided Example	***Practice*** (Exercise 1 p. 548)
The staff senate at a university has representatives from both the classified and professional staff. There are 12 representatives from the 213 classified staff and 7 representatives from the 102 professional staff. If an additional senate position is to be assigned, determine if it should be a classified or professional representative. Use the apportionment criterion to make your decision. To address this question, we will look at the relative unfairness if the senator comes from the classified staff (C) and then compare it to the relative unfairness if the senator is chosen from the professional staff (P). Stage One: Assume 13 senators from classified staff and 7 from professional. The average constituencies are C, $\frac{213}{13} \approx 16.38$ and D, $\frac{102}{7} \approx 14.57$. The relative unfairness is the difference divided by the smaller average constituency; $\frac{16.38-14.57}{14.57} = 0.124$. Stage Two: Assume 12 senators from C and 8 from P. The average constituencies are C, $\frac{213}{12} = 17.75$ and D, $\frac{102}{8} \approx 12.75$. The relative unfairness for this apportionment is $\frac{17.75-12.75}{12.75} = 0.392$. Since the relative unfairness is smaller when the assignment is made to C, we assign the additional senator to the classified staff.	1. The Musician's Guild, with 908 members, has four representatives on the National Arts Advisory Board (NAAB); the Artists' Alliance, with 633 members, has three representatives. If either of the musicians or the artists can be given one additional representative to the NAAB, determine which group should get it by using the following steps: a. Determine the relative unfairness of the apportionment if we give an additional representative to the musicians. b. Determine the relative unfairness of the apportionment if this representative is given to the artists instead. c. Use the apportionment criterion to decide which group deserves the additional representative more.

Objective 2 – Compute Huntington-Hill numbers.

Key Terms
Huntington-Hill numbers

Summary
Huntington-Hill numbers, developed by mathematicians Edward Huntington and Joseph Hill, are used to apportion the U.S. House of Representatives. If a group X has *x* representatives to some organization, it's Huntington-Hill number is $\frac{(\text{number of members of X})^2}{x(x+1)}$.

Notes:

Guided Example	***Practice*** (Exercise 9 p.549)
The staff senate at a university has representatives from both the classified and professional staff. There are 12 representatives from the 213 classified staff and 7 representatives from the 102 professional staff. Compute the Huntington Hill number for the classified staff. The classified staff has 213 members and 12 senate representatives. Their Huntington-Hill number is $\frac{(213)^2}{12(12+1)} = \frac{45,369}{156} \approx 290.83$.	1. The Musician's Guild, with 908 members, has four representatives on the National Arts Advisory Board (NAAB). Compute the Huntington-Hill number for the Musician's Guild.

Objective 3 – Use the Huntington-Hill method to do apportionment.

Key Terms
Huntington-Hill apportionment principle

Summary
Recall that if a group X has x representatives to some organization, it's Huntington-Hill number is $\frac{(\text{number of members of X})^2}{x(x+1)}$. The Huntington-Hill apportionment principle states that if two groups X and Y have already been allotted x and y representatives, respectively, and if an additional representative is to be assigned, it should be given to the group with the larger Huntington-Hill number.

Example: The staff senate at a university has representatives from both the classified and professional staff. There are 12 representatives from the 213 classified staff and 7 representatives from the 102 professional staff. We found the Huntington-Hill number for the classified staff to be 290.83. Verify that the Huntington Hill number for the professional staff is 185.79 and determine which group would be assigned an additional senator if the space was open.

The Huntington-Hill number for the professional staff is $\frac{(102)^2}{7(7+1)} = \frac{10,404}{56} \approx 185.79$. Since the classified staff has a higher Huntington-Hill number, they should be assigned an additional senator. Note that this was the same decision made using the apportionment principle.

The Huntington-Hill principle can be used when making initial apportionment decisions. We will see in the guided example that we start by giving each group one representative, and then use Huntington-Hill numbers to assign additional representation.

Notes:

Guided Example

The student judiciary board at your institution has 9 student representatives. The representatives are assigned by college in proportion to the number of students in the institution. There are 11,000 students in the institution, 5,120 in the College of Arts, 4,120 in the College of Science, and 1,760 in the College of Business and Pre-Professional Programs. Determine the number of representatives from each college using the following table of Huntington-Hill numbers (when computing these numbers, population was expressed in thousands).

Current Number of Representatives	Arts	Sciences	Business and Pre-Professional
1	13.11	8.49	1.55
2	4.37	2.83	0.52
3	2.18	1.41	*
4	1.31	0.85	*
5	0.87	0.57	*

We first allocate one representative from each college, and look at the table to assign representative number four. Arts has the highest Huntington-Hill number (13.11), so they get the next representative and we cross out that number. Now we look at the table

Current Number of Representatives	Arts	Sciences	Business and Pre-Professional
1	~~13.11~~	8.49	1.55
2	4.37	2.83	0.52
3	2.18	1.41	*
4	1.31	0.85	*
5	0.87	0.57	*

and compare the numbers 4.37, 8.49, and 1.55. Sciences has the highest, so we cross that out and look at the following table.

Current Number of Representatives	Arts	Sciences	Business and Pre-Professional
1	~~13.11~~	~~8.49~~	1.55
2	4.37	2.83	0.52
3	2.18	1.41	*
4	1.31	0.85	*
5	0.87	0.57	*

We now look to see which school gets representative number 6. We compare 4.37, 2.83 and 1.55 and assign the representative to Arts. Continuing in this way until we have the nine representatives, we find the assignment would be ASBASASAB.

Practice (Exercise 17 p.549)

1. The electricians', plumbers', and carpenters' unions are sending members to a seven-member labor council. Use the given table of Huntington-Hill numbers to list the order in which the council will be apportioned.

Current Number of Representatives	(E)lectricians	(P)lumbers	(C)arpenters
1	312.5	162.0	480.5
2	104.2	54.0	160.2
3	52.1	27.0	80.1
4	31.3	*	48.1

Section 10.3 Other Paradoxes and Apportionment Methods

Objective 1 – Use standard divisors and quotients to do apportionment.

Objective 2 – Show how the population and new states paradoxes can occur with the Hamilton method.

Objective 3 – Demonstrate the similarities and differences among the Jefferson, Adams, and Webster apportionment methods.

Objective 1 – Use standard divisors and quotients to do apportionment.

Key Terms
Standard divisor
Standard quota
Lower quota
Upper quota
Quota rule

Summary
In previous worksheets, we have discussed the Alabama paradox, but as it turns out, there are other paradoxes that can occur with the Hamilton method. Before addressing those, we introduce some new terminology.

In allocating a group of representatives among several states (groups), we define the standard divisor by the quotient of the total population and the number of representatives being allocated. The standard divisor represents the overall number of persons in the population that are represented by each representative. An individual state's standard quota is defined to be the state's population divided by the standard divisor. The state's standard quota would express the number of representatives that the state deserves. The formulas are $\text{standard divisor} = \dfrac{\text{total population}}{\text{number of representatives being allocated}}$ and $\text{standard quota} = \dfrac{\text{state's population}}{\text{standard divisor}}$. The standard quota often has a fractional part. If we round the standard quota down, we have what is called the lower quota. If we round up, we have the upper quota. If, in making an apportionment, each state is allocated a number of representatives that is between its lower quota and its upper quota, then we say the apportionment satisfies the quota rule.

Now that we have this new language, we can restate Hamilton's apportionment method.

1. Find the standard divisor for the apportionment.
2. Find the lower quota for each state (find the standard quota and round it down). Assign that number of representatives to each state.
3. If there are any representatives left over, assign them to states in order according to the fractional parts of the states' standard quotas.

Notes:

Guided Example	***Practice*** (Exercise 21 p.562)
Consider again the judicial board from a previous worksheet. Find the standard divisor and standard quotas for each group and then apportion the representation on the board. The student judiciary board at your institution has 9 student representatives. The representatives are assigned by college in proportion to the number of students in the institution. There are 11,000 students in the institution, 5,120 in the College of Arts, 4,120 in the College of Science, and 1,760 in the College of Business and Pre-Professional Programs. The standard divisor is $\frac{\text{total population}}{\text{number of representatives being allocated}}$ $= \frac{11,000}{9} \approx 1,222.22.$ The standard quotas are $\frac{\text{college's enrollment}}{\text{standard divisor}}$. For Arts it is $\frac{5,120}{1,222.22} \approx 4.19$, for Sciences $\frac{4,120}{1,222.22} \approx 3.37$, and for Business and Pre-Professional $\frac{1,760}{1,222.22} \approx 1.44$. You should confirm that these are the same numbers we obtained in the first worksheet for this chapter. We allocate 4 representatives to Arts, 3 to Science, and 1 to Business. The ninth representative is given to Business and Pre-Professional, because they have the largest fractional part.	1. A health club instructor has a course load that allows her to teach six two-credit-hour classes. A preregistration survey indicates the following interests: 56 want to take Pilates 29 want to take kickboxing 11 want to take yoga 4 want to take spinning Find the standard divisor and standard quotas for each interest area.

Objective 2 – Show how the population and new states paradoxes can occur with the Hamilton method.

Key Terms
Population paradox
New states paradox

Summary
In the early 1900s, it was discovered that Hamilton's method had serious flaws in addition to the Alabama paradox (see worksheet for section 1). One was that a state with a faster-growing population could lose a representative to a state whose population was growing more slowly. This is called the population paradox; where a state A's population is growing faster than state B's, yet A loses a representative to B.

Another flaw is the new states paradox. The new states paradox occurs when a new state is added and its share of seats is added to the representation, causing a change in allocation of seats previously given to another state.

In each case, we look at the initial apportionment, and then consider the impact of a growing population or an additional state. The fundamental question in this situation is whether other states lose in the presence of growth or presence of another. Keep in mind that although we use the word state, these principles apply to other organizations!

Notes:

Guided Example

A state broadcast organization has 19 representatives from radio and television stations. Does the new states paradox occur if web based groups, with the given population and the appropriate number of additional seats are included in a second Hamilton method apportionment? The appropriate number of representatives to add for the second apportionment is the groups' standard quota, computed using the original two-group standard divisor, rounded down to the nearest integer.

Radio 148
TV 57
Web 41

The standard divisor is $\frac{205}{19} \approx 10.79$. The standard quota for Radio is $\frac{148}{10.79} \approx 13.72$ and for TV is $\frac{57}{10.79} \approx 5.28$. Using the Hamiltonian method, we would assign 13 representatives from radio and 5 from TV, then assign the extra representative from radio, yielding the original a total of 14 from radio.

If we add Web based, they would be allocated $\frac{41}{10.79} \approx 3.80$ or 3, based on the standard divisor.
Now we add the web population to the total and add the 3 additional representatives. The new standard divisor becomes $\frac{246}{22} \approx 11.18$.
Going back to the populations from each group, we find
$\frac{148}{11.18} \approx 13.24$ for radio, $\frac{57}{11.18} \approx 5.10$ for television, and $\frac{41}{11.18} \approx 3.67$ for web based.
The 22 representatives are assigned as follows:
13 for radio
5 for television
4 for web based.

This is an example of the new states paradox because radio has lost representation on the board in the presence of the web based representatives.

Practice (Exercises 25 & 29 pp.562-563)

1. Ten representatives are apportioned amount three states A, B, and C, which have the populations (in thousands) shown in the table. Ten years later, the population of A has increased by 20 thousand and the population of B has increased by 60 thousand, whereas C's population has remained the same. The 10 representatives are reapportioned. Does this example illustrate the population paradox? Explain.

State	Population
A	570
B	2,557
C	6,873

2. Two states A and B have formed a 100-member commission to explore ways in which they can work jointly to improve the business climate in their states. The members will be apportioned to the commission according to the population of the two states. After seeing the success of the commission, a third state asks to join and appoint members to the commission. States A and B agree and will allow state C to add as many members to the commission as they would deserve now using the current standard divisor. The populations of the three states are given below. Does this example illustrate the new states paradox? Explain.

State	Population (thousands)
A	2,000
B	7,770
C	600

Objective 3 – Demonstrate the similarities and differences among the Jefferson, Adams, and Webster apportionment methods.

Key Terms
Modified divisor
Modified quota
Jefferson's method
Adams's method
Webster's method

Summary
In previous examples, we have used the standard quota to determine the number of representatives given to a state or group. Our practice has been to use the whole number portion of the standard quota first, and then go back to look at the fractional parts to determine which group is allocated any unassigned representatives.

We now consider a different way to approach this apportionment using a modified divisor and a modified quota. The modified quota will be used to assign the representatives using one of three methods. In Jefferson's method we consistently round down; with Adam's method we round up, and with Webster's method we round as usual.

The reason for the modified divisor and quota with this method is simple (but potentially difficult to implement). We modify the divisor so that when we determine the quota and use the rounding method identified, we have exactly the right number of assignments for all of the representatives being apportioned.

Determining the modified divisor and modified quota involves trial and error. We will demonstrate this new procedure in the example using Jefferson's method.

Problem Solving: Much of the procedure we will follow is modeled after the work with the Hamilton method. Look to see the similarities and use the strategy *Relate a New Problem to and Older One.*

Notes:

Guided Example	***Practice*** (Exercise 3 p.561)
We return a final time to look at apportioning student representatives on a judiciary board. Recall that the student judiciary board at your institution has 9 student representatives. The representatives are assigned by college in proportion to the number of students in the institution. There are 11,000 students in the institution, 5,120 in the College of Arts, 4,120 in the College of Science, and 1,760 in the College of Business and Pre-Professional Programs. In a previous worksheet we found the standard divisor to be 1,222.22 and the standard quotas to be 4.19 (Arts) 3.37 (Sciences) and 1.44 (Business).	1. Imagine that state A has a population of 19,975 and the total population is 102,000. There are 11 representatives to be apportioned. Find: a. The standard divisor for the apportionment and the standard quota for A. b. The number of representatives that would be given to the state using Jefferson's method. c. The number of representatives that would be given to the state using Adam's method. d. The number of representatives that would be given to the state using Webster's method.

We will now find a modified divisor and modified quotas using the Jefferson's method. Note that in this case, we will round quotas down, and want to modify the quotas so that when we do this, we will have exactly 9 representatives. Since the standard quotas, when rounded down, summed to 8, we need to modified quotas to be larger (so that when we round then down, we get 9). In order to have the modified quotas be larger, we need the modified divisor to be smaller (dividing by a smaller number will give a larger quota). But how much smaller should it be? This is where the trial and error comes in. Let's try a modified divisor of 1,000. It is smaller than the standard divisor of 1,222.22. The modified quotas would be A: $\frac{5,120}{1,000} = 5.12$; S: $\frac{4,120}{1,000} = 4.12$; B: $\frac{1,760}{1,000} = 1.76$. Following Jefferson's rule and rounding down, we would have 5 + 4 + 1 total representatives which is too many! Our standard divisor was too small. Trying another modified divisor, we need it to be larger than 1,000 (but still smaller than 1,222.22). Verify that trying a modified divisor of 1,100 gives modified quotas of 4.65, 3.74, and 1.6. Rounding these down, per Jefferson's method, we are now at 4 + 3 + 1 = 8 representatives. This is too few. This modified divisor is too big! We need a modified divisor between 1,000 and 1,100. Let's try the modified divisor 1,025. The modified quotas would be A: $\frac{5,120}{1,025} = 4.995$; S: $\frac{4,120}{1,025} = 4.02$; B: $\frac{1,760}{1,025} = 1.717$. Rounding down, we have 4, 4, and 1, which sums to 9 (our goal). We assign 4 representatives to Arts 4 to Sciences and 1 to Business and Pre-Professional Programs. If we were using Adams's method, we would go through a similar process, but always rounding up. With Webster's method, we would follow standard rules for rounding.	

Section 10.4 Fair Division

Objective 1 – Be able to describe the differences between discrete and continuous division of items.

Objective 2 – Use the method of sealed bids to make a fair division of items.

Objective 1 – Be able to describe the difference s between discrete and continuous division of items.

Key Terms

Discrete fair division

Continuous fair division

Summary

We will be looking at ways to divide items fairly. Before deciding on a strategy to use, it is important that we look at the type of items being divided! Is it a pizza or Grandma's jewelry? Discrete fair division means that the items being subdivided cannot be divided into fractional parts. Continuous fair division means that the items being distributed can be subdivided into parts and thc division can be done as finely as we want.

Notes:

Guided Example

Identify each situation as dealing with discrete or continuous division of items.

1. Dividing your Grandmother's jewelry among six grand-children.

   This is discrete division. You cannot break her diamond engagement ring into fractional parts.

2. Dividing the remaining pie among the six grandchildren who want seconds.

   This is continuous. Those pieces of pie can be small fractions of the original piece!

Practice (Exercise 1 p.569)

1. Identify each situation as dealing with either discrete or continuous division of items.
   a. Dividing a collection of rare books among five people.
   b. Dividing a chocolate bar among several children.
   c. Apportioning the seats on a city council with representatives chosen from different neighborhoods of the city.

Objective 2 – Use the method of sealed bids to make a fair division of items.

Key Terms
Fair share
Sealed bids

Summary
How can discrete items be shared fairly? We will focus on items contained in an estate. We define a person's fair share of an estate to be the person's estimated value of the estate's value divided by the number of people sharing equally in the estate.

We determine the allocation of the items in the estate (and monies exchanged among the heirs) using the method of sealed bids. The method assumes equal interest in the inheritance.

1. Each person's fair share is determined as follows:
   - Add the person's bids on the various items
   - Divide the total by the number of people sharing in the estate
2. Open the bids. The person bidding the highest amount on an item gets that item. If several people have the same highest bid, then the decision of who gets the item can be based on random selection.
3. If the total value of the items awarded to a person is more than the person's fair share, then the person pays the difference in cash to the estate. If the total value is less than the person's fair share, then the person receives the difference in cash from the estate. If a person does not receive any items, then the person gets his or her fair share in cash from the estate.
4. If, after completing step 3, there is a cash balance in the estate, then it is divided equally among the heirs.

Notes:

Guided Example

Cousins Silvie, Ulla, and Janis need to determine who gets their grandmother's diamond necklace and engagement ring. Each is entitled to $\frac{1}{3}$ of the inheritance. They have submitted sealed bids on the jewelry which are summarized in the table below. Determine who should receive each item and what amount of cash needs to be disbursed as part of a fair division of the items.

	Sylvie	Ulia	Janis
Bid on Ring	$15,000	$14,500	$13,000
Bid on Necklace	$8,000	$10,500	$11,000
Total Value	$23,000	$25,000	$24,000
Estimated Fair Share of Jewelry	$7,666.67	$8,333.33	$8,000

Practice (Exercises 7 & 13 p.570)

1. Three brothers – Ed, Al, and Jerry – want to decide who gets a valuable painting and statue that they have inherited from their grandfather. Complete the table below.

	Ed $(\frac{1}{3})$	Al $(\frac{1}{3})$	Jerry $(\frac{1}{3})$
Bid on painting	$13,000	$8,000	$9,000
Bid on statue	$14,000	$13,000	$15,000
Total value	$27,000	a.	b.
Fair share of estate	c.	d.	e.

2. Consider your answer to the exercise above. Complete the settlement of the estate by competing the fair division accounting of the estate bas indicated in the table.

We see that Sylvie perceives her part of the inheritance as $7,666.67 thus we assume she will be satisfied if she receives this amount. Similarly Ulla and Janis will be satisfied with $8.333.33 and $8,000 respectively.

Since Sylvie had the largest bid on the ring she will receive this. Janis will get the necklace.

Sylvie perceived her portion of the estate as $7,666.67; since she is receiving the ring which she values at $15,000, she owes the estate the difference, which is $7,333.33.

Janis saw her share to be $8,000 but is getting the necklace which she valued at $11,000. She owes the estate $3,000.

Ulla views her portion to be $8333.33, but since she didn't have the highest bid on either item, she doesn't get either. She will receive cash from her cousins!

Sylvie paid the estate $7,333.33 and Janis paid the estate $3,000.00. Based on her valuations, Ulla is owed $8333.33. There is a balance in the estate, 7,333.33 + 3,000.00 − 8,333.33 = $2.000. We divide that equally among the three, so each receives $666.67.

In summary, Sylvie gets the ring and owes $7,333,33 − $666.67 = $6,666.66;
Janis gets the necklace and pays the estate $3,000 − 666.67 = $2,333.33; Ulla gets neither jewel but receives $8,333.33 + $666.67 = $9,000 from the estate.

It is important to note that each cousin receives at least what she perceived her portion to be. It might be in jewelry or in cash, but everyone should be satisfied based on their sealed bid.

	Ed $(\frac{1}{3})$	Al $(\frac{1}{3})$	Jerry $(\frac{1}{3})$
Items obtained with highest bid			
Pays to estate $(+)$ or receives from estate $(-)$			
Division of estate balance ($)			
Summary of cash			

Chapter 10 Answers

<u>Section 10.1</u>

Objective 1 — 1) a. D has 7, C has 4 and B has 1. c. Paradox occurs when assigning the 14th seat; business.

Objective 2 — 1)A; 75, 0.045

<u>Section 10.2</u>

Objective 1 — 1) a. 0.162 b. 0.434 c. musicians

Objective 2 — 1) 41,223.2

Objective 3 — 1) E,P,C,C,E,P,C

<u>Section 10.3</u>

Objective 1 — 1) 16.667; Pilates 3.36, kickboxing 1.74, yoga 0.66, spinning 0.24

Objective 2 — 1) Yes, because A's population has increased faster than B's, but A loses a representative to B in the reapportionment. 2) Yes; when C's representatives are added to the commission, B loses a representative to A.

Objective 3 — 1) a. 9,272.73; 2.15 b. 2 c. 3 d. 2

<u>Section 10.4</u>

Objective 1 — 1) a. discrete b. continuous c. discrete

Objective 2 — 1) a. \$21,000 b. \$24,000 c. \$9,000 d. \$7,000 e. \$8,000 2)

	Ed $\left(\frac{1}{3}\right)$	Al $\left(\frac{1}{3}\right)$	Jerry $\left(\frac{1}{3}\right)$
Items obtained with highest bid	Painting		Statue
Pays to estate (+) or receives from estate (−)	\$4,000 (+)	\$7,000 (−)	\$7,000 (+)
Division of estate balance (\$4,000)	\$1,333.33 (−)	\$1,333.33 (−)	\$1,333.33 (−)
Summary of cash	Pays \$2,666.67	Receives \$8,333.33	Pays \$5,666.67

Chapter 11: Voting

Section 11.1 Voting Methods.

Objective 1 – Use the plurality method to determine the winner of an election.
Objective 2 – Determine winners using the Borda count voting method.
Objective 3 – Use the plurality-with-elimination method to determine the winner of an election.
Objective 4 – Determine the winner of an election using the pairwise comparison method.

Objective 1 – Use the plurality method to determine the winner of an election.

Key Terms
Plurality

Summary
The plurality method is the simplest way to determine the outcome of an election; the person receiving the most votes is declared the winner.

Notes:

Guided Example

There are four candidates for President of a fraternal organization: Evans, Grant, Keith, and Smith. Determine who is the winning candidate if the number of votes for each candidate is given below:

Candidate	Number of votes
Evans	2
Grant	37
Keith	12
Smith	26

Since Grant has more votes than any other candidate, he is deemed the winner.

Practice (Exercise 15 p.585)

1. A small employee-owned internet company is voting on a merger with one of its competitors. The choices are (E)-Mall, (F)irestorm, (S)ecurenet, (T)echenium, and (L)ondram. The employees' votes are summarized in the following preference table. What option is their first choice using the plurality method?

	Number of Ballots				
Preference	15	11	9	10	2
1st	L	F	S	T	L
2nd	F	S	L	E	S
3rd	S	L	F	F	E
4th	E	E	T	L	F
5th	T	T	E	S	T

Objective 2 – Determine winners using the Borda count voting method

Key Terms
Borda count

Summary
Ranked voting methods are becoming increasingly popular. Ranked voting allows voters to not only select their first choice, but also their second, third, and so on. The Borda count method takes into account these rankings and assigns points to the candidates as follows.

If there are k candidates in an election, each voter ranks all candidates on the ballot. Then the first choice is given k points, the second choice is given $k-1$ points, the third choice $k-2$ points, and so on. The candidate who receives the most total points wins the election.

Problem Solving: The computations involved in the Borda count can be cumbersome. Use the *Be Systematic Strategy* to keep your work organized.

Notes:

Guided Example

There are four candidates for President of a fraternal organization: Evans, Grant, Keith, and Smith. Determine who is the winning candidate using the Borda count method with the following preference table:

	Number of Ballots					
Preference	**8**	**12**	**7**	**30**	**18**	**2**
1st	S	K	G	G	S	E
2nd	K	S	K	E	K	G
3rd	G	G	E	S	E	K
4th	E	E	S	K	G	S

Since there are 4 candidates, we assign 4 points for first place votes, 3 for second, 2 for third and 1 for fourth. It is helpful to look at each candidate separately, and count how many first, second, third, and fourth place votes were received. This information is summarized in the following table.

Candidate	1st place votes × 4 points	2nd place votes × 3 points	3rd place votes × 2 points	4th place votes × 1 points	Borda count
Evans	$2 \times 4 = 8$	$30 \times 3 = 90$	$25 \times 2 = 50$	$20 \times 1 = 20$	168
Grant	$37 \times 4 = 148$	$2 \times 3 = 6$	$20 \times 2 = 40$	$18 \times 1 = 18$	212
Keith	$12 \times 4 = 48$	$33 \times 3 = 99$	$2 \times 2 = 4$	$30 \times 1 = 30$	181
Smith	$26 \times 4 = 104$	$12 \times 3 = 36$	$30 \times 2 = 60$	$9 \times 1 = 9$	209

Note that Evans has 2 first place votes, so 8 points for those; 30 second place votes so 90 points; 25 third place votes at 2 points per vote so 50 points; and finally 20 fourth place votes so 20 points for those. Summing, Evans has a Borda count of 168. We do the same for each of the candidates. We see that Grant has the largest Borda count, at 212, and is deemed the winner by this method.

Practice (Exercise 6 p.618)

1. Use the given preference table to determine the winner of the election using the Borda count method.

	Number of Ballots					
Preference	**3**	**5**	**8**	**2**	**6**	**5**
1st	A	B	C	A	B	D
2nd	B	A	D	B	A	C
3rd	C	D	A	C	D	B
4th	D	C	B	D	C	A

Objective 3 – Use the plurality-with-elimination method to determine the winner of an election.

Key Terms
Plurality with elimination

Summary
One of the complaints about the plurality method (where the candidate with the most votes wins) is that the winner has the most votes, but does not necessarily have a majority of the votes. For this reason, the plurality with elimination method is popular. With this method, each voter votes for one candidate. The candidate who receives a majority of votes is declared the winner. If no candidates receives a majority of the votes, then the candidate (or candidates) with the fewest votes is dropped from the ballot and a new election is held. This process continues until a candidate receives a majority of the votes.

When working on these problems, we assume that each voter ranks the candidates, and that this ranking does not change during subsequent rounds of voting.

Notes:

Guided Example

There are four candidates for President of a fraternal organization: Evans, Grant, Keith, and Smith. Determine who is the winning candidate using the plurality with elimination using the following preference table:

	Number of Ballots					
Preference	**8**	**12**	**7**	**30**	**18**	**2**
1st	S	K	G	G	S	E
2nd	K	S	K	E	K	G
3rd	G	G	E	S	E	K
4th	E	E	S	K	G	S

As he noted, Grant has the largest number of first place votes, but with 77 total voters, 37 is not a majority. We eliminate E, who has only 2 first place votes, and look to see who is the winner if E is out of the race.
We reassign the 2 votes for E to candidate G (since G is listed next in the ranking), so G has 39 first place votes. This gives G a majority and he is declared the winner.

Practice (Exercise 17 p.586)

1. Consider again the small internet company voting on a merger with a competitor. The choices are (E)-Mall, (F)irestorm, (S)ecurenet, (T)echenium, and (L)ondram. The employees' votes are summarized in the following preference table. What option is their first choice using the plurality with elimination?

	Number of Ballots				
Preference	**15**	**11**	**9**	**10**	**2**
1st	L	F	S	T	L
2nd	F	S	L	E	S
3rd	S	L	F	F	E
4th	E	E	T	L	F
5th	T	T	E	S	T

Objective 4 –.Determine the winner of an election using the pairwise comparison method.

Key Terms
Pairwise comparison method

Summary
Our strategy in the pairwise comparison method is to look at the possible head-to-head matches, and see who wins more than the other candidates. In the pairwise comparison method, like the plurality with elimination and Borda count method, the voters first rank the candidates and we assume that these rankings would hold throughout the study. If A and B are a pair of candidates, we count how many voters prefer A to B and vice versa. Whichever candidate wins the head-to-head race receives 1 point. If A and B are tied, then each receives $\frac{1}{2}$ point. We do this comparison, assigning points, for each pair of candidates. At the end, the candidate receiving the most points is the winner.

Notes:

Guided Example

Employees at a large company were asked what benefits would most increase their job satisfaction. (F)lex scheduling, (G)ym facilities onsite, or additional (H)olidays were the choices. Using the pairwise comparison method and the preference chart below, determine which additional benefit would be selected.

Preference	20	14	13	10
1st	G	H	F	G
2nd	H	F	H	F
3rd	F	G	G	H

We need to consider all of the head-to-head matchups between the three benefits. We will look at F vs. G, F vs. H, and then G vs. H.

When we look at the matchup between F and G, we essentially cross out the first place votes for H. Doing so results in a table like the one below:

Preference	20	14	13	10
1st	G	~~H~~	F	G
2nd	H	F	H	F
3rd	F	G	G	H

We see that 14 of the votes for H will go to F. This gives F a total of 27 votes and G 30 so G is assigned one point.

Turning to the matchup between F and H, we reassign G's votes; 20 move from G to H, and 10 from G to F. F ends up with 23 and H has 34. H is the winner and is assigned one point.

Finally we look at the matchup between G and H. The 13 votes that F had will go to H, but G wins the matchup 30 to 27 and is assigned one point.

Looking at the three pairwise matches, G has two points and H has one, so G is deemed to be the winner.

Practice (Exercise 12 p. 618)

Use the given preference table to determine who wins the election with the pairwise comparison method.

	Number of Ballots			
Preference	23	47	83	21
1st	A	B	D	C
2nd	B	A	C	B
3rd	C	C	A	A
4th	D	D	B	D

Section 11.2 Defects in Voting Methods.

Objective 1 – Show that the majority criterion is not satisfied by the Borda count method.

Objective 2 – Illustrate how the plurality method violates the Condorcet's criterion.

Objective 3 – Recognize that the pairwise comparison method can fail to satisfy the independence-of-irrelevant-alternatives criterion.

Objective 4 – See how the plurality-with-elimination method fails the monotonicity criterion.

Objective 1 – Show that the majority criterion is not satisfied by the Borda count method.

Key Terms
Majority criterion

Summary
We have discussed several voting methods – and it is understandable to wonder which is best. We now to turn criteria used to evaluate the various methods. The first such criterion is very intuitive; if a majority of the voters rank a candidate as their first choice then that candidate should win the election. This is called the majority criterion.

Plurality and plurality with elimination satisfy the majority criterion, but we will see in the guided example below that it is possible for a candidate to win with the Borda count method when in fact another candidate has a majority of first place votes.

Notes:

Guided Example

1. Using the preference table below, determine the winner using plurality. Does the winner have a majority of the votes?

Preference	20	14	13	10
1st	G	H	F	G
2nd	H	F	H	F
3rd	F	G	G	H

Looking just at first place votes, G has 30, H has 14 and F has 13. G is the winner by plurality. With a total of 57 votes, G's 30 votes represent a majority.

2. Now use the table to determine the Borda count winner.

Recall that with three candidates, we assign 3 points for each 1st place vote, 2 for 2nd, and 1 for 3rd. Verify that The Borda count for G is 117, for H is 118 and for F is 107. H is the Borda count winner.

Since G had a majority of 1st place votes, but was not the Borda count winner, we see an example of the Borda count method violating the majority criterion.

Practice (Exercise 1 p.595)

1. In the preference table, A has a majority of the first-place votes. Who wins the election if we use the Borda count method?

	Number of Ballots			
Preference	12	15	9	13
1st	A	B	C	A
2nd	B	C	B	D
3rd	C	A	D	B
4th	D	D	A	C

Objective 2 – Illustrate how the plurality method violates the Condorcet's criterion.

Key Terms
Condorcet's Criterion

Summary
Condorcet's criterion states that if candidate X can defeat each of the other candidates in a head-to-head vote, then X is the winner of the election. This criterion was proposed by the Marquis de Condorcet, an eighteenth century mathematician, philosopher, and political scientist. It is possible, however, for a winner using the methods we have discussed to violate Condorcet's criterion.

Notes:

Guided Example

Nine players on a baseball team are deciding where to go for dinner after the game. There are three restaurants being considered; A, B and C. The players' preferences are in the table below.

Preference	4	3	2
1st	A	B	C
2nd	B	A	B
3rd	C	C	A

1. What is the winner using plurality?

   Since 4 of the players have A as their top choice, they would select A using plurality.

2. Show that Condorcet's criterion is not satisfied.

   If we look at head-to-head matches, B would beat A and C, so Condorcet would say that B should win. The decision using plurality violates Condorcet's criterion.

Practice (Exercise 7 p.596)

1. Due to a decrease in state funding, a citizen's committee is recommending to the school board ways to reduce expenses. The options are A, reduce sports programs; B, reduce expenditures on art and music programs; C, increase class size; and D, defer maintenance on buildings. Use the preference table to determine the choice that the committee recommends using the plurality-with-elimination method. Show that Condorcet's criterion is not satisfied.

	Number of Ballots					
Preference	9	12	4	5	4	1
1st	C	D	C	A	A	B
2nd	A	A	B	D	B	C
3rd	B	B	D	B	C	A
4th	D	C	A	C	D	D

Objective 3 – Recognize that the pairwise comparison method can fail to satisfy the independence-of-irrelevant-alternatives criterion.

Key Terms
Independence-of-irrelevant-alternatives criterion

Summary
The independence-of-irrelevant-alternatives criterion states that if a candidate X wins an election, X would still win if some non-winners were not present on the ballot. To check to see if this is violated, remove non-winning candidates, and see if the voting method would still find X to be the winner. When using plurality, Borda count or pairwise comparison, it is possible to have a situation where the independence-of-irrelevant-alternatives is violated.

Notes:

Objective 4 – See how the plurality-with-elimination method fails the monotonicity criterion

Key Terms
Monotonicity criterion

Summary
Our final voting criterion is the monotonicity criterion. The monotonicity criterion states that if X wins an election and in a reelection all voters who change their votes only change their votes to favor X, then X also wins the reelection.

This would seem to make sense – and in fact if you are using plurality or the Borda count method, a candidate who wins and gains more support will still be the winning candidate. Using plurality with elimination, however, it is possible to have a candidate gain support, but not win the reelection. This is because the voters changing their allegiance to candidate X may cause a different candidate to be eliminated, which in turn impacts the future stages of the plurality with elimination process.

Notes:

Section 11.3 Weighted Voting Systems

Objective 1 – Be able to represented weighted voting systems numerically.
Objective 2 – Determine the winning coalitions in a weighted voting system.
Objective 3 – Compute the Banzhaf power index of voters in a weighted voting system.

Objective 1 – Be able to represented weighted voting systems numerically.

Key Terms

Weighted voting system
Weights
Quota
Dictator
Veto power

Summary

In the voting systems we have studied thus far, we have assumed that each voter has one vote. In a weighted voting system this is not the case. Each voter is assigned a weight, and in order to pass a resolution a specified sum of weights, called a quota, must be reached. We represent the voting system by first listing the quota, and then the individual weights assigned to each of the voters; {quota : weight of voter 1, weight of voter 2, weight of voter 3, … , weight of voter n}.

Example: Consider the weighted voting system {14: 2, 5, 3, 8}. The voters are assigned weights 2, 5, 3, and 8. In order to pass any resolution, the weight of the voters supporting the resolution must sum to at least 14.

If one of the voter's weights is such that he or she can exercise control, the voter is called a dictator. If the system above were {14: 2, 5, 3, 15} the last voter would be a dictator; that voter alone meets the quota, and the other voters together cannot.

A voter is said to have veto power if that voter can, by her- or himself, prevent a motion from passing.

Notes:

Guided Example	***Practice*** (Exercise 7 p.605)
In the weighted voting system {14: 2, 5, 3, 8} with voters A, B, C, D (in that order) assigned to the weights, identify anyone with veto power. In this voting system, in order to meet the quota you would need voter B and D and one other voter. Without B and D, no resolution would pass, so they both have veto power.	1. Use the weights to represent voters A, B, C, and D, in that order, in the weighted voting system {12 : 1, 3, 5, 7}. Identify a. The quota b. The number and weights of the voters c. The dictators d. Those having veto power.

Objective 2 – Determine the winning coalitions in a weighted voting system.

Key Terms
Coalition
Weight of a coalition
Winning coalition
Critical voter

Summary
Any set of voters who vote the same way is called a coalition. The sum of the weights of the voters in a coalition is called the weight of the coalition. If a coalition has a weight that is greater than or equal to the quota, then that coalition is called a winning coalition.

Problem Solving: If you wish to list the coalitions in a weighted voting system, remember the *Be Systematic* strategy. It will be helpful to look at coalitions consisting of one voter, then those of two, and so on.

What happens to a winning coalition if someone in the coalition leaves? If the coalition is no longer a winning coalition, then that voter is considered a critical voter.

Notes:

Guided Example	***Practice*** (Exercises 23 & 27 p.606)
Three friends are starting up a small internet company. They have set up a weighted voting system based on the amount they have invested in the company. (R)io has 5 votes, (S)ue has 3 votes, and (W)ally has 3 votes. The quota for their voting system is 6. 1. List all winning coalitions First consider coalitions with one person. Since none of the three has a weight of 6 or more, there are none. Look now at the coalitions of two persons. Each of the three coalitions. {R, S}, {R, W}, {S, W} is a winning coalition. We finally look at the coalition {R, S, W} which is also winning. 2. Consider the winning coalitions listed above. For each, determine any critical voters. In the coalitions with 2 voters, all are critical, because if one leaves, the quota will not be met. In the coalition {R, S, W} no voter is critical. Note that if any one voter leaves, the coalition is still winning.	1. The college athletics procedures committee makes policy decisions affecting college athletics programs throughout the state. The committee consists of three (A)dministrators, four (C)oaches in athletics deparments, and three (T) eam captains, and two (N)onathlete students. Assume that each of these four groups votes in a bloc. To make a policy change, a vote of at least 8 is required. Find all the winning coalitions of {A, C, T, N} and state their weights. 2. Determine the critical voters in the winning coalitions above.

Objective 3 – Compute the Banzhaf power index of voters in a weighted voting system.

Key Terms
Banzhaf power index

Summary
In a weighted voting system, a winning coalition is a group of voters whose weights meet or exceed the quota; a critical member of that coalition is a voter who, if no longer in the coalition, would cause it to no longer be a winning one. We measure the power of a voter by the proportion of times that voter is the critical member of a winning coalition. Specifically, the Banzhaf power index for a voter is found using the ratio

$$\frac{\text{the number of times the voter is critical in winning coalitions}}{\text{the total number of times voters are critical in winning coalitions}}.$$

Notes:

Guided Example	***Practice*** (Exercise 41 p.606)
Consider the three friends who are starting up a small internet company. They have set up a weighted voting system based on the amount they have invested in the company. (R)io has 5 votes, (S)ue has 3 votes, and (W)ally has 3 votes. The quota for their voting system is 6. Calculate the Banzhaf power index for each of the friends. We had previously found the winning coalitions to be {R, S}, {R, W}, {S, W} and {R, S, W}. In the coalitions with 2 voters, all are critical, because if one leaves, the quota will not be met. In the coalition {R, S, W} no one voter is critical. Note that if any one voter leaves, the coalition is still winning. There are a total of 6 instances where a voter is critical. Rio is a critical voter in 2 of the 6, so Rio's Banzhaf power index is $\frac{2}{6}$ or $\frac{1}{3}$. Sue is a critical voter in 2 also, so her Banzhaf power index is also $\frac{1}{3}$. Finally, Wally must have Banzhaf power index $\frac{1}{3}$ as well, since all of the Banzhaf power indices must sum to 1. Verify that Wally has a Banzhaf power index of $\frac{1}{3}$.	1. After the 2010 census, Pennsylvania had 20 votes in the Electoral college, North Carolina had 15 and New Mexico had 5. Assume for the moment that these were the only three states in the United States and calculate the Banzhaf power index for each state. Assume that the quota for this weighted voting system is 21.

Section 11.4 The Shapley-Shubik Index

Objective 1 – Determine all the permutations of a set.
Objective 2 – Find the pivotal voters in a coalition.
Objective 3 – Calculate the Shapley-Shubik index for voters in a weighted voting system.

Objective 1 – Determine all the permutations of a set.

Key Terms
Permutation

Summary
We now turn to the study of winning coalitions and the order in which voters joined the coalition. In order to address the questions that will follow, we need to formally define the idea of a permutation of voters. A permutation of the set $\{x_1, x_2, x_3, \ldots, x_n\}$ is an ordering of the elements. We denote a permutation as $\left(x_{i_1} x_{i_2}, x_{i_3}, \ldots, x_{i_n}\right)$, where x_{i_1} is the first element in the permutation, x_{i_2} is the second element in the permutation, and so on.

Problem Solving: Apply the *Splitting Hairs Principle* here; $\{x_1, x_2, x_3, \ldots, x_n\}$ and $\left(x_{i_1} x_{i_2}, x_{i_3}, \ldots, x_{i_n}\right)$ may look similar, but they are quite different! The first represents the set of voters; the second refers to a particular ordering of the voters.

Generally speaking, if we want to order a set with *n* items, there are *n* ways to choose the first element, $n-1$ ways to select the second element, $n-2$ ways to choose the third, and so on. There are $n\times(n-1)\times(n-2)\times\ldots\times 2\times 1$ permutations of these elements. This product is called *n* factorial and is written *n*!

Notes:

Guided Example	***Practice*** (Exercise 1 p.613)
A weighted voting system has voters A, B, and C. List the 3! permutations of this set. Using the *Be Systematic* problem solving strategy, let's first list permutations starting with A. They are (A, B, C) and (A, C, B). Now we turn to permutations starting with B; (B, A, C) and (B, C, A). Finally, those starting with C are (C, A, B) and (C, B, A). There are 6 permutation of the set.	1. Use a tree diagram to find all the permutations of the set {X, Y, Z}

Objective 2 – Find the pivotal voters in a coalition.

Key Terms
Pivotal voter

Summary
As we add voters to a coalition one at a time, the first voter added who makes the coalition a winning coalition is a called a pivotal voter for the coalition.

Problem Solving: When considering pivotal voters, we need to look at the order in which the voters join the coalition. This will cause us to look not just at the voters in the coalition, but the different orderings in which they could join. Use the *Be Systematic strategy* to ensure that you consider all possibilities.

Notes:

Guided Example

Consider again the three friends who are starting up a small internet company. They have set up a weighted voting system based on the amount they have invested in the company. (R)io has 5 votes, (S)ue has 3 votes, and (W)ally has 3 votes. The quota for their voting system is 6. Determine the pivotal voter in the winning coalitions.

Let's look at the order in which they could join. Since there are 3 voters, there are $3! = 6$ permutations to consider. We look at them below, and identify who is the pivotal voter.

Coalition	Sum of weights (until quota is reached)	Pivotal voter
(R, S, W)	$5 + 3 = 8$	S
(R, W, S)	$5 + 3 = 8$	W
(S, R, W)	$3 + 5 = 8$	R
(S, W, R)	$3 + 3 = 6$	W
(W, R, S)	$3 + 5 = 8$	R
(W, S, R)	$3 + 3 = 6$	S

Since the quota is 6, we need two of the three friends to form a winning coalition. In each case, it is the second person entering the coalition who is pivotal.

Practice (Exercise 9 p.613)

1. For the weighted voting system {5: 3, 2, 2} complete the following table, identifying the pivotal voter for each permutation of {A, B, C}.

Sum Weights of Coalition Members	Pivotal Voter
(A, B, C)	
(A, C, B)	
(B, A, C)	
(B, C, A)	
(C, A, B)	
(C, B, A)	

Objective 3 – Calculate the Shapley-Shubik index for voters in a weighted voting system.

Key Terms

Shapley-Shubik index

Summary

We now use the idea of a pivotal voter to define another measure of a voter's power. The Shapley-Shubik index for a voter is defined by the number of times the voter is pivotal in some permutation of the voters divided by the total number of permutations of the voters.

Problem Solving: Based on the way the Shapley-Shubik index is defined, the sum of the index for all voters in the coalition should be 1. Use the *Verify Answers* strategy to ensure that your solution is reasonable.

Notes:

Guided Example	***Practice*** (Exercises 11 p.613)
We return to Rio, Sue, and Wally; three friends starting up an internet company. They have set up a weighted voting system based on the amount they have invested in the company. (R)io has 5 votes, (S)ue has 3 votes, and (W)ally has 3 votes. The quota for their voting system is 6. Determine the Shapley-Shubik index for each voter. In the previous worksheet, we found that in the $3! = 6$ permutations, each of the friends was the pivotal voter in exactly 2 cases. We thus determine that each of the voters has a Shapley-Shubik index of $\frac{2}{6} = \frac{1}{3}$. Even though Rio has more votes than Sue or Wally, she has an equal voting power as measured by this index.	1. Determine the Shapley-Shubik index for the weighted voting system {6: 3, 3, 2}.

Chapter 11 Answers

Section 11.1
Objective 1 1) L (17 votes)
Objective 2 1) B
Objective 3 1) L
Objective 4 1) C

Section 11.2
Objective 1 1) B
Objective 2 1) C wins; however, A defeats all other options head-to-head.

Section 11.3
Objective 1 1) a. 12 b. A, 1 vote; B, 3; C, 5; D, 7 c. no dictator d. C and D have veto power.
Objective 2 1) Winning coalitions and their weights are {A, C, T}, 10; {A, C, N}, 9; {A, T, N}, 8, {C. T, N}, 9; and {A, C, T, N }. 12.
2) {A, C, T, N } has no critical voters; in the other coalitions all voters are critical.
Objective 3 1) Pennsylvania has index 3/5; North Carolina 1/5; New Mexico 1/5.

Section 11.4
Objective 1 1) (X. Y. Z) . (X, Z, Y), (Y, X, Z), (Y, Z, X), (Z, X, Y), (Z, Y, X)
Objective 2 1)

Sum Weights of Coalition Members Until We Reach Quota of 5	Pivotal Voter
(A, B, C) 3 + 2	B
(A, C, B) 3 + 2	C
(B, A, C) 2 + 3	A
(B, C, A) 2 + 2 + 3	A
(C, A, B) 2 + 3	A
(C, B, A) 2 + 2 + 3	A

Objective 3 1) A and B have index $\frac{1}{2}$; C has index 0.

Chapter 12: Counting

Section 12.1 Introduction to Counting Methods

Objective 1 – Count elements in a set systematically.

Objective 2 – Use tree diagrams to represent counting situations graphically.

Objective 3 – Determine the number of branches in a tree without drawing the tree.

Objective 1 – Count elements in a set systematically.

Key Terms

None

Summary

Problem Solving Many problem solving situations require us to count the number of elements in a set. It is helpful to remember our problem solving strategies; in this case the *Be Systematic* strategy is a good choice. You should approach the problem in an organized way and list all of the possible elements.

Example: One of the requirements in your math class is to complete an end of the semester project. Your instructor has presented four different projects; you need to rank your top two choices. Label the projects A, B, C and D and write out the possible choices you can make.

Solving this using the *Be Systematic* strategy, we will start with all choices where A is the top selection. They are AB, AC, AD. Note that AA is not a possible selection. Now list those where B is the top choice. They are BA, BC, BD. Continuing, we find CA, CB, CD and finally DA, DB, DC. Putting these together, we find 12 possible ways to make our top two selections.

Notes:

Guided Example

Your friends (A)my, (B)enito, and (C)am are competing in an obstacle course race. One of the components of the race is climbing over a wall. They need to determine who will climb the wall first, second, and third. List all of the ways your friends could climb the wall.

Using the B*e Systematic* strategy, we will start with Amy and look at the possible orderings if Amy is first. They are

ABC, ACB.

Now consider the orderings where Benito is first

BAC, BCA.

Finally, list those where Cam is first

CAB, CBA.

Putting these together, we find the possible orderings are ABC, ACB, BAC, BCA, CAB, CBA.

Practice (Exercises 1 & 3 p. 626)

In these exercises, you are selecting from the set W = {University of (C)onnecticut, (D)uke, University of (F)lorida, University of (N)orth Carolina} of teams who have won the National Championship in division one men's basketball multiple times since 2000. List all of the ways you can make the selection described.

1. Select two teams without repetition; order is not important. For example DD is not allowed and CD is the same as DC.

2. Select two teams without repetition; order is important. For example DD is not allowed and CD is not the same as DC.

Objective 2 – Use tree diagrams to represent counting situations graphically.

Key Terms
With repetition
Without repetition

Summary
Problem Solving The *Three-way Principle* tells us to view problems graphically. In the case of counting, we can view problems using a tree diagram. In this diagram, we begin and draw branches for all of the choices that can be made in the first stage. This is continued for each stage where a selection is made. When making repeated selections from a group of items, it is important to note if the selection is made with repetition (i.e. you can select the same item twice) or without repetition, where an item is removed from consideration after being selected. These are also called with replacement and without replacement.

Example: A game involves tossing a six-sided die and a quarter. Use a tree diagram to illustrate the possible upward faces when the quarter and die land.

It does not matter which we consider first, the quarter or the die. Let's start with the die. Notice, as shown below, that there are six possible outcomes.

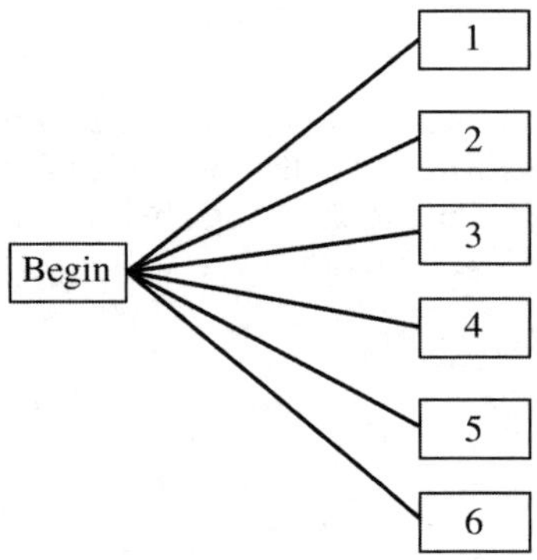

For each of these branches we could find the quarter to land heads side up (H) or tails side up (T). Now we include the branches for the coin:

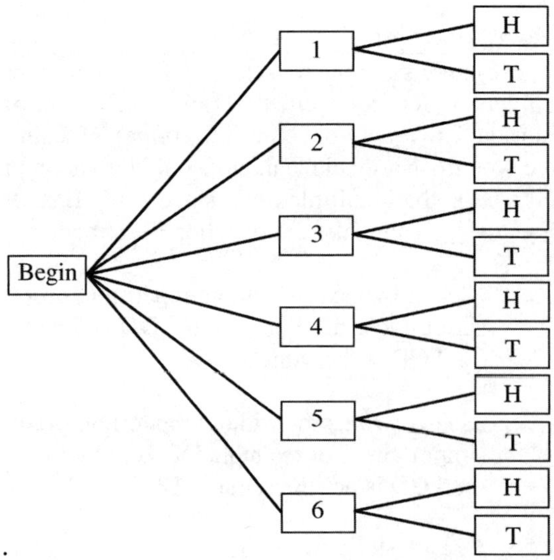

To list the possible outcomes, we start at begin and move across the branches. Starting from the top, we have the following set {1H, 1T, 2H, 2T, 3H, 3T, 4H, 4T, 5H, 5T, 6H, 6T}

Notes:

Guided Example	***Practice*** (Exercises 5 & 7 p. 626)
A restaurant offers their fish of the day (G)rilled or (F)ried and with a choice of one of three seasonings. Draw a tree diagram that illustrates the different ways to order the fish of the day. Draw a tree diagram where the first stage is G or F (grilled or fried) and the second is 1, 2, 3 for the three seasonings. 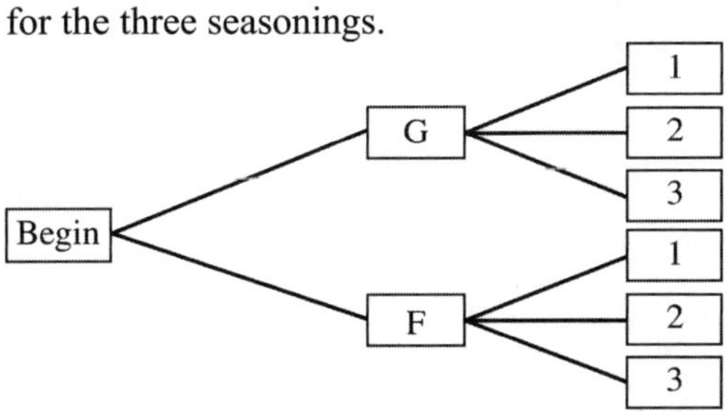 Choices are G1, G2, G3, F1, F2, F3.	Draw a tree diagram that illustrates the different ways to flip a penny, nickel, dime and quarter. Use this diagram to solve the following. 1. In how many ways can we get exactly one head? 2. In how many ways can you get exactly two tails?

Objective 3 – Determine the number of branches in a tree without drawing the tree.

Key Terms
None

Summary
In cases where we don't need a complete listing of the elements in a set, but only need to know how many there are, we can visualize the tree without drawing it. This is particularly helpful when there are a large number of branches at each stage.

Problem Solving A strategy we use when visualizing the tree is *Relate a New Problem to an Older One*. In the previous examples, we saw that the number of branches at each stage was equal to the number of ways to make the selection. We can use this information to help us "see" the tree.

Example: The homecoming committee needs to select a theme and a slogan. There are 4 themes and 3 slogans under consideration. In how many ways can the committee make their selections?

If you visualize the first stage of the tree, there would be 4 branches for the theme selection. For each of these branches, there would be 3 branches in the next stage for slogan. This would result in $4\times3=12$ possibilities for the theme/slogan combination.

Notes:

Guided Example	***Practice*** (Exercise 19 p. 627)
Ryota is an avid runner who exercises daily. He has three pairs of running shoes, five pairs of athletic shorts, and seven shirts. In how many ways can he select his outfit for the daily run? Visualize the tree diagram for a selection of shoes, shorts, and shirt. The first stage would have three branches, one for each of the shoes that could be selected. At the next stage, there would be five branches at each shoe selection for selection of shorts. Finally, for each of the shoe/short selections, there would be seven branches for shirt. We would have $3\times5\times7=105$ total branches, representing the different possible outfits.	1. Imagine that you take six tops, five pairs of pants, and four jackets. How many different outfits are possible from these selections?

Section 12.2 The Fundamental Counting Principle

Objective 1 – Use the fundamental counting principle. **Objective 2** – Use slot diagrams to organize information in counting problems. **Objective 3** – Solve counting problems with special conditions.

Objective 1 – Use the fundamental counting principle.

Key Terms
Fundamental Counting Principle

Summary
The Fundamental Counting Principle (FCP) states that if we want to perform a series of tasks and the first task can be done in *a* ways, the second can be done in *b*, the third can be done in *c*, and so on, then all of the tasks can be done in $a \times b \times c \times ...$ ways.

Problem Solving Using the FCP, we are able to count the number of ways a task can be performed without having to draw a tree diagram. This is just what we did when visualizing the tree diagrams in section 12.1. When we use the FCP, we are using the strategy *Relate a New Problem to an Older One.*

Example: In the game Monopoly, two dice are rolled to determine the number of spaces the player will move their piece. In how many ways can the dice land?

The first die can land with any of 6 faces up; the second can also land with any of 6 sides up. Using the FCP, there are $6 \times 6 = 36$ ways the dice can land.

Notes:

Guided Example	***Practice*** (Exercises 1 & 5 p. 634)
There are fifteen members of the Circle K club on your campus. They are meeting to select a president, vice president, and treasurer. In how many ways can they make the selection from their fifteen members? They could select any one of the 15 members for president. This leaves 14 members who could be selected for vice president, and then 13 for treasurer. Using the FCP, there are $15 \times 14 \times 13 = 2,730$ ways to select the officers.	1. The board of an Internet start-up company has seven members. If one person is to be in charge of marketing and another in charge of research, in how many ways can these two positions be filled? 2. Kevin wants a new paddle board, and will buy a package consisting of a board, custom fin, and paddle. If he can select from ten boards, six fins, and five paddles, how many packages are available for him?

Objective 2 – Use slot diagrams to organize information in counting problems..

Key Terms
Slot diagram

Summary
Problem Solving A slot diagram gives us a systematic way to identify the number of ways we can perform the tasks before we use the fundamental counting principle. When we construct slot diagrams, we are using the *Be Systematic* strategy.

Example: Michael is making trail mix and plans to include nuts, fruit, candy, and seeds. He will select one type of nut from three available; one of four different types of fruit; one type of candy from three choices, and one of two different types of seeds. In how many ways can Michael make the trail mix?

We will construct a slot diagram with 4 spaces $\underline{\textit{nut choice}}\times\underline{\textit{fruit choice}}\times\underline{\textit{candy choice}}\times\underline{\textit{seed choice}}$ and then fill each space with the number of ways each selection can be made. We find that there are $\underline{3}\times\underline{4}\times\underline{3}\times\underline{2}=72$ ways to make the trail mix.

Notes:

Guided Example	***Practice*** (Exercises 13 & 15 p. 635)
Members at a fitness club are assigned an alphanumeric ID. The ID consists of a letter followed by 3 numbers between 0 and 9. The numbers may be repeated. How many different IDs are there? We will construct a slot diagram with 4 blank spaces. $\underline{\textit{letter}}\times\underline{\textit{number}}\times\underline{\textit{number}}\times\underline{\textit{number}}$ Now fill each space with the number of ways each task can be performed and use the FCP to find the solution. $\underline{26}\times\underline{10}\times\underline{10}\times\underline{10}=26{,}000$	Using the digits 0, 1, 2, …, 8, 9, determine how many of each type of four-digit number can be constructed. 1. Zero cannot be used for the first digit; digits may be repeated. 2. The number must be odd and greater than 5,000; digits may be repeated.

Objective 3 – Solve counting problems with special conditions.

Key Terms:
None

Summary
Problem Solving In some counting problems, there are specific conditions that require special consideration. When faced with problems of this type, it is a good idea to *Try a Simpler Version of the Problem* and then work on the problem at hand.

Example: A student ID number consists of 3 numbers and 2 letters. The letters must be next to one another, but are not required to occupy any specific place in the ID. Neither letters nor numbers may be used more than once.

Let's solve a simpler problem! Imagine that the ID number starts with two letters. Then, using a slot diagram, we determine that there would be $\underline{26}\times\underline{25}\times\underline{10}\times\underline{9}\times\underline{8} = 468{,}000$ possible ID numbers.

Now, let's generalize and solve the problem as posed. If the pair of letters can appear in any two adjacent slots, we can generalize and have any of the four following options:

$\underline{letter}\times\underline{letter}\times\underline{number}\times\underline{number}\times\underline{number}$

$\underline{number}\times\underline{letter}\times\underline{letter}\times\underline{number}\times\underline{number}$

$\underline{number}\times\underline{number}\times\underline{letter}\times\underline{letter}\times\underline{number}$

$\underline{number}\times\underline{number}\times\underline{number}\times\underline{letter}\times\underline{letter}$

Notice that there are four different locations for the letters, and for any one of the four, there would be 468,000 possible ID numbers. Thus there are $4\times26\times25\times10\times9\times8 = 4\times468{,}000 = 1{,}872{,}000$ possibilities.

Notes:

Guided Example	***Practice*** (Exercises 35 & 37 p. 636)
The five starters for your basketball team are lining up at center court for the National Anthem. They may line up in any order, but the two seniors will stand next to one another. In how many ways can the five line up? If the seniors are in the two left most spots, we would have $\underline{senior}\times\underline{senior}\times\underline{\quad}\times\underline{\quad}\times\underline{\quad}$ Since there are 2 seniors and 3 underclassmen, there would be $\underline{2}\times\underline{1}\times\underline{3}\times\underline{2}\times\underline{1} = 12$ ways to arrange them. Looking at the big picture, there are four places to arrange the seniors, $\underline{senior}\times\underline{senior}\times\underline{\quad}\times\underline{\quad}\times\underline{\quad}$ $\underline{\quad}\times\underline{senior}\times\underline{senior}\times\underline{\quad}\times\underline{\quad}$ $\underline{\quad}\times\underline{\quad}\times\underline{senior}\times\underline{senior}\times\underline{\quad}$ $\underline{\quad}\times\underline{\quad}\times\underline{\quad}\times\underline{senior}\times\underline{senior}$ Putting this all together, there are $4\times2\times1\times3\times2\times1 = 4\times12 = 48$ ways for them to line up.	Assume that we wish to seat three men and three women-Alex, Bonnie, Carl, Daria, Edith, and Frank-in a row of six chairs. Answer the following using the fundamental counting principle. Where appropriate, first identify the separate tasks involved in making the arrangements. 1. In how many ways can we seat these people with no restrictions? 2. In how many ways can we seat these people if all the men must sit together and all of the women must sit together?

Section 12.3 Permutations and Combinations

Objective 1 – Calculate the number of permutations of n objects taken r at a time.
Objective 2 – Use factorial notation to represent the number of permutations of a set of objects.
Objective 3 – Calculate the number of combinations of n objects taken r at a time
Objective 4 – Apply the theory of permutations and combinations to solve counting problems.

Objective 1 – Calculate the number of permutations of n objects taken r at a time.

Key Terms
Permutation

Summary
A permutation is an ordering of distinct objects. We define $P(n,r)$ to be the number of ways we can select r different objects from a set of n and arrange them in order. The selections are made without replacement, and order of selection is important!

Problem Solving To determine the number of permutations, we will *Relate a New Problem to an Older One* and use the FCP or a slot diagram to find the solution.

Example: The DJ of the college radio station has time to play 3 more songs and will select them from a group of nine. How many playlists are possible? Express your answer using $P(n,r)$ notation.

We will use a slot diagram with
number of ways so select first song × *number of ways to select second* × *number of ways to select third*
There are 9 songs, so the first selection can be made in 9 ways. The second selection is from the 8 remaining, and then the final selection from the 7 that are left. Thus we have $\underline{9}\times\underline{8}\times\underline{7}=504$ ways to select the 3 songs on the playlist $P(9,3)=504$.

Notes:

Guided Example	***Practice*** (Exercises 15 & 17 p. 647)
There are six cars in a NASCAR race. Determine how many ways first, second, and third prizes can be awarded and use $P(n,r)$ notation to express your answer. We will use a slot diagram with the number of ways to fill the following: *First place* × *second place* × *third place* $\underline{6}\times\underline{5}\times\underline{4}=120$ $P(6,3)=120$	Find the number of permutations of 1. Eight objects taken three at a time. 2. Ten objects taken eight at a time.

Objective 2 – Use factorial notation to represent the number of permutations of a set of objects.

Key Terms
Factorial

Summary
The number of ways to arrange n items in order is given by n!, read n factorial. To compute n!, we multiply n by all counting numbers less than n. The formula is $n! = n\cdot(n-1)\cdot(n-2)\cdot(n-3)\cdot...\cdot 2\cdot 1$. We can use factorial notation to give a formula for the number of permutations of n items taken r at a time. We will see that

$$P(n,r) = \frac{n!}{(n-r)!}$$

Example: In the previous section, we computed $P(9,3)$ using a slot diagram. Now we will use the definition

$$P(n,r) = \frac{n!}{(n-r)!}.$$

We substitute $n = 9$ and $r = 3$ to find $P(9,3) = \frac{9!}{(9-3)!} = \frac{9!}{6!} = \frac{9\cdot 8\cdot 7\cdot \cancel{6}\cdot \cancel{5}\cdot \cancel{4}\cdot \cancel{3}\cdot \cancel{2}\cdot \cancel{1}}{\cancel{6}\cdot \cancel{5}\cdot \cancel{4}\cdot \cancel{3}\cdot \cancel{2}\cdot \cancel{1}} = 9\cdot 8\cdot 7$ which is 504.
When using this definition, there will often be terms in the numerator which will cancel with those in the denominator.

Notes:

Guided Example

1. Calculate $(7-2)!$

When solving this remember to perform operations inside the parentheses first. Thus we have
$(7-2)! = 5! = 5\cdot 4\cdot 3\cdot 2\cdot 1 = 120$

2. Calculate $P(5,2)$

Using the definition, we compute
$$P(5,2) = \frac{5!}{(5-2)!} = \frac{5!}{3!} = \frac{5\cdot 4\cdot \cancel{3}\cdot \cancel{2}\cdot \cancel{1}}{\cancel{3}\cdot \cancel{2}\cdot \cancel{1}} = 20$$

Practice (Exercises 3 and 9 p. 647)

1. Calculate $(8-5)!$

2. Calculate $P(6,2)$.

Objective 3 – Calculate the number of combinations of *n* objects taken *r* at a time.

Key Terms

Combination

Summary

If we are interested in finding the number of ways to select *r* items from *n* where order of selection is not important, then we do not use permutations. In this case, we count the number of combinations. We can use factorials to find the number of combinations of *r* items from *n*. The definition is

$$C(n,r)=\frac{P(n,r)}{r!}=\frac{n!}{r!\cdot(n-r)!}$$

Example: You are living in a suite with 5 roommates. In how many ways can you select two of your roommates to join you in a team for a three on three basketball tournament?

We use $C(n,r)$ because the order in which you select the players is not important. Substituting $n=5$ and $r=2$, we have $C(5,2)=\frac{5!}{2!\cdot(5-2)!}=\frac{5!}{2!\cdot 3!}=\frac{5\cdot 4\cdot \cancel{3}\cdot \cancel{2}\cdot \cancel{1}}{2\cdot 1\cdot \cancel{3}\cdot \cancel{2}\cdot \cancel{1}}=10$.

Notes:

Guided Example

Find the number of combinations of seven objects taken 2 at a time.

$$C(7,2)=\frac{7!}{2!(7-2)!}=\frac{7!}{2!5!}$$
$$=\frac{7\cdot 6\cdot \cancel{5}\cdot \cancel{4}\cdot \cancel{3}\cdot \cancel{2}\cdot \cancel{1}}{2\cdot 1\cdot \cancel{5}\cdot \cancel{4}\cdot \cancel{3}\cdot \cancel{2}\cdot \cancel{1}}$$
$$=21$$

Practice (Exercises 19 & 21 p. 647)

Find the number of combinations of

1. Eight objects taken 3 at a time.
2. Ten objects taken eight at a time.

Objective 4 – Apply the theory of permutations and combinations to solve counting problems.

Key Terms
None

Summary
Often we combine counting methods when solving application problems. In some cases, we will use permutations and combinations to solve individual parts, and then the fundamental counting principle to complete the problem.

Problem Solving When solving these problems, it is good to employ the *Be Systemic* strategy. Think about the problem in an organized way. If the problem involves several parts, identify each and determine if selection of items is with or without replacement. If selection is without replacement, then you might use combinations or permutations. Which one fits each piece of the problem? The factor that determines your answer is order. If order of selection is important, then you should use permutations, if order of selection is not important you should use combinations.

Example: The panel for a disciplinary committee consists of four persons selected from six students and eight faculty. If the committee must have an equal number of students and faculty, in how many ways could the committee be selected?

Using the *Be Systematic* strategy, we should think of the two decisions that need to be made; we must select two students from the six available and two faculty from eight. There is no indication that order is important in making the selections, so we will use combinations. The first stage is to determine the number of ways to select the students. Since there are six students and we need two, the number of ways to select the students is $C(6,2)=15$. Similarly, the number of ways to select the faculty, which represents the second stage of the problem, also uses combinations and is $C(8,2)=28$. Our final step is to use the fundamental counting principle to incorporate the number of ways to select both the students and the faculty. The committee could be made in $15\times28=420$ ways.

Notes:

Guided Example	***Practice*** (Exercises 55 & 57 pp. 648- 649)
Rand Smith is making a four digit password using the letters in his name. Letters may not be repeated. The first two letters will be selected from his first name, and the last two will come from his last name. How many passwords does he have to select from? Since order is important in passwords, we will use permutations. We will be systematic and first determine the number of ways to select two letters from Rand's first name. This selection can be done in $P(4,2)=12$ ways. Next we select two letters from the five in his last name. This can be done in $P(5,2)=20$ ways. Using the fundamental counting principle, we determine that there are $12\times20=240$ different passwords.	1. A password for a computer consists of three different letters of the alphabet followed by four different digits from 0 to 9. How many different passwords are possible? 2. HazMat, Inc. is sending a group of eight people to attend a seminar on toxic waste disposal. They will send two of eight engineers and three of nine crew supervisors, and the remainder will be chosen from the five senior managers. In how many ways can these choices be made?

Section 12.4 Counting and Gambling

Objective 1 – Use the fundamental counting principle to analyze counting problems that occur in gambling.
Objective 2 – Use the theory of permutations and combinations to compute the number of ways to draw various card hands.

Objective 1 – Use the fundamental counting principle to analyze counting problems that occur in gambling.

Key Terms
none

Summary
The slot machine is a classic casino game. A player pulls a lever (or pushes a button), the wheels spin, and depending on the arrangement of the wheels, the player may receive a payoff. We will look at the ways the wheels can land.

Problem Solving To determine the number of ways the wheel can land, we will *Be Systematic* and use the FCP or a slot diagram to find the solution.

Example: Imagine that a slot machine has three wheels with symbols as in the figure below. When the player pulls the lever, each wheel spins separately. In how many ways can the three wheels stop?

Wheel 1	Wheel 2	Wheel 3
✈	📖	📖
✂	🔔	✂
❀	☺	📖
🔔	🔔	☺
✂	✂	✈
✈	✈	☺
🔔	📖	📖
☺	☺	❀
📖	✈	❀
✂	🔔	🔔

Since each of the wheels has 10 spaces, using the fundamental counting principle we determine the solution

$$\underline{\textit{number of ways first wheel can stop}} \times \underline{\textit{number of ways second wheel can stop}} \times \underline{\textit{number of ways third wheel can stop}}$$

$= \underline{10} \times \underline{10} \times \underline{10}$

$= 1{,}000$

Notes:

Guided Example	***Practice*** (Exercises 1 & 3 p. 655)
Use the slot machine figure on the previous page to determine the number of ways the wheel could land with bells on all three wheels. We again use the fundamental counting principle. Since the first wheel has 2 bells, the second has 3 bells, and the third has one, the number of ways all three could land with bells is $\underline{2}\times\underline{3}\times\underline{1}=6$	Use Figure 12.15 on page 651 of the text to answer the following. 1. In how many ways can we obtain cherries on only the first two wheels? 2. In how many ways can we obtain oranges on all three wheels?

Objective 2 – Use the theory of permutations and combinations to compute the number of ways to draw various card hands.

Key Terms
Poker hands: royal flush, straight flush, four of a kind, full house, flush, straight, three of a kind, two pairs, one pair

Summary
We can use combinations to determine the number of ways various poker hands can be dealt. We use combinations rather than permutations because the order in which the cards are dealt is not important. After all, many card players organize the cards in their hand to facilitate play.

Problem Solving To determine the number of ways the cards can be dealt, we will *Be Systematic* and break the question into subproblems.

Example: In how many ways can a player be dealt a hand with three Aces and two other cards (not aces and not the same rank)? Assume the cards are dealt from a fair deck of 52 cards.

We will solve this problem using combinations and the fundamental counting principle.
Our first subproblem is to determine the number of ways to be dealt three Aces. Since there are four aces in the deck we use combinations to determine the number of ways to select three Aces from four. Use the combinations formula to verify $C(4,3)=4$.
In the second subproblem, we need to find the number of ways to select the two other cards. The fourth card can be chosen from the 12 remaining ranks in 48 ways, and then the fifth card (which must be a different rank from the others) can be chosen in 44 ways. Thus the last two cards can be selected in $\frac{48\times44}{2}=1,056$ ways. Note that we divide by two because order is not important, and if we had used 48×44 we would be counting queen of hearts and ten of clubs as a different outcome than ten of clubs and queen of hearts.
Putting the two parts of the problem together, there are $4\times1,056=4,224$ ways to be dealt a hand with three Aces and two cards of different rank.

Notes:

Guided Example	***Practice*** (Exercises 7 & 9 p. 655)
In how many ways can we construct a five-card poker hand that contains a pair of kings and a pair of queens? We will consider a series of subproblems and then use the FCP to determine the final solution. We can choose the pair of kings in $C(4,2)=6$ ways and the queens in $C(4,2)=6$ ways. The two pairs can be selected in $6\times 6=36$ ways. The fifth card can be any one of the 44 remaining cards (we use 44 and not 48 here, because of the 48 remaining cards, 4 are either kings or queens so cannot be the fifth card or we would have a full house). Now, using the FCP we find that there are $36\times 44=1,584$ ways.	1. In how many ways can we obtain a royal flush (10, J, Q, K, and A, all in the same suit)? 2. To construct a flush, we first select a suit and then choose five cards, without regard to order, from that suit. a. In how many ways can we choose the suit? b. In how many ways can we choose the five cards? c. In how many ways can we construct a flush?

Chapter 12 Answers

Section 12.1

Objective 1	1) CD, CF, CN, DF, DN, FN 2) CD, CF, CN, DC, DF, DN, FC, FD, FN, NC, ND, NF
Objective 2	1) 4 2) 6
Objective 3	1) 120

Section 12.2

Objective 1	1) 42 2) 300
Objective 2	1) 9,000 2) 2,500
Objective 3	1) 720 2) 72

Section 12.3

Objective 1	1) 336 2) 1,814,400
Objective 2	1) 6 2) 30
Objective 3	1) 56 2) 45
Objective 4	1) $P(26,3)\times P(10,4) = 78,624,000$ 2) 23,520

Section 12.4

Objective 1	1) 120 2) 20
Objective 2	1) 4 2) a. 4 b. 1,287 c. 5,148 - 40 = 5,108

Chapter 13: Probability

Section 13.1 The Basics of Probability Theory

Objective 1 – Calculate probabilities by counting outcomes in a sample space.

Objective 2 – Use counting formulas to compute probabilities.

Objective 3 – Perform computations using the relationship between probability and odds.

Objective 4 – Use probability theory to investigate genetic diseases.

Objective 1 – Calculate probabilities by counting outcomes in a sample space.

Key Terms

Experiment
Outcomes
Sample space
Event
Probability
Empirical assignment

Summary

An experiment is any observation of random phenomenon. The different possible results are called outcomes; the set of all possible outcomes for an experiment is called a sample space. An event is a subset of the sample space.

Example: Imagine we toss a fair coin and a fair six-sided die. What are the possible outcomes of the experiment? What outcomes are included in the event "die lands with the five dot side up?"

There are 12 outcomes for this experiment. They are {(H, 1), (H, 2), (H, 3), (H, 4), (H, 5), (H, 6), (T, 1), (T, 2), (T, 3), (T, 4), (T, 5), (T, 6)}. The outcomes included in the event "die lands with the five dot side up" are {(H, 5), (T, 5)}.

The probability of an outcome in a sample space is a number between 0 and 1 inclusive. The sum of the probabilities of all the outcomes in the sample space must be 1. The probability of an event E, written $P(E)$, is defined as the sum of the probabilities of the outcomes that make up E.

If E is an event and we perform an experiment several times, then we estimate the probability of E as follows: $P(E) = \dfrac{\text{number of times } E \text{ occurs}}{\text{the number of times the experiment is performed}}$. The ratio is sometimes called the relative frequency of E.

Notes:

Guided Example	***Practice*** (Exercises 9 & 27 p.672-673)
Consider again the experiment where a fair coin and fair die are tossed. Define event E to be the event that the die lands with the five dot side up. Determine $P(E)$. The listing of the elements in the sample space, above had 12 elements. There were two elements in the outcome set "die lands with five dot side up." Using the relative frequency approach, the probability of E can be found $P(E) = \frac{\text{number of times } E \text{ occurs}}{\text{the number of times the experiment is performed}}$ $= \frac{2}{12} = \frac{1}{6}.$	1. We are rolling two four-sided dice having the numbers 1, 2, 3, and 4 on their faces. Outcomes in the sample space are pairs such as (1, 3) and (4, 4). a. How many elements are in the sample space? b. Express the event "the total showing is even" as a set. c. What is the probability that the total showing is even? d. What is the probability that the total showing is greater than six? 2. In a given year, 2,048,861 males and 1,951,379 females were born in the United States. If a child is selected randomly from this group, what is the probability that it is a female?

Objective 2 – Use counting formulas to compute probabilities.

Key Terms
Basic properties of probability

Summary
If E is an event in a sample space S with all equally likely outcomes, then the probability of E is given by the formula $P(E)=\frac{n(E)}{n(S)}$ where $n(E)$ is the number of ways event E can occur, and $n(S)$ is the number of outcomes in the sample space.

The following basic properties of probability are always true. Assume that S is a sample space for some experiment and that E is an event in S.

$1.\ 0 \le P(E) \le 1 \qquad 2.\ P(\varnothing)=0 \qquad 3.\ P(S)=1.$

Problem Solving: Remember the *Always Principle.*; the basic rules of probability must always hold.

Notes:

Guided Example

We are playing a card game where we will draw a 2-card hand randomly from a standard 52-card deck. What is the probability that our two card hand includes the ace of hearts and the king of diamonds?

We assume that all 2-card hands are equally likely. We define E to be the event that we are dealt the ace of hearts and king of diamonds and will use the formula $P(E)=\frac{n(E)}{n(S)}$. Since E is one particular 2-card hand, $n(E) = 1$. The number of ways to select 2 cards from a standard 52 card hand is found using combinations; using our work from chapter 12 verify that $C(52,2)=1,326$.

Putting these together we find $P(E)=\frac{1}{1,326}$.

Practice (Exercise 13 p.672)

1. An experimenter testing for extrasensory perception has five cards with pictures of a (s)tar, a (c)ircle, (w)iggly lines, a (d)ollar sign, and a (h)eart. She selects two cards without replacement. Outcomes in the sample space are represented by pairs such as (s,d) and (h,c).

   a. How many elements are in this sample space?
   b. Express the event "a star appears on one of the cards" as a set.
   c. What is the probability that a star appears on one of the cards?
   d. What is the probability that a heart does not appear?

Objective 3 – Perform computations using the relationship between probability and odds.

Key Terms
Odds

Summary
We often use the word odds to express the notion of probability. When we do this, we usually state the odds against something happening. If the outcomes of a sample space are equally likely, then the odds against an event *E* are simply the number of outcomes that are against *E* compared with the number of outcomes in favor of *E* occurring. We write these odds as $n(E'):n(E)$, where *E'* is the complement of event *E* (*E'* consists of all elements in the sample space that are not in *E*). We can compute the odds against E using probabilities; the odds against *E* are $\frac{P(E')}{P(E)}$. This expression of odds will be in the form $\frac{a}{b}$ which is equivalent to $a:b$ or *a* to *b*.

Notes:

Guided Example	***Practice*** (Exercises 15 & 17 p.673)
The odds against a horse winning the Kentucky Derby are 5:1. What is the probability that the horse will win the race? The odds against are 5 to 1, which means that there are 5 outcomes favoring the horse losing the race, 1 outcome favoring the horse winning the race, and 6 outcomes total. Using our rules for probabilities, we find the probability of the horse winning to be $\frac{1}{6}$.	In the exercises, find a) the probability of the given event. b) the odds against the given event. 1. A total of nine shows when we roll two fair dice. 2. We draw a heart when we select one card randomly from a standard 52-card deck.

Objective 4 – Use probability theory to investigate genetic diseases.

Key Terms

Gene

Punnett Square

Summary

The use of probability in the study of genetics was proposed by Austrian monk Gregor Mendel when cross-breeding plants. He noticed that a characteristic of the plant would disappear in a generation of offspring but then reappear at a later time. He theorized that the plants had some hidden factor that was transmitted to later generations. We now call that factor a gene. Mendel was a pioneer in the area of genetics, which we use to determine the possibility of a disease being transmitted to a child even when not present in the parent. As we will see in the guided example, the possible gene combinations for an offspring can be organized in a table called a Punnett Square.

Notes:

Guided Example

Sickle cell anemia is a serious inherited disease. A person with two sickle-cell genes will have the disease, but a person with only one sickle-cell gene will be a carrier of the gene. If a parent who is a carrier and a parent who has no sickle cell gene have a child, what is the probability that the child will be a carrier?

We denote the gene for sickle-cell anemia with *s* and normal gene with *n*. The Punnett square below indicates the possible combinations for a child.

		Second Parent	
		n	*n*
First Parent	*s*	*sn* carrier	*sn* carrier
	n	*nn* normal	*nn* normal

In two of the four boxes, the child would have one sickle-cell gene and would be a carrier. Therefore, the probability that the child is a carrier is $\frac{2}{4} = \frac{1}{2}$.

Practice (Exercise 35 p.673)

1. One parent who has sickle-cell anemia and one parent who is a carrier have a child. Find the probability that the child is a carrier of sickle-cell anemia. Construct a Punnett square as done in the guided example.

Section 13.2 Complements and Unions of Events.

Objective 1 – Use the relationship between the probability of an event and its complement to simplify problems.

Objective 2 – Calculate the probability of the union of two events.

Objective 3 – Use complement and union formulas to compute the probability of an event.

Objective 1 – Use the relationship between the probability of an event and its complement to simplify problems.

Key Terms
Complement

Summary
Sometimes we need to compute the probability of an event which has many possible outcomes. Trying to determine how many outcomes in the sample space are in the event can be difficult, but determining which outcomes in the sample space which are not included in the event may be easier. We can solve the simpler problem and then use the complement rule below to determine our answer.

Recall that the complement of an event *E*, denoted *E'*, includes the outcomes in the sample space which are not included in the event *E*. Since an element in the sample space occurs with probability 1, we have $P(E') = 1 - P(E)$.

Example: If two fair dice are tossed, the probability that the sum on the upward faces is 5 is $\frac{4}{36} = \frac{1}{9}$. What is the probability that the sum on the upward faces is not 5?

If we define the event *E* to be the sum of the upward faces is 5, $P(E) = \frac{1}{9}$. The probability that the sum on the two faces is not 5 is found using the complement rule, $P(E') = 1 - P(E) = 1 - \frac{1}{9} = \frac{8}{9}$.

Notes:

Guided Example	***Practice*** (Exercise 3 & 7 p.683)
Three fair dice are tossed. What is the probability that at least one of the dice lands with the two dot (also called pip) face up? First we will use our counting rules to determine the number of ways the dice could land. Using the Fundamental Counting Principle, we have $6 \cdot 6 \cdot 6 = 216$ ways. If we were to compute the probability of at least one die with the two pip side up directly, there would be many options to consider. We would need to think about the number of ways to have the dice land with exactly one die having the two pip side up, the number of ways to have two dice landing with the two pip side up, and the number of ways to have all three with the two pip side up. It is much easier to consider the complement to solve the problem. The complement would be that none of the dice land with the two pip side up. For each die, there are 5 sides that do not have two pips. There are $5 \cdot 5 \cdot 5 = 125$ ways for all dice to land with a side other than the two pip side up. We use the complement rule and find *P*(at least one die lands with two pip side up) $= 1 - \frac{125}{216} = \frac{91}{216}$	1. If there is a 1 in 1,000 chance that you will pick the numbers correctly in tonight's' lottery, what is the probability that you will not pick the numbers correctly? 2. If five coins are flipped, what is the probability of obtaining at least one head?

Objective 2 – Calculate the probability of the union of two events.

Key Terms
Union
Mutually exclusive

Summary
Given two events *A* and *B*, if we are interested in the event that *A* or *B* or both occurs, then we use the rule for unions. $P(A \cup B) = P(A) + P(B) - P(A \cap B)$. We subtract $P(A \cap B)$ because the elements in $A \cap B$ are added twice, once with *A* and again with *B*.

Example: At a coffee shop, 75% of the customers purchase (C)offee, 42% purchase a (B)agel, and 23% purchase coffee and a bagel. Find the probability that a randomly selected customer purchases a coffee or a bagel.

Based on the information provided, we know that $P(C) = 0.75$, $P(B) = 0.42$, and $P(C \cap B) = 0.23$. Using the formula above, $P(C \cup B) = P(C) + P(B) - P(C \cap B) = 0.75 + 0.42 - 0.23 = 0.94$.

If *A* and *B* do not share any outcomes, we call them mutually exclusive events. In this case, $A \cap B = \varnothing$ and $P(A \cap B) = 0$. The rule for unions simplifies to $P(A \cup B) = P(A) + P(B)$.

Notes:

Guided Example	***Practice*** (Exercises 9 & 15 p.683)
Use the rule $P(A \cup B) = P(A) + P(B) - P(A \cap B)$ and the fact $P(A) = 0.38$, $P(B) = .045$ and $P(A \cup B) = 0.67$ to find $P(A \cap B)$. We will substitute the given information into the formula and use algebra to solve for the unknown probability. $0.67 = 0.38 + 0.45 - P(A \cap B)$. This simplifies to $0.67 = 0.83 - P(A \cap B)$ and we find $P(A \cap B) = 0.83 - 0.67 = 0.16$.	1. If a single card is drawn from a standard 52 card deck, what is the probability that it is either a five or a red card? 2. Given events *A* and *B* with $P(A \cup B) = 0.70$, $P(A) = 0.40$, $P(A \cap B) = 0.25$, find $P(B)$.

Objective 3 – Use complement and union formulas to compute the probability of an event.

Key Terms
None

Summary
Given two events A and B, we can use the complement formula and the formula for unions to compute the probability that neither A nor B occurs.

Notes:

Guided Example	***Practice*** (Exercises 29 & 31 p.684)
At a coffee shop, 75% of the customers purchase (C)offee, 42% purchase a (B)agel, and 23% purchase coffee and a bagel. Find the probability that a randomly selected customer purchases neither a coffee nor a bagel. In the previous worksheet we found $P(C \cup B) = 0.94$. The event "purchases neither a coffee nor a bagel" is the complement of $C \cup B$ so we find the solution by subtracting $P(C \cup B)$ from 1. The probability that someone purchases neither a coffee nor a bagel is 0.06. **Problem Solving:** It is helpful to use a Venn Diagram to visualize the sample space. We use the *Draw Pictures* strategy. In the Venn diagram below, the shaded area represents the event purchases neither coffee nor bagel. 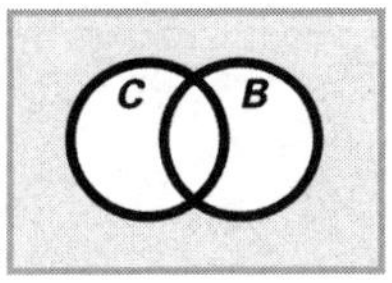	1. If we draw a card from a standard 52 card deck, what is the probability that the card is neither a heart nor a queen? 2. Your car's windshield wipers have not been working properly. The probability that you need a new motor is 0.55, the probability that you need a new switch is 0.4, and the probability that you need both is 0.15. What is the probability that you need neither a new motor nor a new switch?

Section 13.3 Conditional Probability and Intersections of Events

Objective 1 – Be able to compute conditional probabilities.
Objective 2 – Calculate the probability of the intersection of two events.
Objective 3 – Use probability trees to compute conditional probabilities.
Objective 4 – Be able to determine when events are dependent and independent.

Objective 1 – Be able to compute conditional probabilities.

Key Terms
Conditional probability

Summary
In previous worksheets, when computing probabilities the denominator has represented the number of outcomes in the sample space. We now turn to problems where we limit the possible outcomes based on some given information.

When we compute the probability of an event *F* assuming that the event *E* has already occurred, we call this the conditional probability of *F* given *E*. We denote this probability as $P(F \mid E)$. We read $P(F \mid E)$ as "the probability of *F*, given that E has occurred," or "the probability of *F* given *E*."

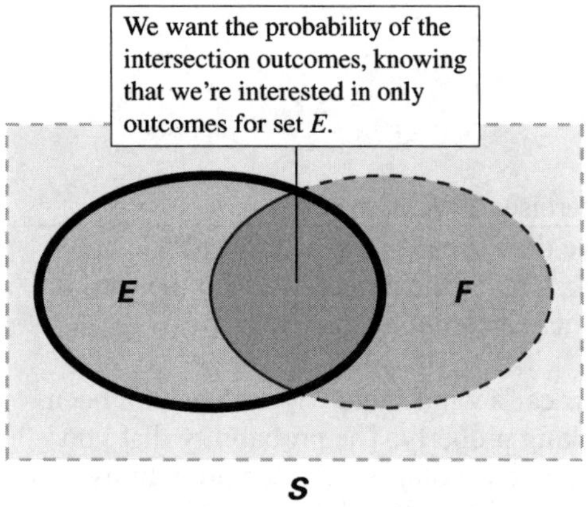

If the outcomes in the sample space are equally likely, we can compute $P(F \mid E)$ using the formula $P(F \mid E) = \dfrac{n(E \cap F)}{n(E)}$. Recall that $n(E)$ represents the number of outcomes in the event *E*. It is important that we divide by the number of outcomes in the event which is given, or known to have occurred.

Example: A card is to be drawn from a standard 52-card deck. What is the probability that the card is a king, given that the card is a heart?

We will use *K* for the event "card is a king" and *H* for "card is a heart". The questions is asking us to compute $P(K \mid H)$. We know that there are 13 hearts in the deck, so $n(H) = 13$. The numerator $n(H \cap K) = 1$ since the only card in the deck which is a heart and a king is the king of hearts. Putting this information together, we find $P(K \mid H) = \dfrac{n(H \cap K)}{n(H)} = \dfrac{1}{13}$.

The general rule for computing $P(F \mid E)$ is $P(F \mid E) = \dfrac{P(E \cap F)}{P(E)}$.

Notes:

Guided Example	***Practice*** (Exercises 3& 7, pp.697-698)
At a coffee shop, 75% of the customers purchase (C)offee, 42% purchase a (B)agel, and 23% purchase coffee and a bagel. Find the probability that a randomly selected customer purchases a bagel, given that they purchase coffee. We use the general rule, $P(B \mid C) = \frac{P(B \cap C)}{P(C)} = \frac{0.23}{0.75} \approx 0.307$. Note that we divide by $P(C)$ because purchasing coffee is the given event.	1. We are rolling two fair dice. First compute $P(F)$ and then $P(F \mid E)$. E – a three shows on at least one of the dice. F – the total is less than 5. 2. We are drawing a single card from a standard 52-card deck. Find $P(\text{seven} \mid \text{nonface card})$.

Objective 2 – Calculate the probability of the intersection of two events.

Key Terms
None

Summary
We can determine a general formula for the intersection of two events using the general rule for computing conditional probability. Recall that $P(F \mid E) = \dfrac{P(E \cap F)}{P(E)}$. Multiply each side of the equation by $P(E)$ to obtain the rule for computing the probability of the intersection of two events; $P(E \cap F) = P(E)P(F \mid E)$.

Problem Solving: As we will see in the guided example and exercises, many of these problems involve sequential selection from a group of items. It is very important that we identify if the selection is done with replacement or without replacement. We use the *Splitting Hairs* principle to ensure that we are answering the question correctly.

Notes:

Guided Examples	***Practice*** (Exercise 17 p.698)
A jar contains 5 blue marbles, 3 red marbles, and 2 yellow marbles. Two marbles are selected at random. Find the probability that both marbles are blue if the marbles are drawn a. without replacement We will define B_1 to be the event first marble is blue and B_2 the event that the second marble is blue. Since 5 of the 10 marbles are blue, $P(B_1)=\frac{5}{10}=\frac{1}{2}$. After the first marble is drawn, it is not replaced, so there are 9 marbles in the jar. The conditional probability $P(B_2 \mid B_1)$ is the probability that the second marble is blue given that the first one was. If the first marble selected was blue, then there are 4 remaining, so $P(B_2 \mid B_1)=\frac{4}{9}$. Using the rule for computing the intersection we obtain $P(B_1 \cap B_2)=P(B_1)P(B_2 \mid B_1)=\frac{1}{\cancel{2}}\cdot\frac{\cancel{4}^{2}}{9}=\frac{2}{9}$ b. with replacement The probability of B_1 does not change, but we need to consider the impact of replacing the marble on $P(B_2 \mid B_1)$. Since the first marble drawn is replaced, there are 10 marbles, including 5 blue when making the second selection. Thus $P(B_2 \mid B_1)=\frac{5}{10}=\frac{1}{2}$. Substituting this new value into the formula yields $P(B_1 \cap B_2)=P(B_1)P(B_2 \mid B_1)=\frac{1}{2}\cdot\frac{1}{2}=\frac{1}{4}$.	1. We draw 2 cards from a standard 52-card deck. Find the probability that we draw a face card followed by a nonface card a. assuming that the cards are drawn without replacement. b. assuming that the cards are drawn with replacement.

Objective 3 – Use probability trees to compute conditional probabilities.

Key Terms
Probability tree

Summary
We can represent an experiment that happens in stages with a tree whose branches represent the outcomes of the experiment. We calculate the probability of an outcome by multiplying the probabilities found along the branch representing that outcome. We call these tree diagrams probability trees.

Problem Solving: In the guided example below, we use the *Draw Pictures* strategy to organize the information in the problem.

Notes:

Guided Example

A call to a customer service help line initially goes to one of three units; 25% to unit *X*, 25% to unit *Y* and 50% to unit *Z*. 6% of the calls answered by *X*, 8% of the calls answered by *Y*, and 15% of the calls answered by *Z* need to be transferred to a manager. Draw a probability tree to illustrate this situation and use the tree to determine the probability that a randomly selected call to the help line needs to be transferred to a manager.

We will let *T* denote the event the call needs to be transferred, and *X, Y, Z* represent the three units. We draw branches to each of the units, and on the branches indicate the probability that a call would be answered by each unit. From each of the call unit branches, we have two branches, one for transferred (*T*) and one for no transfer needed, which is *T'*.

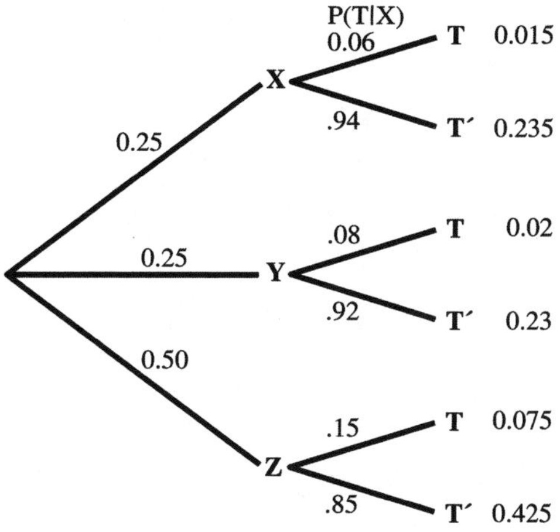

In order to find the probability that a call needs to be transferred, we look at each of the branches which ends at a *T*. The call could go to *X* and need a transfer (probability is 0.015), the call could go to *Y* and need a transfer (probability is 0.02) or it could go to *Z* and need a transfer (probability is 0.075). To find the probability of a transfer to a manager, we sum these three probabilities and find $0.015 + 0.02 + 0.075 = 0.11$.

Practice (Exercises 59 & 61 p.699)

Imagine that you are taking part in a study to test a new cold medicine. Although you don't know exactly what drug you are taking, the probability that it is drug *A* is 0.10 and the probability that it is drug *B* is 0.20 and the probability that it is drug *C* is 0.70. From past clinical trials the probabilities that these drugs will improve your condition are: *A* (0.30), *B* (0.60), *C* (0.70).

1. Draw a probability tree to illustrate this drug trial situation.

2. What is the probability that you will improve?

Objective 4 – Be able to determine when events are dependent and independent.

Key Terms
Independent
Dependent

Summary
We have defined the conditional probability of *F* given *E*, $P(F \mid E)$, to be the probability that *F* occurs given that we know *E* has occurred. Does event *E* impact the chance that *F* will occur? To answer this question we define a new concept; statistical independence. Events *E* and *F* are independent if $P(F \mid E) = P(F)$. If $P(F \mid E) \neq P(F)$ then *E* and *F* are dependent.

Notes:

Guided Example	***Practice*** (Exercises 37 & 39 p.698)
In a previous worksheet, we considered purchases made at a coffee shop. Recall that 75% of the customers purchase (*C*)offee, 42% purchase a (*B*)agel, and 23% purchase coffee and a bagel. We found *P*(*B*|*C*) to be 0.307, indicating that 30.7% of customers ordered a bagel, given that they ordered coffee. Are *B* and *C* independent? Since *P*(*B*) = 0.42 which is not the same as *P*(*B*|*C*), *B* and *C* are dependent. It appears that if someone orders coffee, they are less likely to order a bagel.	In each of the following exercises an experiment and two events are given. Determine if the events are independent or dependent. 1. A card is drawn randomly from a standard 52-card deck; "the card is red" and "the card is a face card." 2. Two dice (one red, one green) are rolled; "the total is greater than nine" and "the total is even."

Section 13.4 Expected Value

Objective 1 – Be able to compute the expected value to predict the long-term results of repeating an experiment.
Objective 2 – Calculate the expected value of lotteries and games of chance.
Objective 3 – Use expected value to solve applied problems.

Objective 1 – Be able to compute the expected value to predict the long-term results of repeating an experiment.

Key Terms
Expected value

Summary
We can use probability to predict what will happen in a situation over the long run. When taking the standardized test, will you earn more points if you guess or leave the answer blank? Is the lottery game you are playing fair? The answer to either of these questions could be answered by determining the expected value. The expected value uses the possible outcomes of an experiment and the associated probabilities to predict an expected outcome.

Assume that an experiment has outcomes numbered 1 to n with probabilities $P_1, P_2, P_3, \ldots, P_n$. Each outcome has a numerical value associated with it and these are labeled $V_1, V_2, V_3, \ldots, V_n$. The expected value of the experiment is $P_1V_1 + P_2V_2 + P_3V_3 + \ldots + P_nV_n$.

Notes:

Guided Example

Even with electronic records, doctor's offices use a lot of paper. An office manager collected data on the number of packages of paper used each day and estimated the probabilities in the following table. Calculate the expected value.

Outcome	Probability
0	0.05
1	0.25
2	0.35
3	0.15
4	0.15
5	0.05

To compute the expected value, we multiply each outcome by the associated probability and sum the results. Doing this yields

$$(0)(0.05)+(1)(0.25)+(2)(0.35)+(3)(0.15)$$
$$+(4)(0.15)+(5)(0.05)=2.25.$$

Over the long run, this office uses 2.25 packages of paper per day. It is good to check that our answer is reasonable; it should be a value within the range of the table (in other words, between 0 and 5).

Practice (Exercise 1 p.707)

1. In the table below, we give the probabilities and values associated with five outcomes of an experiment. Calculate the expected value.

Outcome	Probability	Value
A	0.1	4
B	0.3	6
C	0.4	−2
D	0.15	−4
E	0.05	8

Objective 2 – Calculate the expected value of lotteries and games of chance.

Key Terms
Fair game

Summary
We can use expected value to predict long term winnings (or more likely, losses) when playing games of chance. The expected value tells us how much we would expect to win (or lose) over the long run if a game were played repeatedly. If the expected value of the game is 0, the game is called a fair game. A game in which the expected value is not 0 is called an unfair game.

Notes:

Guided Example	***Practice*** (Exercise 9 p.707)
Fifty tickets are sold at \$100 apiece for a raffle. The winner receives \$1,000. Find the expected value of the game. If it is not a fair game, indicate the ticket price that would make it fair. There are two outcomes for a person purchasing a ticket. They could win, in which case they have a net win of $\$1,000-\$100=\$900$ or they could not win and have a loss, which we indicate as $-\$100$. The probability associated with winning is $\frac{1}{50}=0.02$ and the probability of losing is 0.98. The expected value of the game is $(0.02)(900)+(0.98)(-100)=-80.$ The expected value indicates that for each \$100 ticket, we expect to lose \$80. For the game to be fair, we should pay \$80 less for the ticket, or \$20. We verify that if the ticket cost \$20, the game is fair by changing 900 to 980 and $-\$100$ to $-\$20$ in the formula above and obtain an expected value of 0. $(0.02)(980)+(0.98)(-20)=0.$	In the exercise below, we describe a lottery. First calculate the expected value of the lottery and determine if it is a fair game. If the game is not fair, determine a price for playing the game that would make it fair. 1. The Daily Number lottery costs \$1 to play. You must pick three digits in order from 0 to 9 and duplicates are allowed. If you win, the prize is \$600.

Objective 3 – Use expected value to solve applied problems.

Key Terms
None

Summary
You can use expected value to determine the long run behavior of many different activities. One relevant question to students involves guessing on multiple choice questions. As you will see in the guided example, in cases where there are penalties for wrong answers, as on many standardized tests, you should know if it is better in the long run to guess if you don't know the answer.

Notes:

Guided Example	***Practice*** (Exercise 19 p.708)
You are taking a standardized test where there are multiple choice questions with five choices. You will earn 1 point for a correct answer, lose $\frac{1}{3}$ point for an incorrect answer, and neither receive not lose points if you leave the answer blank. If you encounter a problem where you do not know the answer, are you better off leaving the answer blank or guessing? If you leave the answer blank you will receive 0 points. If you guess, you have a $\frac{1}{5}$ chance of earning 1 point and a $\frac{4}{5}$ chance of losing $\frac{1}{3}$ point. The expected value of guessing is $\left(\frac{1}{5}\right)(1)+\left(\frac{4}{5}\right)\left(-\frac{1}{3}\right)=\frac{1}{5}-\frac{4}{15}=-\frac{1}{15}$. Over the long run, you will lose points if guessing is your strategy (for each 15 questions you guess on, you will expect to lose 1 point).	1. A student is taking the GRE, consisting of several multiple-choice questions. One point is awarded for each correct answer. Questions left blank neither receive nor lose points. If there are four options for each question and the student is penalized $\frac{1}{4}$ point for each wrong answer, is it in the student's best interest to guess? Explain.

Section 13.5 Binomial Experiments

Objective 1 – Be able to compute binomial probabilities.

Objective 2 – Use binomial probability to solve applied problems.

Objective 1 – Be able to compute binomial probabilities.

Key Terms

Binomial experiment

$B(n,k;p)$

Summary

There is a class of experiments in probability theory that have similar characteristics and we provide a formula that can be used to compute probabilities for these experiments. The experiments are similar in that they consist of repeating an activity a certain number of times, where each repetition (also called a trial) results in one of two possible outcomes. A sequence of trials is called a binomial experiment if it has the following properties:

1. The experiment is performed for a fixed number of trials (n trials).
2. Each trial has only two outcomes, often denoted success and failure.
3. The probability of success is the same for each trial.
4. The trials are independent.

If we identify an experiment as being a binomial experiment, we can compute the probability of exactly k successes in n trials using the formula $C(n,k)(p)^k(1-p)^{n-k}$ where p is the probability of success on any trial. We will denote this probability by $B(n,k;p)$.

Notes:

Guided Example	***Practice*** (Exercise 9 p.715)
A binomial experiment involves $n = 6$ trials with probability of success 0.20 on each trial. What is the probability of exactly 1 success in the 6 trials? Using our $B(n,k;p)$ notation, we are looking for $B(6,1;0.2)$ Substituting into the formula $C(n,k)(p)^k(1-p)^{n-k}$ we find $B(6,1,0.2) = C(6,1)(0.2)^1(1-0.2)^{6-1}$ $= 6(0.2)(0.8)^5$ $= 0.393216$ or approximately 0.393.	1. Compute the value of $B\left(5,3;\frac{1}{4}\right)$.

Objective 2 – Use binomial probability to solve applied problems.

Key Terms
None

Summary
There are many applications where binomial probabilities can be applied. In order to use the binomial probability formula, you should first check to see that the experiment satisfies all of the required properties.

Notes:

Guided Example	***Practice*** (Exercise 19 p.715)
There have been reports that 20% of bottled water brands contain tap water. If you were to independently sample 10 water brands, what is the probability that none of them would contain tap water? Use this to determine the probability that at least 1 of the 10 contains tap water. First we will check the properties. We have a fixed number of trials (10), and in each case there are two outcomes. In this example, we are counting the number of bottles that contain tap water, so a "success" is defined as the bottle containing tap water. We are assuming independent sampling (which would require us to sample with replacement) and a fixed probability of success equal to 0.2. To find the probability that none of the bottles contain tap water, we need $B(10,0;0.2)$. Using the formula, we obtain $B(10,0;0.2) = C(10,0)(0.2)^0(1-0.2)^{10-0}$ $= 1(1)(0.8)^{10} \approx 0.107.$ Finding at least one bottle with tap water is the complement of finding no bottles containing tap water. We can use the complement rule; the probability of at least one of the bottles containing tap water is $1-0.107=0.893$.	Chase Utley currently has a 0.250 batting average. If we ignore all other factors, what is the probability that if he comes to bat five times he gets exactly two hits in today's game? Assume the at bats are independent.

Chapter 13 Answers

Section 13.1

Objective 1 — 1) a. 16 b {(1, 1), (1, 3), (2, 2), (2, 4), (3, 1), (3, 3), (4, 2), (4, 4)} c. 0.5 d) 0.1875 2) 0.49

Objective 2 — 1) a. 20 b. { (s, c), (s, w), (s, d), (s, h), (c, s), (w, s), (d, s), (h, s)} c. $\frac{2}{5}$ d. $\frac{3}{5}$.

Objective 3 — 1) a. $\frac{1}{9}$ b. 8 to 1 2) a. $\frac{1}{4}$ b. 3 to 1.

Objective 4 — 1) The probability that the child is a carrier is $\left(\frac{1}{2}\right)$

		Second Parent	
		s	*n*
First	*s*	*ss*	*sn*
Parent	s	*ss*	*sn*

Section 13.2

Objective 1 — 1) $\frac{999}{1,000}$ 2) $\frac{31}{32}$

Objective 2 — 1) $\frac{7}{13}$ 2) 0.55

Objective 3 — 1) $\frac{9}{13}$ 2) 0.2

Section 13.3

Objective 1 — 1) $\frac{1}{6}$, $\frac{2}{11}$ 2) $\frac{1}{10}$

Objective 2 — 1) a. $\frac{40}{221}$ b. $\frac{30}{169}$

Objective 3 — 1) 2) 0.64

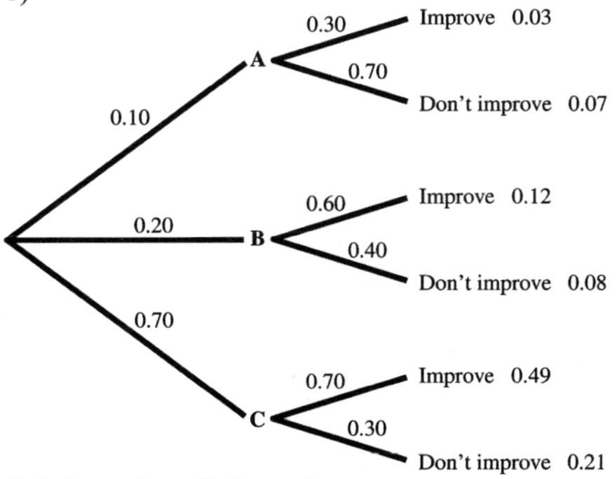

Objective 4 — 1) independent 2) dependent

Section 13.4

Objective 1 — 1) 1.2

Objective 2 — 1) −\$0.40; \$0.60

Objective 3 — 1) Student should guess; expected value is $\frac{1}{16}$

Section 13.5

Objective 1 — 1) 0.0879

Objective 2 — 1) 0.2637

Chapter 14: Descriptive Statistics

Section 14.1 Descriptive Statistics

Objective 1 – Explain the difference between a sample and a population.

Objective 2 – Organize data in a frequency table.

Objective 3 – Use a variety of methods to represent data visually.

Objective 4 – Use stem-and-leaf displays to compare data.

Objective 1 – Explain the difference between a sample and a population.

Key Terms

Statistics
Data
Sample
Population
Bias

Summary

The study of statistics involves making sense of information called data. The set of all data that we are interested in describing is called the population. Sometimes we are not able to access all of the data in a population; it may be too expensive, time consuming, or simply not reasonable. In such cases we take a sample from the population. It is important that the sample data be representative of the population from which the sample is selected. If it is not, selection bias may occur, which means that the sample does not accurately reflect the population as a whole. There are other types of biases that can impact our results. One is leading question bias, where the way a question is asked impacts the response.

Notes:

Objective 2 – Organize data in a frequency table.

Key Terms
Qualitative data
Quantitative data
Frequency distribution
Frequency
Relative frequencies
Frequency table

Summary
We refer to a collection of information as data. Data can be qualitative or quantitative. Qualitative data consists of information about a characteristic such as eye color or gender, whereas quantitative data gives a meaningful numerical measurement.

In order to organize qualitative data, we construct a frequency distribution, and list the count, or frequency with which each value occurs. We could also compute the relative frequencies, where we divide the frequency by the number of observations in the data set.

Example: Your school's basketball coach is collecting data on semester averages (in letter grades) for players on the team who are receiving scholarships. The data is {B, C, B, A, D, B, C, A, B, B, C, D, A}. Organize this data in a frequency table, and include a column for relative frequencies.

To construct the table, we assign a row for each of the grades that occur in the set. We report the number of times each grade occurs, and then divide by the total number (13) to compute the relative frequencies.

Grade	Frequency	Relative Frequency
A	3	$\frac{3}{13} \approx 0.231$
B	5	0.385
C	3	0.231
D	2	0.154

When data is quantitative, we can construct a frequency table with rows for each value, but often there are so many values that the list would be very long. Another option is to group values into classes and then determine the number of data values in each class. If we group the values, it is important that each data value falls in exactly one class, and that each class has the same number of possible values.

Example: The number of passengers per car on an Amtrak Acela train is reported in the set below. Construct a frequency table and a relative frequency table.
{38, 39, 39, 32, 40, 38, 30, 37, 40, 28, 38, 33, 30. 37, 40, 37, 40, 38, 20, 30}

There are 20 values in this data set, and while they are similar, it would be good to group them as they range from 20 to 40. We will group them 20 – 24, 25 – 29, 30 – 34, 35 – 39, and 40 – 44. Note that each value will fit in only one group (there are no overlaps) and that each group includes 5 possible values.

Number of Passengers	Frequency	Relative Frequency
20 – 24	1	0.05
25 – 29	1	0.05
30 – 34	5	0.25
35 – 39	9	0.45
40 – 44	4	0.20

Objective 3 – Use a variety of methods to represent data visually.

Key Terms
Bar graph
Histogram

Summary
By presenting data graphically, we can observe patterns more easily. A bar graph is one way to visualize a frequency distribution for qualitative data. We list the possible classes or groups on the horizontal axis and the frequencies on the vertical axis.

Example: Use the basketball grade data from the previous worksheet to construct a bar graph. The data is {B, C, B, A, D, B, C, A, B, B, C, D, A}.

We obtained the frequency table below; the graph provides a visual representation of the data.

Grade	Frequency	Relative Frequency
A	3	$\frac{3}{13} \approx 0.231$
B	5	0.385
C	3	0.231
D	2	0.154

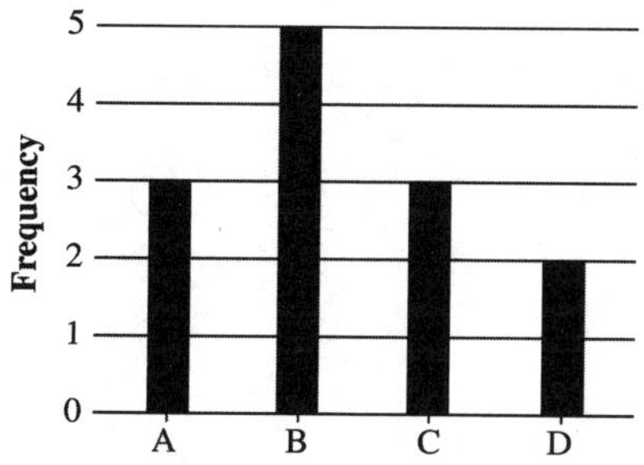

We see that more athletes earned a B than any other grade, and notice that there are the same number of A grades as C grades.

When data is quantitative, we use a graph called a histogram to visually represent the frequency distributions. If the data is discrete (takes on only specific values, but not arbitrary values between them) we can make a class for each possible value, but as we found in the previous worksheet, it is often helpful to group the data into classes. When the data is continuous (can take on values between stated values, and we cannot list all of them), we must group the data into classes. When constructing such a graph for continuous data, it is helpful to state where the first class will start and the class width (the amount you add to the starting point to determine the starting value for the next class).

Notes:

Guided Example

Construct a histogram for the data below using a class width of 2 starting the first class at 10.

{10.2, 13.5,14.7, 11.3, 16.5, 18, 12.1, 11.6, 10.8, 13.8. 19.5, 16.6, 12.5, 11.7, 19.9}

First we will construct a frequency table. We start the first class at 10 and the next at 12 since the class width is 2. In order to be clear about where a value of 12 would be placed, we use the notation 12 +.

Classes	Frequency
10+ – 12	5
12+ – 14	4
14+ – 16	1
16+ – 18	2
18+ – 20	3

To construct the histogram, we plot the possible values of the variable on the horizontal axis, and the frequencies on the vertical axis. It is important to note that the adjacent bars touch.

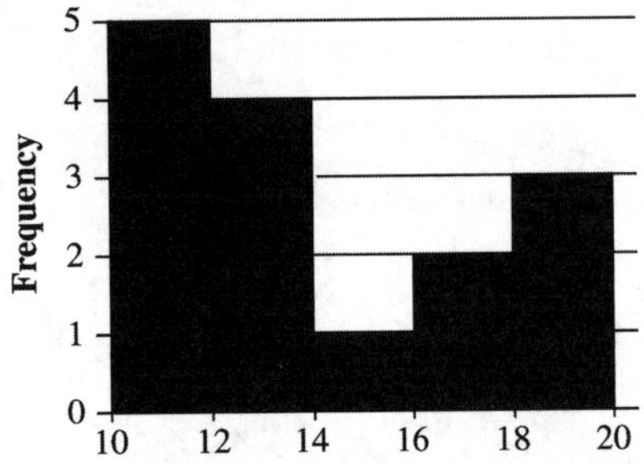

Practice (Exercises 3 & 7 p.729)

1. Construct a histogram for the data in the table below, which gives mileage ratings for 38 domestic cars.

MPG	19	20	21	22	23	24	25	26	27	28	29	30
Frequency	2	5	3	2	1	8	1	2	3	2	7	2

2. Group the data below using a class width 2, starting at 64.5, and construct a histogram. The data represents the heights (in inches) of the players in four Eastern Conference teams of the WNBA for a recent season. Consider the data as a single set of data.

   Chicago Sky: 69, 71, 74, 74, 78, 68, 68, 71, 75, 69, 76, 68, 69, 74, 65, 73

   NY Liberty: 72, 67, 72, 73, 75, 77, 73, 69, 67, 75, 76, 76, 73, 74, 69, 70

   Washington Mystics: 68, 72, 75, 68, 70, 69, 68, 74, 74, 69, 73, 76, 76, 80, 73, 78

   Atlanta Dream: 72, 77, 71, 80, 69, 73, 69, 75, 66, 68, 77, 74, 76, 73, 72, 75

Objective 4 – Use stem-and-leaf displays to compare data.

Key Terms

Stem-and-leaf display

Summary

A stem-and-leaf display is an effective way to display data quickly. To construct a stem-and-leaf display, separate the digits in your data value into two groups based on their place value. The stem must have the larger place value. List the stem values on the left side of a vertical bar and then write all leaf values associated with each stem to the right of the bar.

Notes:

Guided Example

Construct a stem-and-leaf display for the data below.

{10.2, 13.5,14.7, 11.3, 16.5, 18, 12.1, 11.6, 10.8, 13.8. 19.5, 16.6, 12.5, 11.7, 19.9}

We will use the ones place for the stem unit and the tenths place for the leaves. As a reminder numbers greater than 10, say 13 represents 13 ones. We write the possible stem values on the left side of the vertical bar (10 – 19) and then on the line that is appropriate, we write the leaf value. For example, 11.3 goes on the line for a stem of 11, and we have a 3 in the leaf area of the display.

Ones	Tenths
10	28
11	367
12	15
13	58
14	7
15	
16	56
17	
18	0
19	59

If we were comparing this data with another set, you could put the leaves for the second set of data on the other side of the stems, as you will see in the exercise.

Practice (Exercise 9 p.730)

1. Represent the two sets of data below on a combined stem-and-leaf display.

A: 29, 32, 34, 43, 47, 43, 22, 38, 42, 39, 37, 33, 42, 18, 22, 39, 21, 26, 18, 43

B: 32, 38, 22, 39, 21, 26, 28, 16, 13, 20, 21, 29, 22, 24, 33, 47, 23, 22, 18, 33

Section 14.2 Measures of Central Tendency

Objective 1 – Compute the mean, median, and mode of distributions.

Objective 2 – Find the five-number summary of a distribution.

Objective 3 – Apply measures of central tendency to compare data.

Objective 1 – Compute the mean, median, and mode of distributions.

Key Terms
Measures of central tendency
Mean
Median
Mode

Summary
Statisticians are often interested in describing the average of a set of data. There are several different measures of central tendency; we will focus on the mean, median, and mode.

The mean of a set of data is the sum of the data divided by the number of data points in the set. If a data set contains *n* values, the sample mean, $\bar{x}$, is $\bar{x} = \frac{\sum x}{n}$. We will assume that our data sets represent a sample unless otherwise stated. If we have population data, the mean is denoted by the lower case Greek letter μ.

The median of a data set is the middle number. It is important that we first order the data (increasing or decreasing) before we attempt to find the median. If there is an odd number of values in the set, then the median is in the middle position; if there are an even number of values, the median is the mean of the two middle values.

The mode is the data point (or points) which occurs most frequently. It is possible for a data set to have multiple modes or no mode.

Example: Find the mean, median, and mode of the set $\{3, 2, 10, 4, 2, 15\}$.

To find the mean we find the sum of the data, $\sum x = 36$, and divide by $n = 6$ to obtain $\bar{x} = \frac{\sum x}{n} = \frac{36}{6} = 6$. In order to find the median, we need to put the data in order. Since there are an even number of data values, the median will be the mean of the 3rd and 4th values. The ordered data set is $\{2, 2, 3, 4, 10, 15\}$ and the median is $\frac{3+4}{2} = 3.5$. The mode of the data set is 2 because the value 2 occurs more frequently than any other value.

Problem Solving: Using the *Analogies Principle* can help us remember what the median represents. The median strip in a roadway is the middle portion just as the median is the middle number in the data set.

In the guided example, we will find the mean and median for data which is described in a frequency table. We use a frequency table to compute the mean of a data set as follows:

1. Write all products $x \cdot f$ of the values and their frequencies in a new column of the table.
2. Represent the sum of the products you calculated in step 1 by $\sum(x \cdot f)$.
3. Denote the sum of the frequencies, which is equal to *n,* by $\sum f$.
4. The mean is found by dividing the quantities in steps 2 & 3; $\bar{x} = \frac{\sum(x \cdot f)}{\sum f}$.

Notes:

Guided Example

Compute the mean, median, and mode for the data in the frequency table below.

x	f
4	2
7	4
8	7
9	5
11	1
13	1

Let's start with the mean. We start by finding the products $x \cdot f$ in step 1 above. They are in the third column of the table below.

x	f	xf
4	2	8
7	4	28
8	7	56
9	5	45
11	1	11
13	1	13

Next we sum the products and find, $\sum(x \cdot f) = 8 + 28 + 56 + 45 + 11 + 13 = 161$.

The sum of the frequencies, $\sum f = 20$ gives us the value for *n*. We find the mean using the formula $\bar{x} = \frac{\sum(x \cdot f)}{\sum f} = \frac{161}{20} = 8.05$.

The median would be the mean of the 10th and 11th values $\left(\text{since } n = \sum f = 20\right)$. The 10th and 11th values would both be 8 since the first six values are less than 8, and then the next seven data values, values 7 – 13 are all equal to 8. The median is 8.

The mode is the value which occurs most frequently. For this data set, that is 8 since it has the highest frequency.

Practice (Exercise 3 & 11 p.743)

1. Find the mean, median, and mode for the data $\{4, 6, 4, 6, 7, 9, 3, 9, 10, 11\}$.

2. Use the graph below to find the mean, median, and mode of the distribution.

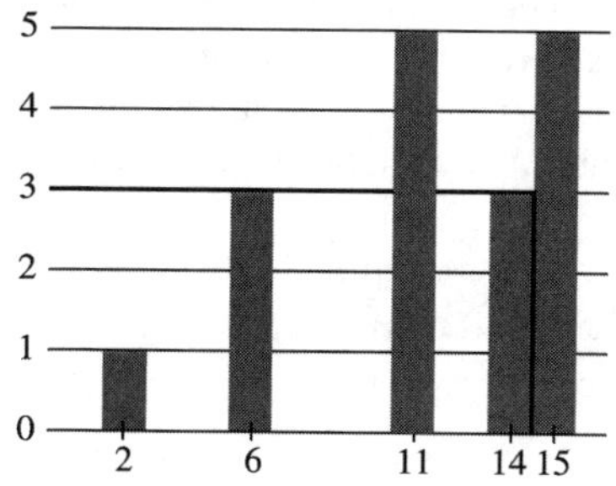

Objective 2 – Find the five-number summary of a distribution.

Key Terms
Five number summary
Box plot

Summary
The median cuts a data set into two halves. The set of numbers below the median is called the lower half and the set above the median is called the upper half. The median of the lower half is called the first quartile and is indicated by Q_1; the third quartile, denoted by Q_3 is the median of the upper half. The five number summary of a set of data values consists of the minimum value, Q_1, median, Q_3, and the maximum value.

The five number summary is used to construct a box-and-whisker plot, which provides a useful way to visualize the distribution of the data.

The steps below outline the process for constructing a box-and-whisker plot.

1. Determine the values for the five-number summary
2. Draw a horizontal axis from the minimum value to the maximum value.
3. Draw a rectangular box over the axis from Q_1 to Q_3. The height of the box is not important.
4. Draw a vertical line segment in the box where the median is located.
5. Draw a horizontal segment extended from the edges of the box (the quartiles) to the minimum and maximum values. These are the whiskers.

Notes:

Guided Example

Given the data below, find the five-number summary and construct a box-and-whisker plot.
{10.2, 13.5,14.7, 11.3, 16.5, 18, 12.1, 11.6, 10.8, 13.8. 19.5, 16.6, 12.5, 11.7, 19.9}

In order to find the five-number summary, we must write the data in order. We can then find the median.

The ordered data is {10.2, 10.8, 11.3, 11.6, 11.7, 12.1, 12.5, 13.5, 13.8, 14.7, 16.5, 16.6, 18, 19.5, 19.9}. There are 15 data points, so the median is the value in the 8th spot, which is 13.5.
The lower half of the data is {10.2, 10.8, 11.3, 11.6, 11.7, 12.1.12.5}. The middle number of the lower half, 11.6, is Q_1.

The upper half of the data is {13.8, 14.7, 16.5, 16.6, 18, 19.5, 19.9}. The middle number of the upper half, Q_3, is 16.6.

The five-number summary is 10.2, 11.6, 13.5, 16.6, 19.9.

To construct the box plot, we draw a line segment from 10 to 20. It is very important that we have scale on this segment so that we locate the values of the five-number summary in the correct relative locations.

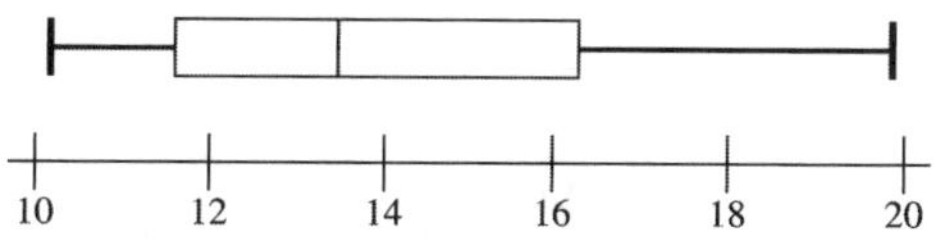

Note that the box spans from Q_1 to Q_3. The median is not in the middle of the box because the value of the median, 13.5, is closer to 11.6 than it is to 16.6. The whiskers extend to the minimum and maximum values. The box-and-whisker plot gives insight into the spread of the data, which we address in a future worksheet.

Practice (Exercise 17 p.744)

1. Give the five-number summary for the data and draw a box-and-whisker plot.
{11, 23, 25, 17, 26, 31, 45, 18, 41, 26, 31, 33, 48, 44, 53}

Objective 3 – Apply measures of central tendency to compare data.

Key Terms
None

Summary
Imagine you are at a coffee shop near campus and you collect data on the net worth of those in the shop. The mean and median will likely be similar. Then the patron with the highest net worth walks out and Taylor Swift walks in the shop. What impact would her net worth have on the median? Nothing! Her net worth, which is likely much more than anyone else's, does not impact the middle number. What about the mean? Her income will change the mean a great deal because it will be part of the sum. In cases where there are extreme values, the median may be a better measure of central tendency to describe the data.

Notes:

Guided Example

A real estate website provides the following prices for recently sold homes (in thousands): {\$452, \$219, \$358, \$1,599, \$375}. Find the mean and median and suggest which most accurately describes the center of the data.

The ordered data is {\$219, \$358, \$375, \$452, \$1,599}. The median of the five data values is the one in the middle position, \$375.

The mean of the data set is

$\bar{x} = \frac{\sum x}{n} = \frac{3,003}{5} = 600.6$ or \$600.6. This is larger than 4 of the 5 data values, and is influenced by the one very high value. The median is more appropriate as a measure of center.

Practice (Exercise 25 p.744)

1. The table below gives the gross domestic receipts for the top 10 movies of 2015. Find the mean, median, and mode for the data. Explain which measure you believe best describes a typical value in the distribution.

Movie Title	Total Gross (\$ millions)
Avengers: Age of Ultron	459.0
Cinderella	201.2
Furious 7	353.0
Inside Out	356.5
Jurassic World	652.3
Minions	336.0
Spectre	200.0
Star Wars: The Force Awakens	936.7
The Hunger Games: Mockingjay — Part 2	281.7
The Martian	228.4

Data from www.BoxOfficeMojo.com

Section 14.3 Measures of Dispersion

Objective 1 – Compute the range of a data set.

Objective 2 – Explain how the standard deviation measures the spread of a distribution.

Objective 3 – Use the coefficient of variation to compute the relative variability of different distributions.

Objective 1 – Compute the range of a data set.

Key Terms
Range

Summary
The measures of center we have discussed give us an idea of the center, or middle of the data set. This is very useful information, but it does not bring into consideration the variability in the values. The simplest measure of dispersion, or variability, is the range. The range of a data set is the difference between the largest and smallest data values in the set. Before computing the range, be sure to put the data values in order!

Example: Compute the range of the data set $\{10.2, 13.5, 14.7, 11.3, 16.5, 18, 12.1, 11.6, 10.8, 13.8. 19.5, 16.6, 12.5, 11.7, 19.9\}$.

We first put the data in order and then subtract the minimum value from the maximum value. The ordered data is $\{10.2, 10.8, 11.3, 11.6, 11.7, 12.1, 12.5, 13.5, 13.8, 14.7, 16.5, 16.6, 18, 19.5, 19.9\}$. The range is $19.9-10.2=9.7$.

Problem Solving: The range is the length of the box-and-whisker plot. We *Relate a Newer problem to an Older One* and use the range to help us describe the information included in the box-and-whisker plot.

Notes:

Objective 2 – Explain how the standard deviation measures the spread of a distribution.

Key Terms
Deviation from the mean
Standard deviation
Variance

Summary
A better way to find the spread of data is to use the standard deviation, a measure based on calculating the distance each value is from the mean. We denote the standard deviation of a sample of *n* data values by *s*, which is defined using the formula $s = \sqrt{\dfrac{\sum\left(x-\bar{x}\right)^2}{n-1}}$. We compute the standard deviation in four steps:

1. Compute the mean of the data set, $\bar{x}$.
2. Find the deviations from the mean for each value of *x*, $x-\bar{x}$ and square the result yielding $\left(x-\bar{x}\right)^2$.
3. Add the squares found in step 2 and divide this sum by $n-1$. This quantity $\dfrac{\sum\left(x-\bar{x}\right)^2}{n-1}$ is called the variance.
4. Compute the square root of the number found in steps 3. This is the standard deviation.

It is less common to compute the standard deviation for a set of population data. We represent the standard deviation of a population by σ rather than s, and we calculate it by the formula $\sigma = \sqrt{\dfrac{\sum\left(x-\bar{x}\right)^2}{n}}$. For population data, the denominator in step 3 is *n* rather than $n-1$.

Example: Calculate the standard deviation of the data set {3, 2, 10, 4, 2, 15}.

1. In a previous worksheet, we found the sample mean $\bar{x}$, to be 6.
2. In step 2 it is helpful to use a table to organize your work.

x	$x-\bar{x}$	$(x-\bar{x})^2$
3	−3	$(-3)^2 = 9$
2	−4	$(-4)^2 = 16$
10	4	$(4)^2 = 16$
4	−2	$(-2)^2 = 4$
2	−4	$(-4)^2 = 16$
15	9	$(9)^2 = 81$

3. In step 3, we sum the squares in the far right column of the table and divide by $n-1=5$. We have $\dfrac{\sum\left(x-\bar{x}\right)^2}{n-1} = \dfrac{142}{5} = 28.4$.
4. The square root of the quantity found in step 3, 5.3, is our standard deviation.

To find the standard deviation of data summarized in a frequency table, use $s = \sqrt{\dfrac{\sum\left(x-\bar{x}\right)^2 f}{n-1}}$. We demonstrate this in the guided example.

Notes:

Guided Example

Find the standard deviation of the data describe in the following frequency table.

x	f
4	2
7	4
8	7
9	5
11	1
13	1

In a previous worksheet we found the mean to be 8.05. We will round this to 8 for purposes of calculation (it is preferable not to round at this stage, but as we find in this example it will not have a great impact).

In the table below, we compute $x-\bar{x}$, $\left(x-\bar{x}\right)^2$, and $\left(x-\bar{x}\right)^2 f$ for each of the values.

x	f	$x-\bar{x}$	$(x-\bar{x})^2$	$(x-\bar{x})^2 f$
4	2	-4	$(-4)^2 = 16$	$16 \cdot 2 = 32$
7	4	-1	$(-1)^2 = 1$	$1 \cdot 4 = 4$
8	7	0	0	0
9	5	1	$(1)^2 = 1$	5
11	1	3	$(3)^2 = 9$	9
13	1	5	$(5)^2 = 25$	25

The sum of the far right column is 75, so we find $s = \sqrt{\dfrac{75}{19}} = \sqrt{3.947} \approx 1.99$.

Note that if we had used the exact value for $\bar{x}$ the sum $\sum\left(x-\bar{x}\right)^2 f$ would have been 74.95, not 75, but the standard deviation, correct to the hundredths place, would be unchanged.

Practice (Exercises 25 & 19 pp. 756-757.)

1. The table gives the annual incomes for eight families, in thousands of dollars. Find the number of standard deviations family H's income is from the mean.

Family	A	B	C	D	E	F	G	H
Annual Income (in \$ thousands)	47	48	50	49	51	47	49	51

2. The following are the scores of 20 people who took a paramedics licensing test. Find the mean and standard deviation for these data.

Score	72	73	78	84	86	93
Frequency	7	1	3	6	2	1

Objective 3 – Use the coefficient of variation to compute the relative variability of different distributions.

Key Terms
Coefficient of Variation

Summary
The standard deviation provides us with information about the dispersion in the distribution. The size of the standard deviation, however, is tied to the size of the data values. In order to compare variability in different distributions, we use the coefficient of variation.

For a set of data with mean $\bar{x}$ and standard deviation s, we define the coefficient of variation, denoted CV as $CV = \frac{s}{\bar{x}} \cdot 100\%$. The coefficient of variation expresses the size of the standard deviation as a percentage of the size of the mean and is not dependent on the units of the original distribution. When comparing two sets of data, the one with the largest coefficient of variation, relative speaking, has more variability.

Notes:

Guided Example	***Practice*** (Exercise 33 p.757)
Compute the coefficient of variation for the following sets of data and indicate what you can conclude from the results. For athletes on the men's crew team, the mean weight is 162 pounds with a standard deviation, $s = 12$ pounds. For athletes on the men's lacrosse team, the mean weight is 205 pounds with standard deviation $s = 14$ pounds. The coefficient of variation for the weights on the crew team is $CV = \frac{12}{162} \cdot 100\% \approx 7.4\%$. For the lacrosse team, the coefficient of variation is $CV = \frac{14}{205} \cdot 100\% \approx 6.8\%$. Although the lacrosse team has a higher mean weight and larger standard deviation, the relative variability is slightly less than that of those on the crew team.	1. As the manager of the human resources department at your firm, you are investigating the number of sick days employees took during the last calendar year. For employees in the marketing area, the average number of sick days is 4.2 with standard deviation 2; for those in research and design the mean is 5.6 with standard deviation 3. Compute and compare the coefficient of variation for each set of data.

Section 14.4 The Normal Distribution

Objective 1 – Describe the basic properties of the normal distribution.
Objective 2 – Relate the area under a normal curve to z-scores.
Objective 3 – Make conversions between raw scores and z-scores.
Objective 4 – Use the normal distribution to solve applied problems.

Objective 1 – Describe the basic properties of the normal distribution.

Key Terms
Normal distribution
68-95-99.7 rule

Summary
The normal distribution is the most common distribution in statistics and describes many real-life data sets. The properties of a normal distribution are listed below:

1. A normal curve is bell shaped.
2. The curve is symmetric with respect to its mean, where the curve reaches its highest point.
3. The mean, median, and mode of the distribution are the same.
4. The total area under the curve is 1.
5. Roughly 68% of the data values are within 1 standard deviation from the mean, 95% of the data values are within 2 standard deviations from the mean, and 99.7% of the data values are within 3 standard deviations from the mean.

The last property is often referred to as the 68-95-99.7 rule, and we can use it to make conclusions about a normal distribution using only the mean and standard deviation. The figure below gives a visual description of a normal distribution.

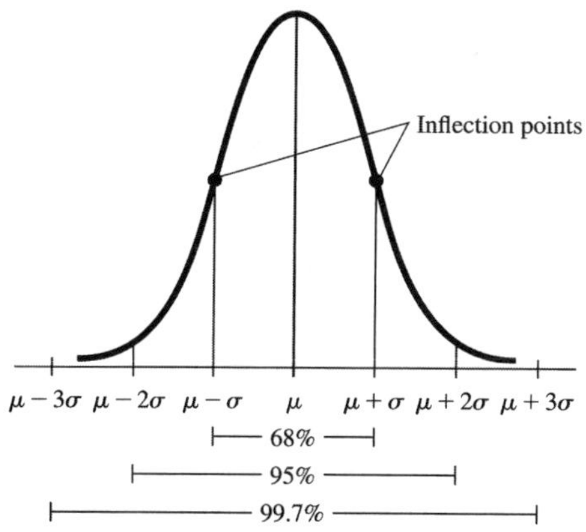

We assume that the normal distribution is describing an entire population, so we use μ for the mean and σ for the standard deviation.

Notes:

Guided Example	***Practice*** (Exercises 7, 9, & 11 p.769)
The annual rainfall in a city has a normal distribution with mean 67 inches and standard deviation 20 inches. Use the 68-95-99.7 rule to find the percentage of years the annual rainfall 1. Between 47 and 87 inches. In order to use the 68-95-99.7 rule, we need to know how many standard deviations 47 and 87 are from 67. Since $87 = 67 + 20 = \mu + \sigma$ and $47 = 67 - 20 = \mu - \sigma$ the interval described is one standard deviation from either side of the mean. Using the 68-95-99.7 rule we would expect 68% of the years to have rainfall between 47 and 87 inches. 2. Above 107 inches. 107 is 40 inches above the mean, which is two standard deviations. Since we want the percentage of years with rainfall about that number, we use the fact that 95% of the years would have rainfall within two standard deviations from the mean (so between 27 and 107) and the fact that the curve is symmetric to conclude that $\frac{100\% - 95\%}{2}$ or 2.5% of the years would have rainfall above 107 inches.	Assume that a distribution has a mean of 12 and a standard deviation of 3. Use the 68-95-99.7 rule to find the percent of values in the distribution: 1. Below 9. 2. Above 6. 3. Between 15 and 18.

Objective 2 – Relate the area under a normal curve to *z*-scores.

Key Terms
z-scores

Summary
The 68-95-99.7 rule enables us to discuss the percentage of values in a normal distribution in several key intervals, but if we are interested in intervals other than 1, 2 and 3 standard deviations from the mean we need a different strategy. We will find these by using the standard normal curve areas table. The standard normal distribution is one with mean 0 and standard deviation 1. The table gives the area under the normal curve between the mean and a number called a *z*-score. A *z*-score represents the number of standard deviations a data value is from the mean. The table has only positive valued *z*-scores, but we will use the symmetry property of the normal distribution to answer questions when the *z*-score is negative.

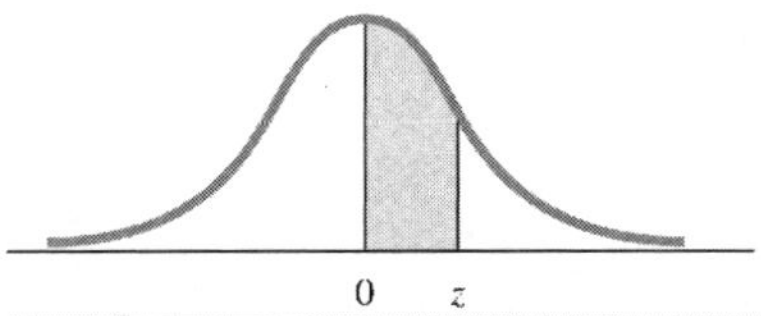

z	.00	.01	.02	.03	.04	.05	.06	.07	.08	.09
.0	.0000	.0040	.0080	.0120	.0160	.0199	.0239	.0279	.0319	.0359
.1	.0398	.0438	.0478	.0517	.0557	.0596	.0636	.0675	.0714	.0753
.2	.0793	.0832	.0871	.0910	.0948	.0987	.1026	.1064	.1103	.1141
.3	.1179	.1217	.1255	.1293	.1331	.1368	.1406	.1443	.1480	.1517
.4	.1554	.1591	.1628	.1664	.1700	.1736	.1772	.1808	.1844	.1879
.5	.1915	.1950	.1985	.2019	.2054	.2088	.2123	.2157	.2190	.2224
.6	.2257	.2291	.2324	.2357	.2389	.2422	.2454	.2486	.2517	.2549
.7	.2580	.2611	.2642	.2673	.2704	.2734	.2764	.2794	.2823	.2852
.8	.2881	.2910	.2939	.2967	.2995	.3023	.3051	.3078	.3106	.3133
.9	.3159	.3186	.3212	.3238	.3264	.3289	.3315	.3340	.3365	.3389
1.0	.3413	.3438	.3461	.3485	.3508	.3531	.3554	.3577	.3599	.3621
1.1	.3643	.3665	.3686	.3708	.3729	.3749	.3770	.3790	.3810	.3830
1.2	.3849	.3869	.3888	.3907	.3925	.3944	.3962	.3980	.3997	.4015
1.3	.4032	.4049	.4066	.4082	.4099	.4115	.4131	.4147	.4162	.4177
1.4	.4192	.4207	.4222	.4236	.4251	.4265	.4279	.4292	.4306	.4319
1.5	.4332	.4345	.4357	.4370	.4382	.4394	.4406	.4418	.4429	.4441
1.6	.4452	.4463	.4474	.4484	.4495	.4505	.4515	.4525	.4535	.4545
1.7	.4554	.4564	.4573	.4582	.4591	.4599	.4608	.4616	.4625	.4633
1.8	.4641	.4649	.4656	.4664	.4671	.4678	.4686	.4693	.4699	.4706
1.9	.4713	.4719	.4726	.4732	.4738	.4744	.4750	.4756	.4761	.4767
2.0	.4772	.4778	.4783	.4788	.4793	.4798	.4803	.4808	.4812	.4817
2.1	.4821	.4826	.4830	.4834	.4838	.4842	.4846	.4850	.4854	.4857
2.2	.4861	.4864	.4868	.4871	.4875	.4878	.4881	.4884	.4887	.4890
2.3	.4893	.4896	.4898	.4901	.4904	.4906	.4909	.4911	.4913	.4916
2.4	.4918	.4920	.4922	.4925	.4927	.4929	.4931	.4932	.4934	.4936
2.5	.4938	.4940	.4941	.4943	.4945	.4946	.4948	.4949	.4951	.4952
2.6	.4953	.4955	.4956	.4957	.4959	.4960	.4961	.4962	.4963	.4964
2.7	.4965	.4966	.4967	.4968	.4969	.4970	.4971	.4972	.4973	.4974
2.8	.4974	.4975	.4976	.4977	.4977	.4978	.4979	.4979	.4980	.4981
2.9	.4981	.4982	.4982	.4983	.4984	.4984	.4985	.4985	.4986	.4986
3.0	.4987	.4987	.4987	.4988	.4988	.4989	.4989	.4989	.4990	.4990

Source: Anders Hald, *Statistical Tables and Formulas*, 1952.

Problem Solving: It is very helpful to sketch the area you are trying to find. Use the *Draw Pictures Strategy* when finding normal curve areas.

Notes:

Guided Example

Use the standard normal table to find the percentage of the area under the curve that is indicated.

1. Between $z = 0$ and $z = 2.31$

   Let's sketch the standard normal distribution curve. We are looking for the area between $z = 0$ and $z = 2.31$ as indicated below

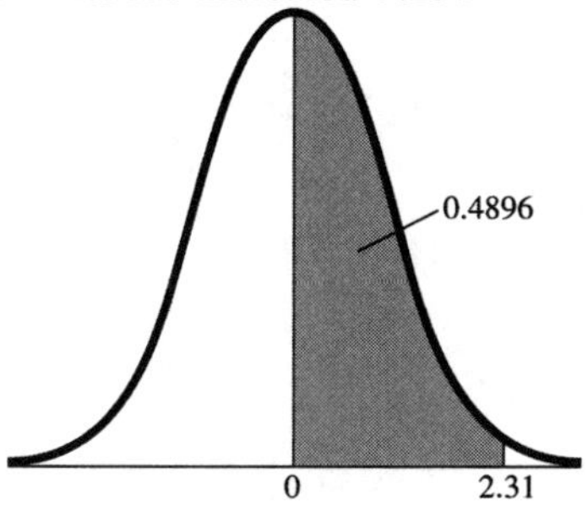

   The area, 0.4896 is found by looking at the table in the row for $z = 2.3$ and the column for 0.01. This tells us that 48.96% of the area is between the indicated values.

z	.00	.01
2.3	.4893	.4896

   If the question had asked you to find the area between $z = 0$ and $z = -2.31$ the answer would be the same, because the curve is symmetric.

2. Between $z = 1.45$ and $z = 2.31$

   We again sketch a normal curve. We found the area from $z = 0$ to $z = 2.31$ in the previous example; to find the area between $z = 1.45$ and $z = 2.31$ we need to subtract the area from $z = 0$ to $z = 1.45$ as indicated in the sketch.

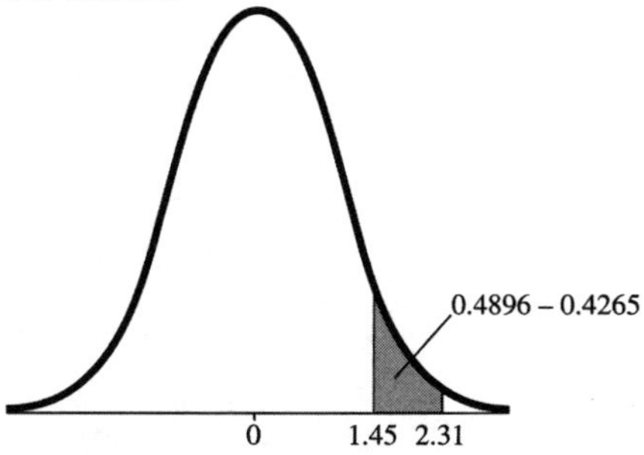

   The desired area under the curve is $0.4896 - 0.4265 = 0.0631$, or 6.31%.

Practice (Exercises 15, 17 & 21 p.770)

Use the normal table to find the percent of the area under the standard normal curve that is specified.

1. Between $z = 0$ and $z = 1.23$
2. Between $z = 0$ and $z = -0.75$
3. Between $z = -0.38$ and $z = -0.76$

Objective 3 – Make conversions between raw scores and z-scores.

Key Terms

z-score
Raw score

Summary

The normal distribution has many real-life applications. In order to use the standard normal table for normal distributions that aren't centered at 0 with standard deviation 1, we need to convert from the original distribution to the standard normal by computing a *z*-score.

Assume a normal distribution has a mean of μ and a standard deviation of σ. We use the equation $z = \dfrac{x-\mu}{\sigma}$ to convert a value x in the nonstandard normal distribution to a *z*-score.

Notes:

Guided Example	***Practice*** (Exercises 35 & 43 p.770)
A normal distribution has mean 32 and standard deviation 4. 1. What is the *z*-score that corresponds to a raw score of 25.68? Substitute $x = 25.68$, $\mu = 32$ and $\sigma = 4$ into the formula $z = \dfrac{x-\mu}{\sigma}$. We obtain $z = \dfrac{25.68-32}{4} = -1.58$. 2. A *z*-score of 2.31 corresponds to what raw score? Using the same formula, we substitute in $z = 2.31$, $\mu = 32$ and $\sigma = 4$ and solve for x. We find $2.31 = \dfrac{x-32}{4}$; solving this equation yields $x = (2.31)(4) + 32 = 41.24$.	1. Convert the raw score $x = 87$ to a *z*-score. The distribution has mean 80, standard deviation 5. 2. Find the raw score corresponding to a *z*-score of -0.45. The distribution has mean 35 and standard deviation 3.

Objective 4 – Use the normal distribution to solve applied problems.

Key Terms
None

Summary
We can now use the z-scores obtained from raw scores and the standard normal table to determine the percentage of a distribution in a given interval. You are always encouraged to sketch the curve when doing these problems to guide your work. Identify the raw score(s) of interest, convert them to z-scores, and then use the table to find the area.

Some problems give us a percentage of interest, and then ask that we find the raw score. Again, a sketch is a good first step. We use the normal table in reverse (start by looking in the body of the table for the area you need, and then identify the appropriate z-score). The final step of the problem is to use the z-score formula, $z = \frac{x-\mu}{\sigma}$, with the z-score you found in the table, to find x.

Notes:

Guided Example

Replacement times for tablets have a normal distribution with a mean of 7.1 years and a standard deviation of 1.4 years.

1. What percentage of tablets would you expect to have a replacement time of less than 8 years?

   To answer this question we will find the z-scores corresponding to a raw score of 8 years and then use the standard normal table to find the area.

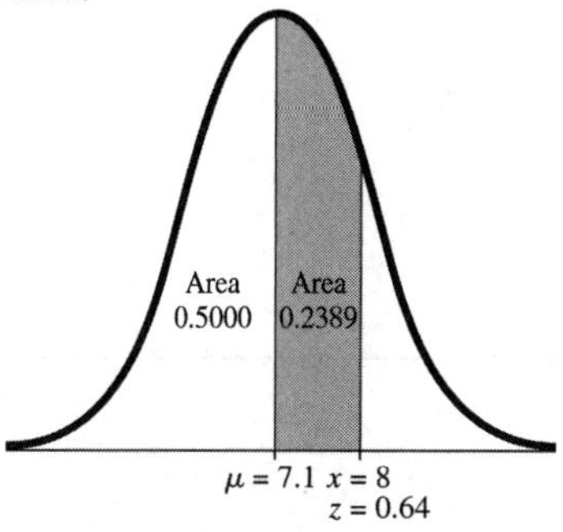

Practice (Exercises 53 & 63 pp.770-771)

1. Assume that the distribution of 21-year old women's heart rates at rest has a mean of 68 beats per minute and a standard deviation of 4 beats per minute. If 200 women are examined, how many would you expect to have a heart rate of less than 70?

2. If the distribution of race times for a cross country race has a mean of 85 minutes and a standard deviation of 9 minutes, what is the cutoff time for a runner to finish in the top 20% of the runners?

The z-score is $z = \frac{8-7.1}{1.4} \approx 0.64$. We round our z-scores to the hundredths place. Looking at the table for $z = 0.64$ we find an area of 0.2389 between $z = 0$ and $z = 0.64$. We must not forget, however, the values that are less than the mean. Because a normal curve is symmetric, another 50% of the values lie below 7.1. The percentage of tables that would have a replacement time less than 8 years is 23.89% + 50% = 73.89%.

2. What is the cutoff time that would show the lifetime that only 15% of tablets would exceed?

In this case, we would like to find the value which cuts off the top 15% of the distribution, as indicated in the sketch.

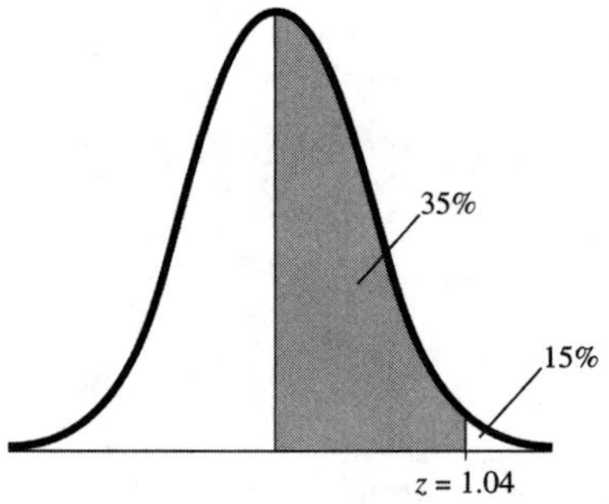

We use the standard normal table to find the z-score which has area 0.3500 between it and $z = 0$. The closest area in the table is 0.3508 which corresponds to a z-score of $z = 1.04$. This tells us that the desired value lies 1.04 standard deviations above the mean. We can find the exact value by substituting into the z-score formula and solving for x.

Doing this, we find $1.04 = \frac{x-7.1}{1.4}$.

Multiplying both sides by 1.4 and adding 7.1 to each side yields $x = (1.04)(1.4) + 7.1 = 8.556$.

Fifteen percent of tablets would be expected to last longer than 8.556 years.

Section 14.5 Linear Correlation

Objective 1 – Construct a scatterplot to show the relationship between two variables.
Objective 2 – Explain the properties of the linear correlation coefficient.
Objective 3 – Use linear regression to find the line of best fit for a set of data points.

Objective 1 – Construct a scatterplot to show the relationship between two variables.

Key Terms
Scatterplot

Summary
To this point, we have considered data for a single variable. We now look to see if two sets of data are related in any way. As a first step in answering this question, we construct a scatterplot. A scatterplot is a graph of ordered pairs.

Example: A dieting app assigns points for exercise activities. Eight persons using this app reported the number of points earned for exercise and the number of pounds lost for a given week. Construct a scatterplot for the data. Does there appear to be a relationship between the two variables?

Exercise Points, x	Pounds Lost, y
2	0.5
15	4
5	1
7	2
10	2.5
8	1
12	2
15	3

We construct the scatterplot by plotting the ordered pair (points earned, pounds lost) for each of the participants. The graph below shows that as the number of points earned increases, the pounds lost also appears to increase. This relationship does not suggest that higher exercise points causes larger pounds lost. There are other factors to be considered, notably what the participants are eating!

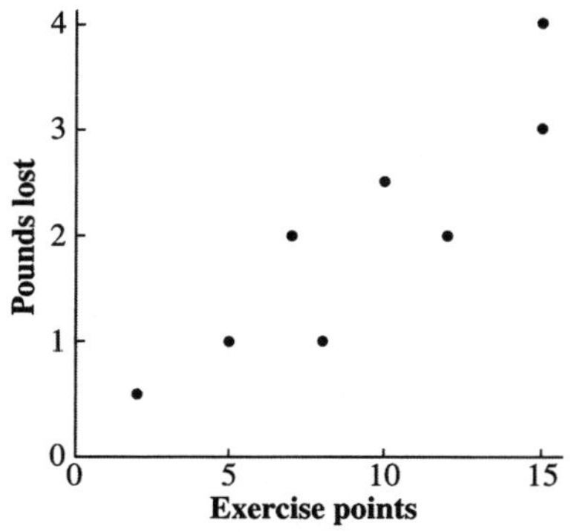

Notes:

Objective 2 – Explain the properties of the linear correlation coefficient.

Key Terms
Linear correlation coefficient
Significant linear correlation

Summary
In the last worksheet, we noticed that as the number of exercise points increases, the pounds lost also increases. The points do not fall on a line, but they appear to have a somewhat linear trend. The linear correlation coefficient measures to what degree the points lie along a straight line.

The linear correlation coefficient for a set of ordered pairs (*x*, *y*) is given by the formula
$r = \dfrac{n\sum xy - \left(\sum x\right)\left(\sum y\right)}{\sqrt{n\left(\sum x^2\right) - \left(\sum x\right)^2}\sqrt{n\left(\sum y^2\right) - \left(\sum y\right)^2}}$. The linear correlation coefficient has the following properties.

1. r is a number between -1 and 1, inclusive. If $r = 1$ or $r = -1$, then all of the points lie on a straight line.
2. If r is positive, there is positive linear correlation between the variables, meaning that an increase in x is associated with an increase in y. If r is negative, there is negative linear correlation between the variables, indicating that an increase in x is associated with a decrease in y. Values of r closer to 1 or -1 indicate a stronger relationship, and that the points lie closer to a straight line.
3. If r is close to 0, then there is little linear correlation between the variables.

In property 2 we stated that if the correlation coefficient is close to 1 or -1, the relationship is stronger. But how close must the values be to 1 or -1? We use the table below to determine if the linear correlation between the variables is significant. To use the table, select the row appropriate for your sample size. Compare the absolute value of r, $|r|$, with the value in the columns to the right. If $|r|$ exceeds the number in the column labeled $\alpha = 0.05$, we can be 95% confident that there is significant linear correlation between the variables. That is, there is only a 5% chance that we will be incorrect if we conclude that there is a linear correlation between the variables. In a similar fashion, if $|r|$ exceeds the number in the column labeled $\alpha = 0.01$, we can be 99% confident that there is significant linear correlation between the variables, and there is only a 1% chance that we will be incorrect if we conclude that there is significant linear correlation between the variables.

n	$\alpha = 0.05$	$\alpha = 0.01$
4	0.950	0.999
5	0.878	0.959
6	0.811	0.917
7	0.754	0.875
8	0.707	0.834
9	0.666	0.798
10	0.632	0.765
11	0.602	0.735
12	0.576	0.708
13	0.553	0.684
14	0.532	0.661
15	0.514	0.641
16	0.497	0.623
17	0.482	0.606
18	0.468	0.590
19	0.456	0.575
20	0.444	0.561

Problem Solving: Computing the linear correlation coefficient involves a lot of steps. Use the *Be Systematic* strategy and *Order Principle.* It helps to first compute all of the sums in a systematic way, and then be sure to perform calculations with the correct order of operations.

Notes:

Guided Example

Compute the linear correlation coefficient for the data in the table below. Determine if we can be 95% or 99% confident that there is significant linear correlation between the exercise points earned and pounds lost.

Exercise Points, x	Pounds Lost, y
2	0.5
15	4
5	1
7	2
10	2.5
8	1
12	2
15	3

In order to compute r we need to find the following sums:

$\sum x$, the sum of the data in the x column

$\sum y$, the sum of the data in the y column

$\sum x^2$, the sum of the squares of the data in the x column

$\sum y^2$, the sum of the squares of the data in the y column

$\sum xy$, the sum of the products, $x \cdot y$.

We use the table to organize this information; the sums are in the last row.

Practice (Exercise 15 p.779)

1. The following table lists the number of years of education past high school and the annual income of five people. Determine the linear correlation coefficient and also determine whether we can be 95% or 99% confident that there is significant linear correlation between the two variables.

Years of Education Past High School	Annual Income (in $ thousands)
0	23
1	22
2	27
2	28
5	35

x	y	x^2	y^2	xy
2	0.5	4	0.25	1
15	4	225	16	60
5	1	25	1	5
7	2	49	4	14
10	2.5	100	6.25	25
8	1	64	1	8
12	2	144	4	24
15	3	225	9	45
$\Sigma x = 74$	$\Sigma y = 16$	$\Sigma x^2 = 836$	$\Sigma y^2 = 41.5$	$\Sigma xy = 182$

Now we substitute the sums into the formula for the linear correlation coefficient.

$$r = \frac{n\sum xy - \left(\sum x\right)\left(\sum y\right)}{\sqrt{n\left(\sum x^2\right) - \left(\sum x\right)^2}\sqrt{n\left(\sum y^2\right) - \left(\sum y\right)^2}}$$

$$= \frac{8(182) - (74)(16)}{\sqrt{8(836) - (74)^2}\sqrt{8(41.5) - (16)^2}}$$

$$= \frac{272}{\sqrt{1,212}\sqrt{76}} \approx 0.896$$

Our linear correlation coefficient is positive indicating positive linear correlation. Comparing $|r|$ to the values in the table above for $n = 8$, we can be 99% confident that there is a significant linear correlation because $|r|$ exceeds 0.834.

Objective 3 – Use linear regression to find the line of best fit for a set of data points.

Key Terms
Line of best fit

Summary
We found that there was a significant linear correlation between exercise points and pounds lost. We will now find the line that best models the data. This is called the line of best fit. In general, the line does not pass through all of the data points, it is the line that minimizes the sum of the squares of all the vertical distances from the data points to the line, as shown below.

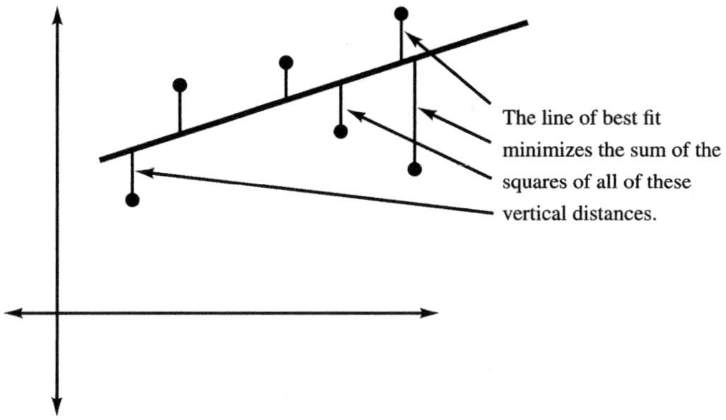

The line of best fit for a set of data points (*x*, *y*) is of the form $y = mx + b$, where $m = \dfrac{n\sum xy - \left(\sum x\right)\left(\sum y\right)}{n\left(\sum x^2\right) - \left(\sum x\right)^2}$

and $b = \dfrac{\sum y - m\left(\sum x\right)}{n}$.

Notes:

Guided Example

Find the line of best fit for the data on exercise points and pounds lost, below.

Exercise Points, x	Pounds Lost, y
2	0.5
15	4
5	1
7	2
10	2.5
8	1
12	2
15	3

We found the necessary sums in the last worksheet; they are $\sum x = 74$, $\sum y = 16$, $\sum x^2 = 836$, $\sum y^2 = 41.5$, $\sum xy = 182$.

Substituting into the formula for the slope, we have

$$m = \frac{n\sum xy - \left(\sum x\right)\left(\sum y\right)}{n\left(\sum x^2\right) - \left(\sum x\right)^2}$$

$$= \frac{8(182) - (74)(16)}{8(836) - (74)^2}$$

$= \dfrac{272}{1{,}212} \approx 0.224$. We then find the y-intercept;. $b = \dfrac{\sum y - m\left(\sum x\right)}{n}$. Substituting, we find $b = \dfrac{16 - 0.224(74)}{8} = -0.072$. The line of best fit is $y = 0.224x - 0.072$. We can interpret the slope and y-intercept. A person with no exercise points would expect to lose -0.072 pounds (so gain pounds); for each additional exercise point earned, a person is expected to lose 0.224 pounds.

Practice (Exercise 19 p.780)

1. Find the line of best fit for the data below.

Years of Education Past High School	Annual Income (in $ thousands)
0	23
1	22
2	27
2	28
5	35

Chapter 14 Answers

Section 14.1

Objective 3 1)

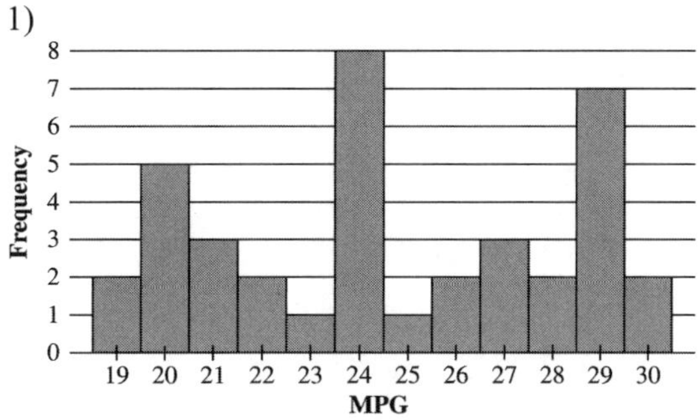

2)

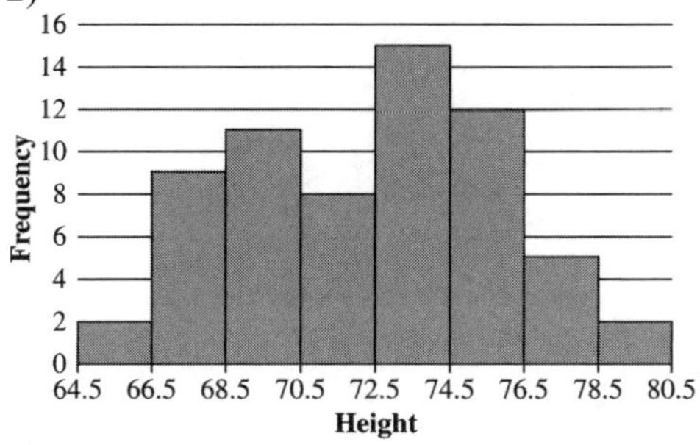

Objective 4 1)

A		B
8 8	1	3 6 8
9 6 2 2 1	2	0 1 1 2 2 2 3 4 6 8 9
9 9 8 7 4 3 2	3	2 3 3 8 9
7 3 3 3 2 2	4	7

Section 14.2

Objective 1 1) mean 6.9; median 6.5; modes 4, 6 and 9 2) mean 11.29; median 11; modes 11 and 15.

Objective 2 1) 11, 23, 31, 44, 53

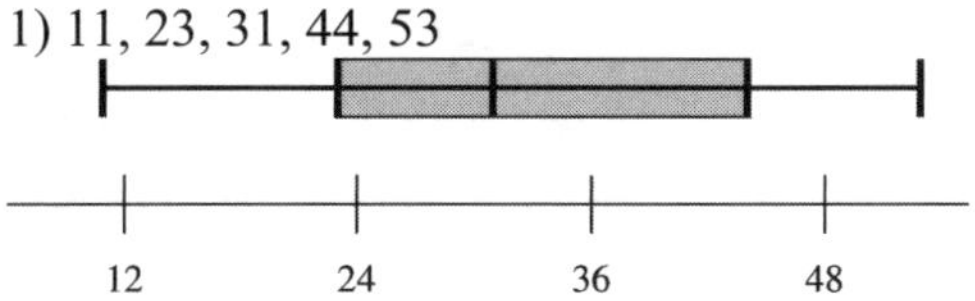

Objective 3 1) mean 400.48; median 344.5; no mode. The median is most representative.

Section 14.3

Objective 2 1) Family H's income is 1.25 standard deviations above the mean. 2) 79; 6.59

Objective 3 1) *CV* for marketing is 47.6%; *CV* for R&D is 53.6%. R&D has more relative variability.

Section 14.4

Objective 1 1) 16% 2) 97.5% 3) 13.5%

Objective 2 1) 39.07% 2) 27.34% 3) 12.84%

Objective 3 1) 1.4 2) 33.65

Objective 4 1) $69.15\% \times 200 = 138.3$ so we expect 138. 2) 92.56 min.

Section 14.5

Objective 2 1) $r = 0.96$; we can be 99% confident that there is positive linear correlation.

Objective 3 1) $y = 2.64x + 21.71$

Chapter 1: Problem Solving

Objective 1 – Percent conversions.
Objective 2 – Translate algebraic expressions to word phrases and word phrases to algebraic expressions.
Objective 3 – Evaluate an algebraic expression.
Objective 4 – Evaluate a formula.

Objective 1 – Percent Conversions

Key Terms

None

Summary

To change a percent to the equivalent fraction, drop the % sign from the given percent and place the number over 100. Then simplify the resulting fraction into lowest terms if possible.

$$23\% = \frac{23}{100}$$

To change a percent to the equivalent decimal, drop the % sign and move the decimal point two places to the left.

$$23\% = 0.23$$

To change a decimal to the equivalent percent, move the decimal point two places to the right and insert a % sign after the number.

$$0.23 = 23\%$$

To change a fraction to the equivalent percent, first change the fraction to a decimal by division. Then change the decimal to a percent as stated above.

$$\frac{23}{100} = 0.23 = 23\%$$

Notes:

Guided Example	***Practice***
1. Write 12% as a fraction. Drop the % sign and write 12 over 100. $\frac{12}{100}$ Reduce the fraction. $\frac{12}{100} = \frac{\cancel{4} \cdot 3}{\cancel{4} \cdot 25} = \frac{3}{25}$	1. Write 24% as a fraction.
2. Write 15% as a decimal. Drop the % sign and move the decimal point 2 places to the left. $15\% = 0.\underline{15}\cancel{\%} = 0.15$	2. Write 32% as a decimal.
3. Write 0.236 as percent. Move the decimal point two places to the right and insert a % sign after the number. $0.236 = \underline{23}.6\% = 23.6\%$	3. Write 0.615 as a percent.
4. Write $\frac{4}{5}$ as a percent. Change the fraction to a decimal by division. $\frac{4}{5} = 0.8$ Now change the decimal to a percent. $0.8 = \underline{80}\% = 80\%$	4. Write $\frac{3}{8}$ as a percent.

Objective 2 – Translate algebraic expressions to word phrases and word phrases to algebraic expressions.

Key Terms

Variable:

Constant:

Algebraic Expressions:

Terms:

Summary
Translating between algebraic expressions and word phrases involves taking statements that are in one form, either algebraic or word phrases, and rewriting them into the other form. In doing translation, we look for key words that have algebraic meaning. For example, plus or more means "+", minus or less means "-", quotient and ratio means division, etc.

Notes:

Guided Example	***Practice***
Write each algebraic phrase as a word phrase.	Write each algebraic phrase as a word phrase.
1. $6x$ Since the number and variable are together there is an implied multiplication: 6 times x	1. $\frac{2}{3}y$
2. $\frac{z}{3}$ This is a fraction so is it division: z divided by 3	2. $\frac{m}{-2}$
3. $t-(-6)$ Subtraction means difference: the difference between t and -6	3. $m-7$
4. $\frac{2}{3}y$ Multiplication by a fraction means "of": $\frac{2}{3}$ of y	4. $\frac{3}{5}x$

5. $x + a$

The "+" sign means sum:
the sum of x and a

6. $2(x + y)$

The "+" means sum and the 2 in front of the parenthesis implies multiplication:
2 times the sum of x and y

Translate each word phrase into an algebraic expression.

7. x decreased by 3

"decreased by" means subtraction
$x - 3$

8. the quotient of -2 and y

"quotient" means division
$\frac{-2}{y}$

9. $\frac{1}{3}$ of z

"of" means multiplication
$\frac{1}{3}z$

10. The difference between t and the product of 2 and y

Difference means subtraction and product means multiplication
$t - 2y$

11. 6 times and sum of a and b

"times" means multiplication and "sum" means addition:
$6(a + b)$

5. $y + t$

6. $7(a + b)$

Translate each word phrase into an algebraic expression.

7. the sum of x and $-y$

8. the quotient of x and -3

9. $\frac{3}{5}$ of x

10. 6 less the product of 3 and y

11. -2 times the quotient of a and b

Objective 3 – Evaluate an algebraic expression.

Key Terms

Variable:

Summary
To evaluate an algebraic expression means that we replace the variables that are used in the expression with numbers. Then we do the mathematical computation using the correct order of operations. First evaluate any parenthesis. Then evaluate any multiplication or division from left to right. Then evaluate any addition or subtraction from left to right.

Notes:

Guided Example	***Practice***
Evaluate each expression for the given values.	Evaluate each expression for the given value(s).
1. $6-x$ for $x=-3$ Replace x with $x=-3$ $6-x=6-(-3)=6+3=9$	1. $10+x$ for $x=-8$
2. $-5ab$ for $a=2, b=-3$ Replace a with 2 and b with -3 $-5ab=-5(2)(-3)=30$	2. $2xy$ for $x=3, y=-4$

Objective 4 – Evaluate a formula.

Key Terms

Variable:

Summary

To evaluate a formula means that we analyze the application problem and replace the variables that are used in the formula with numbers given in the application. Then we do the mathematical computation using the correct order of operations.

Notes:

Guided Example	***Practice***
Evaluate each expression for the given values.	Evaluate each expression for the given value(s).
1. The volume of a box of given length, *l*, width, *w*, and height, *h*, is $V = l \cdot h \cdot w$. Find the volume of a box whose length is 8 in, width is 6 in and height is 4 in. Replace *l* with 8 in, *w* with 6 in and *h* with 4 in. $V = l \cdot h \cdot w = (8\text{in})(6\text{in})(4\text{in}) = 192\ \text{in}^3$	1. Find the volume of a large container that is 7 ft long, 4 ft wide and 5 ft high. $V = l \cdot h \cdot w$
2. The formula for the perimeter of a rectangle is $P = 2(l + w)$. Find the perimeter of a rectangular field that is 200 ft long and 150 ft wide. Replace *l* with 200 and *w* with 150 $P = 2(l + w) = 2(200 + 150)$ $= 2(350) = 700$ ft	2. Find the perimeter of a rectangular garden that is 3 yd long and 5.5 yd wide. $P = 2(l + w)$

Chapter 1 Answers

Objective 1 1. $\frac{6}{25}$ 2. 0.32 3. 61.5% 4. 37.5%

Objective 2 1. $\frac{2}{3}$ times y 2. m divided by -2 3. The difference between m and 7

4. $\frac{3}{5}$ of x 5. The sum of y and t 6. 7 times the sum of a and b

7. $x-y$ 8. $\frac{x}{-3}$ 9. $\frac{3}{5}x$ 10. $6-3y$ 11. $-2\cdot\frac{a}{b}$

Objective 3 1. 2 2. −24

Objective 4 1. 140 ft^3 2. 17 yd

Chapter 2: Set Theory

Objective 1 – Exponents and order of operations.
Objective 2 – Compare real numbers.
Objective 3 – Solve linear equations using the addition property.

Objective 1 – Exponents and order of operations.

Key Terms

Exponent

Summary
When a number has an exponent, that number is multiplied by itself the number of time equal to the exponent.

$$a^n = \underbrace{a \cdot a \cdot a \cdots a}_{\text{n times}}$$

To calculate a numeric expression correctly it is important that we do the mathematics of the expression in the correct order. First we calculate the inside of any parenthesis or other grouping symbols, from left to right. Next we evaluate any exponents from left to right. Continue from left to right with any multiplication and/or division. Perform the multiplication and division in the order they appear in the expression. Lastly, do the addition and/or subtraction in the order they appear from left to right.

Notes:

Guided Example	***Practice***
1. Calculate 3^6. Since the 3 has an exponent of 6, multiply the 3 by itself 6 times: $3^6 = 3 \cdot 3 \cdot 3 \cdot 3 \cdot 3 \cdot 3 = 729$	1. Calculate 4^3.
2. Calculate $2^3 \cdot 4^2$ The 2 has an exponent of 3 and the 4 has an exponent of 2. Multiply the 2 by itself 3 times and multiply the 4 by itself 2 times and then multiply these two together. $2^3 \cdot 4^2 = 2 \cdot 2 \cdot 2 \cdot 4 \cdot 4 = 8 \cdot 16 = 128$	2. Calculate $3^4 \cdot 5^2$.

3. Calculate $30-24\div 8\cdot 4$ First do the multiplication and division from left to right as they appear. $30-24\div 8\cdot 4=30-3\cdot 4$ $=30-12$ Now do the subtraction. $30-12=18$	3. Calculate $12\div 3+2\cdot 4$.
4. Calculate $2+3\cdot(5+2^3)$ First do the inside of the parenthesis, $2+3\cdot(5+2^3)=2+3\cdot(5+8)$ $=2+3\cdot(13)$ Next we do the multiplication, $2+3\cdot(13)=2+39$ Next we do the addition, $2+39=41$	4. Calculate $40-3\cdot(4+2^3)$.

Objective 2 – Compare real numbers.

Key Terms

Equal to

Not equal to

Less than

Less than or equal

Greater than

Greater than or equal

Summary
As you move from left to right on a number line the numbers increase in value. When you compare two numbers, the number to the right of another number on the number line is the greater of the two. The number to the left of another number on the number line is the lesser of the two. An easy way to remember which symbol to use is to always point the symbol to the smaller of the two numbers. If a number is equal to itself, include the equal sign in the inequality. Therefore a number is less than or equal to itself or greater than or equal to itself.

Notes:

Guided Example	***Practice***
Insert $<, >,$ or $=$ in the blank to make a true statement. 1. -6 ___ -3 Since -6 is to the left of -3 on the number line, then it is less than. So, $-6 < -3$ 2. 5 ___ -7 Since 5 is to the right of -7 on the number line, then it is greater than. So, $5 > -7$ 3. -2 ___ -5 Since -2 is to the right of -5 on the number line, then it is greater than. So, $-2 > -5$	Insert $<, >,$ or $=$ in the blank to make a true statement. 1. -5 ___ -8 2. 7 ___ -8 3. -9 ___ -6

Objective 3 – Solve linear equations using the addition property.

Key Terms

Addition property

Additive inverse

Linear equation in one variable

Summary
The key idea in solving a linear equation is finding the value of the variable that makes the equation a true statement. To do this, isolate the variable on one side of the equal sign and the answer will be on the other side of the equal sign. The addition property is sometimes used to isolate the variable.

Both sides of the equation must be manipulated in the same way so that the equality is not effected. For example, if we add a number to one side, that same number must be added to the other side or else the equation will not be equal. This is called the addition property. To isolate the variable a quantity may need to be 'removed' from the side that has the variable. Therefore, we can add the additive inverse of the number to both sides. A general linear equation might be in the form $x+a=b$. To isolate the x on the left side, add the additive inverse of *a*, *-a*, to both sides:

$$x+a+(-a)=b+(-a)$$
$$x+0=b-a$$
$$x=b-a$$

The x has been isolated and the solution is the arithmetic expression on the opposite side.

Notes:

Guided Example	***Practice***
Solve the following linear equations. 1. $x+5=16$ Add the additive inverse of 5 to both sides and simplify: $x+5+(-5)=16+(-5)$ $x+0=11$ $x=11$ The solution is 11. 2. $y-6=10$ Add the additive inverse of -6 to both sides and simplify: $y-6+(6)=10+(6)$ $y+0=16$ $y=16$ The solution is 16.	Solve the following linear equations. 1. $t+6=32$ 2. $m-10=-5$

3. $a - 7 = -2$ Add the additive inverse of -7 to both sides and simplify: $a - 7 + (7) = -2 + (7)$ $a + 0 = 5$ $a = 5$ The solution is 5.	3. $-2 + w = -8$

Chapter 2 Answers

Objective 1	1. 64	2. 2025	3. 12	4. 4
Objective 2	1. >	2. >	3. <	
Objective 3	1. 26	2. 5	3. −6	

Chapter 5: Numeration Systems

Objective 1 – Decimals.
Objective 2 – Add real numbers.
Objective 3 – Subtract real numbers.
Objective 4 – Multiply real numbers.
Objective 5 – Divide real numbers.

Objective 1 – Decimals.

Key Terms

Decimal number

Summary

A decimal number contains a whole part that comes before the decimal and a fractional part that follows the decimal. Place value in the decimal part follows a similar pattern as the whole number part. The following chart shows the place values for several positions on each side of the decimal point.

Hundreds	Tens	Ones	.	Tenths	Hundredths	Thousandths
100	10	1	And	$\frac{1}{10}$	$\frac{1}{100}$	$\frac{1}{1000}$

Knowing the place value is important in order to properly round a decimal number. To round a decimal number to a given place, follow these rules:

1. Mark the place to which you are rounding with an underscore.
2. Look at the digit following the underscored digit. If this digit is 5 or higher, add 1 to the underscored digit; if it is less than 5, do not change the underscored digit.
3. Drop all the digits on the right of the underscored digit.

Notes:

Guided Example	***Practice***
1. Round 23.792 to the nearest hundredth. Underline the hundredth place. 23.792 Since the next digit after the 9 is less than 5 do not change the 9 and drop all digits after the 9. Thus, the rounded number is: 23.79	1. Round 75.57 to the nearest tenth.
2. Round 74.56 to the nearest tenth. Underline the tenth place. 74.56 Since the next digit after the 5 is more than 5 add 1 to the 5 to make 6 and drop all digits after the 6. Thus, the rounded number is: 74.6	2. Round 56.714 to the nearest hundredth.
3. Round 51.698 to the nearest hundredth. Underline the hundredth place. 51.698 Since the next digit after the 9 is more than 5 add 1 to the 9. Since the hundredths place has a 9 and adding 1 results in 10 this makes the tenths place and hundredths place 70 and then and drop all digits after the 9. Thus, the rounded number is: 51.70	3. Round 45.96 to the nearest tenth.

Objective 2 – Add real numbers.

Key Terms

Additive inverse

Additive Identity Property

Summary

To add two real numbers, use a number line to understand the concept. Start on the number line at the first number. If the second number being added is positive, move to the right the number of positions indicated by the second number. If the second number is negative, move to the left the number of positions indicated by the second number.

Add $3+(-5)$. Since the second number is negative, we start at 3 on the number line and move to the left 5 positions. The results is -2.

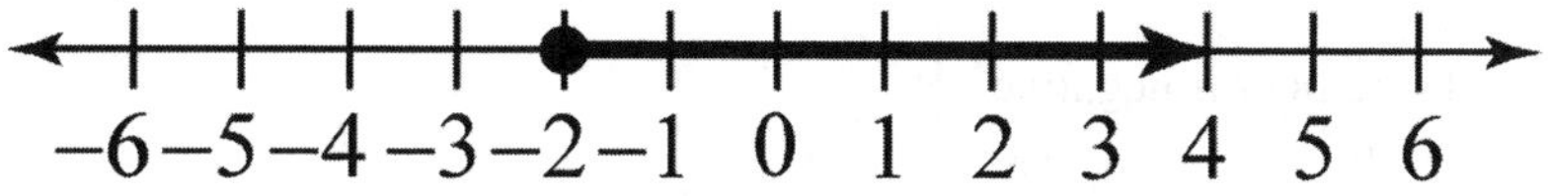

Add $-2+6$. Since the second number is positive, we start at -2 on the number line and move to the right 6 positions. The results is 4.

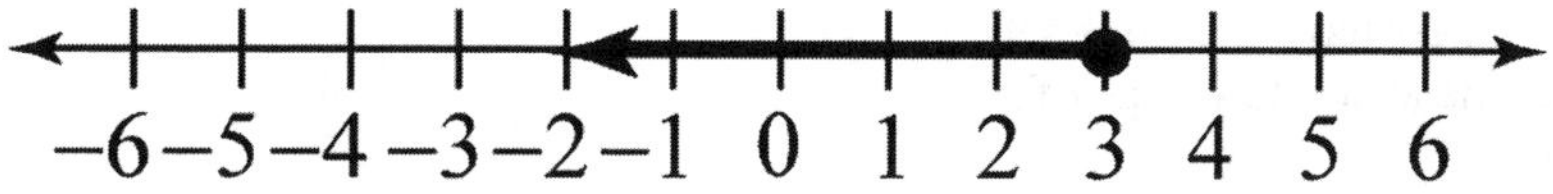

Two numbers that have opposite signs only are called additive inverses and their sum is zero: $a+(-a)=0$. Every number has an additive inverse.

Another property of addition of real numbers is the Additive Identity Property. If zero is added to any real number the results is not changed; $a+0=a$.

All of these statements lead to the general rules for adding real numbers.

1. If the numbers being added have the same sign, add their absolute values and keep their sign.
2. If the numbers being added have different signs, find the difference in the absolute values (subtract the smaller absolute value from the larger absolute value) and take the sign of the one with the larger absolute value.

Notes:

Guided Example	***Practice***
Add the following real numbers.	Add the following real numbers.
1. $23+(-16)$ Since the numbers have different signs we take the difference in the absolute values of the numbers $23-16=7$ The larger of the number is positive so the answer is 7.	1. $-35+19$
2. $15+(-21)$ Since the numbers have different signs we take the difference in the absolute values of the numbers $21-15=6$ The larger of the number is negative so the answer is -6	2. $27+(-14)$
3. $-\frac{1}{6}+\left(-\frac{5}{6}\right)$ Since the numbers have the same sign we take the absolute values of the numbers. $\frac{1}{6}+\left(\frac{5}{6}\right)=\frac{6}{6}=1$ Since both numbers are negative the answer is -1	3. $-\frac{3}{5}+\left(-\frac{7}{5}\right)$
4. $-9.1+7.2$ Since the numbers have different signs we take the difference in the absolute values of the numbers $9.1-7.2=1.9$ The larger of the number is negative so the answer is -1.9	4. $-10.5+17.3$

Objective 3 – Subtract real numbers.

Key Terms

NONE

Summary
To subtract real numbers, two previous topics will be used: addition and finding the opposite. Adding two real numbers depends on the signs of the two numbers. Subtraction will involve changing the operation to addition and finding the inverse of the number being subtracted.

To subtract two real numbers:

1. Change the operation to addition.
2. Change the number being subtracted to its opposite (sign).
3. Perform the addition by following the rules for adding two real numbers.

Thus, $a - b = a + (-b)$

Notes:

Guided Example	***Practice***
Subtract the following real numbers.	Subtract the following real numbers.
1. $16-(-4)$ Change the operation to addition. Also change the number being subtracted to its opposite. Since the number being subtracted is -4 change it to 4. Therefore the problem becomes: $16-(-4)=16+(4)=20$	1. $15-(-23)$
2. $-12-8$ Change the operation to addition. Also change the number being subtracted to its opposite. Since the number being subtracted is 8 change it to -8. Therefore the problem becomes: $-12+(-8)=-20$	2. $-45-12$
3. $-4-(-8)$ Change the operation to addition. Also change the number being subtracted to its opposite. Since the number being subtracted is -8 change it to 8. Therefore the problem becomes: $-4+8=4$	3. $34-23$

Objective 4 – Multiply real numbers.

Key Terms

Multiplicative Identity Property

Multiplicative Property of Zero

Summary
There are basic rules to follow when multiplying real numbers. From basic arithmetic we know that multiplying two positive numbers results in a positive number. We can think about multiplication of a positive times a negative in the real world as owing multiple people money. If you owe 3 people 10 dollars (-10) then your debt is $3(-10)=-30$, that is, you owe 30 dollars. So a positive number times a negative number results in a negative number. The only other combination is multiplying two negative numbers. The results of multiplying two negative numbers is a positive number. This leads to the following rules for multiplication of real numbers:

1. Multiplying two real numbers with like signs results in a positive number.
2. Multiplying two real numbers with unlike signs results in a negative number.

There are two special properties of multiplication:

1. If we multiply a real number by 1 the result is the original number. This is called the Multiplicative Identity Property; $1 \cdot a = a$
2. If we multiply a real number by 0 the result is 0. This is called the Multiplicative Property of Zero; $0 \cdot a = 0$

Notes:

Guided Example	***Practice***
Multiply the following real numbers. 1. $(-2)(-4)$ Since the numbers have the same sign we multiply the numbers and the result will be positive: $(-2)(-4)=8$	Multiply the following real numbers. 1. $(6)(-3)$
2. $(3)(-7)$ Since the numbers have opposite signs we multiply the numbers and the result will be negative: $(3)(-7)=-21$	2. $(8)(5)$
3. $(5)(8)$ Since the numbers have the same sign we multiply the numbers and the result will be positive: $(5)(8)=40$	3. $(-9)(7)$

4. $(8)(0)$ Multiplying a number by 0 results in 0 so: $(8)(0) = 0$ 5. $(-14)(1)$ Multiplying a number by 1 results in the same number so: $(-14)(1) = -14$	4. $(-9)(0)$ 5. $(15)(1)$

Objective 5 – Divide real numbers.

Key Terms

NONE

Summary

Since division is related to multiplication, division of real numbers basically follows the same rules as multiplication. We can use the following rules:

To divide two real numbers:

1. Divide the absolute value of the two real numbers.
2. If the two numbers have like signs the result is a positive number.
3. If the two numbers have unlike signs the result is a negative number.

Notes:

Guided Example	***Practice***
Divide the following real numbers.	Divide the following real numbers.
1. $35 \div (-7)$ Divide the absolute values of the numbers. Since the numbers have unlike signs the result will be negative. Therefore: $35 \div (-7) = -5$	1. $-25 \div 5$
2. $-12 \div 6$ Divide the absolute values of the numbers. Since the numbers have unlike signs the result will be negative. Therefore: $-12 \div 6 = -2$	2. $60 \div (-12)$
3. $\frac{24}{6}$ Divide the absolute values of the numbers. Since the numbers have like signs the result will be positive. Therefore: $\frac{24}{6} = 4$	3. $\frac{-48}{-4}$
4. $-56 \div (-8)$ Divide the absolute values of the numbers. Since the numbers have like signs the result will be positive. Therefore: $-56 \div (-8) = 7$	4. $27 \div 9$

Chapter 5 Answers

Objective 1	1. 75.6	2. 56.71	3. 46.0		
Objective 2	1. −16	2. 13	3. −2	4. 6.8	
Objective 3	1. 38	2. −57	3. 11		
Objective 4	1. −18	2. 40	3. −63	4. 0	5. 15
Objective 5	1. −5	2. −5	3. 12	4. 3	

Chapter 6: Number Theory and the Real Number System

Objective 1 – Square roots.
Objective 2 – Solve applied problems involving algebraic expressions or formulas.

Objective 1 – Square roots.

Key Terms

Square root

Exponent

Radical

Radicand

Factors

Product

Perfect Square

Approximate symbol

Square root symbol

Summary
The square root of a number is a factor we can multiply by itself to get the number. For example, 3 time 3 is 9 so the square root of 9 is 3. We can also say that $3^2 = 9$ so the square root of 9 is 3. The notation is: $\sqrt{a \cdot a} = \sqrt{a^2} = |a|$. Note that the absolute value of *a* is used because we want a single answer to the problem. There is no square root of a negative number in the real number system since any number times itself is always positive. The square root sign is also called a radical sign and the number under the radical sign is called the radicand.

Sometimes a number is not a perfect square, that is, there is no factor you can multiply by itself to get the number. We can calculate this square root is two ways. First, we can factor the radicand and see if there are any perfect square factors. If so, we can take the square root of the perfect square factors. For example 12 is not a perfect square but we can factor 12 into 4 times 3. Since 4 is a perfect square then:

$$\sqrt{12} = \sqrt{4 \cdot 3} = \sqrt{4}\sqrt{3} = 2\sqrt{3}$$

Warning: This does not work for addition. You cannot take the square root of terms. $\sqrt{4+9} \neq \sqrt{4} + \sqrt{9}$

The other way to calculate the square root of a number that is not a perfect square is by using a calculator and finding the approximation. Round to the decimal place given in the problem. We use the approximation symbol, $\approx$, to indicate that the answer is not exact.

Notes:

Guided Example

Find the exact square root of the following.

1. $\sqrt{36}$
   Since 36 is a perfect square, then:
   $\sqrt{36} = \sqrt{6 \cdot 6} = \sqrt{6^2} = 6$

2. $\sqrt{75}$
   75 is not a perfect square. Factor 75 to determine if there is a perfect square factor.
   $\sqrt{75} = \sqrt{25 \cdot 3} = \sqrt{25} \cdot \sqrt{3} = 5 \cdot \sqrt{3}$

3. $\sqrt{9} + \sqrt{16}$
   This is the addition of two separate square roots:
   $\sqrt{9} + \sqrt{16} = 3 + 4 = 7$

4. $\sqrt{9+16}$
   This is the square root of addition so the addition must be done first:
   $\sqrt{9+16} = \sqrt{25} = 5$

5. $\sqrt{\frac{9}{16}}$
   To find the square root of a fraction, take the square root of the numerator and denominator separately:
   $\sqrt{\frac{9}{16}} = \frac{\sqrt{9}}{\sqrt{16}} = \frac{3}{4}$

Find the approximate value of the following:

6. $\sqrt{20}$ (round to the nearest hundredth)
   Use a calculator to obtain:
   $\sqrt{20} \approx 4.47$

7. $\sqrt{500}$ (round to the nearest tenth)
   Use a calculator to obtain:
   $\sqrt{500} \approx 22.4$

Practice

Find the exact square root of the following.

1. $\sqrt{64}$

2. $\sqrt{24}$

3. $\sqrt{25} - \sqrt{4}$

4. $\sqrt{114-50}$

5. $\sqrt{\frac{4}{25}}$

Find the approximate value of the following:

6. $\sqrt{48}$ (round to the nearest hundredth)

7. $\sqrt{800}$ (round to the nearest hundredth)

Objective 2 – Solve applied problems involving algebraic expressions or formulas.

Key Terms

NONE

Summary

To evaluate an expression or formula for an applied problem involves analyzing the problem for the missing quantities and determining which variables are given and which variable needs to be calculated.

Reading the problem carefully is very important. Make a list of the variables. Then determine which variable needs to be computed. Find the value of all of the other variables in the problem. Substitute the given values and compute the needed value using the correct order of operations.

Notes:

Guided Example	***Practice***
1. The volume of a box is calculated by multiplying the width by the length by the height. Find the volume of a box that is 60 mm wide, 85 mm long, and 70 mm high. The formula for the volume of a box is: $V = w \cdot l \cdot h$. The width, *w*, of the box is 60 mm. The length, *l*, of the box is 85 mm and the height, *h*, of the box is 70 mm. Substitute these values into the formula and calculate the volume: $V = w \cdot l \cdot h$ $= (60 \text{ mm})(85 \text{ mm})(70 \text{ mm})$ $= 357{,}000 \text{ mm}^3$	1. A farmer wishes to install fencing around his rectangular field. The field is 2500 feet by 4500 feet. Find the amount of fencing the farmer will need. The formula for perimeter is $P = 2l + 2w$
2. The formula for calculating simple interest is $I = Prt$, where *I* is the interest (in dollars), *P* is the principal (amount invested in dollars), *r* is the annual interest rate (expressed as a decimal), and *t* is the time (in years) that the principal has been on deposit. Find the amount of simple interest on a principal deposit amount of \$4500 for 4 years at a 2% annual interest rate. *P* is the principal so $P = 4500$. The annual rate is, $r = 0.02$, and *t* is the number of years so $t = 4$. $I = Prt$ $= (4500)(0.02)(4)$ $= \$360.00$	2. Find the amount of simple interest earned on a deposit amount of \$7000 for 6 years at an interest rate of 2.5%.

3. A gardener wishes to calculate the area of her circular garden. The garden has a radius of 12 feet. The formula for the area of a circle is $A = \pi r^2$. Find the approximate area of the circle. Use $\pi \approx 3.14$ and round to the nearest tenth. The radius is 12 feet so $r = 12$ ft. So the approximate area is: $A \approx \pi r^2$ $\approx (3.14)(12 \text{ ft})^2$ $\approx 452.2 \text{ ft}^2$	3. Calculate the approximate volume of a cylinder that has a height of 12 inches and a radius of 2 inches. The volume of a cylinder is given by: $V = \pi r^2 h$. Use $\pi \approx 3.14$ and round to the nearest tenth.

Chapter 6 Answers

Objective 1	1. 8	2. $2\sqrt{6}$	3. 3	4. 8	5. $\frac{2}{5}$
	6. 6.93	7. 28.28			
Objective 2	1. 14,000 ft	2. \$1,050	3. 150.7 in^3		

Chapter 7: Algebraic Models

Objective 1 – Solve a linear inequality using the addition and multiplication properties of inequalities.
Objective 2 – Find the slope of a line that passes through two given points.
Objective 3 – Determine whether the slope of a given line is positive, negative, zero, or undefined.
Objective 4 – Solve a proportion.
Objective 5 – Solve quadratic equations by using the square root property.

Objective 1 – Solve a linear inequality using the addition and multiplication properties of inequalities.

Key Terms

Addition Property of Inequalities

Multiplication Property of Inequalities

Summary
The solution to an inequality is any value of the variable that makes the inequality true. Solving an inequality means finding all solution to the inequality. We can use the addition and multiplication properties to solve the inequality. Similar to the Addition Property of Equality, the same value can be added to both sides of an inequality without changing the solution set. Thus the following is called the Addition Property of Inequalities:

For any real numbers, *a*, *b*, and *c*;

1. If $a < b$, then $a + c < b + c$
2. If $a > b$, then $a + c > b + c$

This also works for $\leq, \geq$.

When a number is added or subtracted to both sides of an inequality, do not change the direction of the inequality.

Multiplication of both sides of an inequality may affect the direction of the inequality. If you multiply by a negative number, the direction of the inequality is changes. The following is the Multiplication Property of Inequalities:

For any real numbers *a*, *b*, and *c*, $c \neq 0$;

1. If *c* is positive and $a < b$, then $ac < bc$
2. If c is positive and $a > b$, then $ac > bc$
3. If *c* is negative and $a < b$, then $ac > bc$ (reverse the inequality)
4. If *c* is negative and $a > b$, then $ac < bc$ (reverse the inequality)

The same holds true for $\leq, \geq$.

To solve an inequality, isolate the variable on one side of the inequality.

Notes:

Guided Example	***Practice***
Solve the following inequalities. 1. $3x+2<-15$ The goal is to isolate the x. The first step is to subtract 2 from both sides. Subtraction does not change the direction of the inequality. $3x+2<-15$ $3x+2-2<-15-2$ $3x<-17$ Next, multiply both sides by the reciprocal of 3 which is $\frac{1}{3}$. We do not multiply by a negative number so we do not change the sign of the inequality. $3x<-17$ $\frac{1}{3}\cdot 3x<\frac{1}{3}\cdot(-17)$ $x<-\frac{17}{3}$	Solve the following inequalities. 1. $-5x-7>60$
2. $-6m-3>-30$ The goal is to isolate the m. The first step is to add 3 from both sides. Addition does not change the direction of the inequality. $-6m-3>-30$ $-6m-3+3>-30+3$ $-6m>-27$ Next, multiply both sides by the reciprocal of -6 which is $-\frac{1}{6}$. We multiply by a negative number so we must change the sign of the inequality. Also, reduce the answer. $-6m>-27$ $\left(-\frac{1}{6}\right)\cdot(-6)m<\left(-\frac{1}{6}\right)\cdot(-27)$ $m<\frac{27}{6}$ $m<\frac{9}{2}$	2. $4x+5<-24$

3. $4y-3(2y-1)<9$ Follow the order of operations to eliminate the parenthesis and combine the like terms. $4y-3(2y-1)<9$ $4y-6y+3<9$ $-2y+3<9$ Next subtract 3 from both sides and multiply by the reciprocal of -2 which is $-\frac{1}{2}$. We multiply by a negative number so we must change the sign of the inequality. $-2y+3<9$ $-2y+3-3<9-3$ $-2y<6$ $\left(-\frac{1}{2}\right)\cdot(-2)y>\left(-\frac{1}{2}\right)\cdot 6$ $y>-3$	3. $5w-4(2w-1)>10$

Objective 2 – Find the slope of a line that passes through two given points.

Key Terms

Slope

Summary

The slope of a line is also called the rate of change. The slope can tell us if the line is going up or down or if it is vertical or horizontal. Slope can be used to tell us how fast a quantity is changing such as the speed of an object or the rate of change of a cell phone bill. To calculate the slope of a straight line, two points are needed. Label the two points (x_1, y_1) and (x_2, y_2) to represent the coordinates of the points. The slope is represented by m. The formula for calculating the slope is:

$$m = \frac{\text{change is y values}}{\text{change in x values}} = \frac{y_2 - y_1}{x_2 - x_1} \text{ where } x_1 \neq x_2$$

The change is y values goes in the numerator and the change in the x values goes in the denominator. Make sure you use the coordinates in the same order, $y_2 - y_1$ and $x_2 - x_1$, in the numerator and denominator.

Notes:

Guided Example	***Practice***
Find the slope of the line through the following sets of points. Then graph the line. 1. $(2,1),(-2,-3)$ Let $(2,1)$ represent (x_1, y_1) and $(-2,-3)$ represent (x_2, y_2). The slope is: $m = \frac{y_2 - y_1}{x_2 - x_1} = \frac{(-3)-(1)}{(-2)-(2)} = \frac{-4}{-4} = 1$ 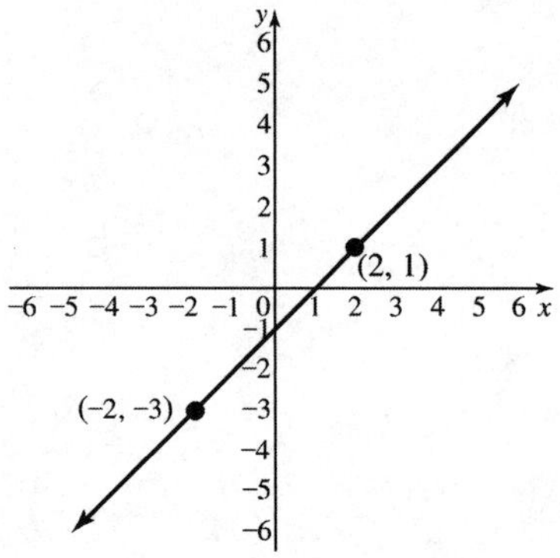	Find the slope of the line through the following sets of points. Then graph the line. 1. $(-3,-2),(-5,6)$

2. $(-1,3),(4,-1)$ Let $(-1,3)$ represent (x_1, y_1) and $(4,-1)$ represent (x_2, y_2). The slope is: $m = \frac{y_2 - y_1}{x_2 - x_1} = \frac{(-1)-(3)}{(4)-(-1)} = \frac{-4}{5} = -\frac{4}{5}$ 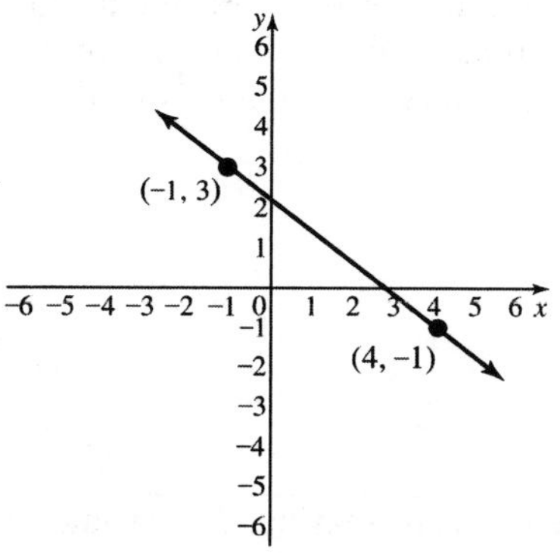 	2. $(4,4),(-7,-1)$

Objective 3 – Determine whether the slope of a given line is positive, negative, zero, or undefined.

Key Terms

NONE

Summary

If the y values get larger as the x values get larger, the slope is positive and the line is said to be increasing. If the y values get smaller as the x values get larger, the slope is negative and the line is said to be decreasing. If the y values are the same for all x values, then the numerator is zero thus the slope is zero and the line is horizontal. If the x values are the same for all y values, then the denominator is zero. The denominator of a fraction cannot be zero so the slope is undefined. Since the x values are the same the line is vertical.

Notes:

Guided Example	***Practice***
Determine if the line is increasing, decreasing, vertical or horizontal.	Determine if the line is increasing, decreasing, vertical or horizontal.
1. 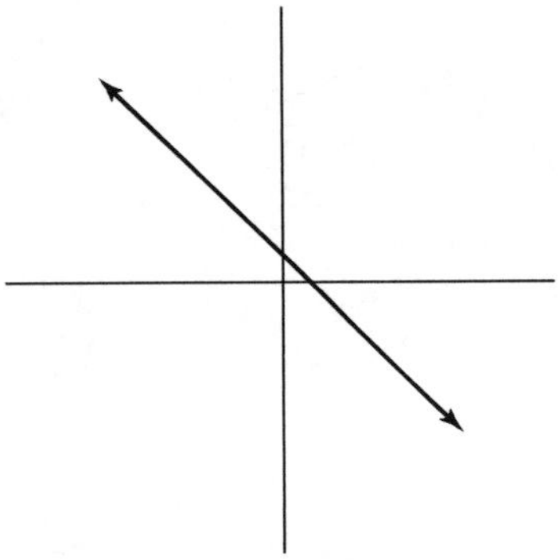 Since the y values are getting smaller as the x values are getting larger the slope is negative and the line is decreasing.	1. 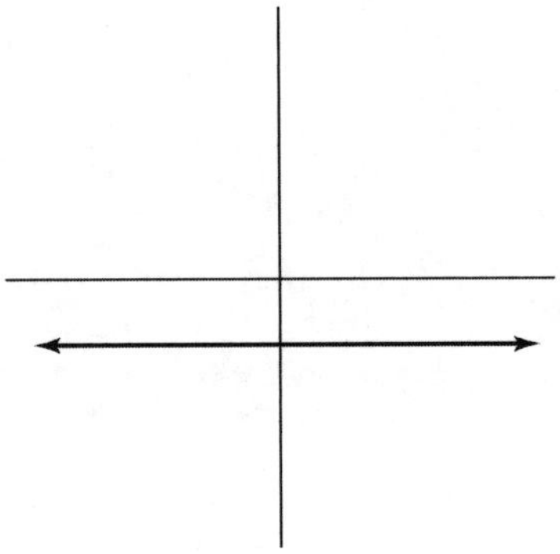
2. 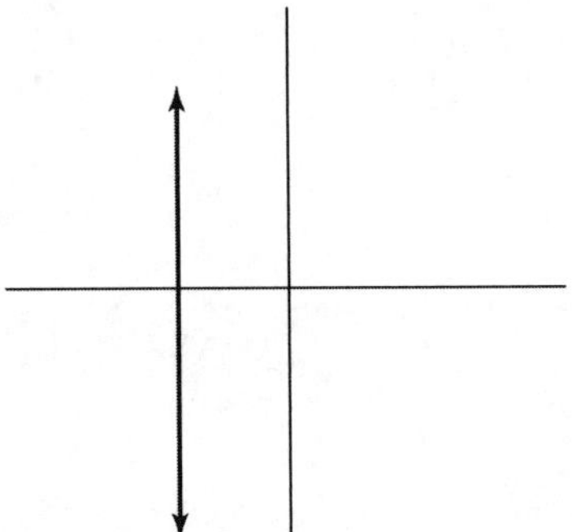 Since the x values are the same for all y values the slope is undefined and the line is vertical.	2. 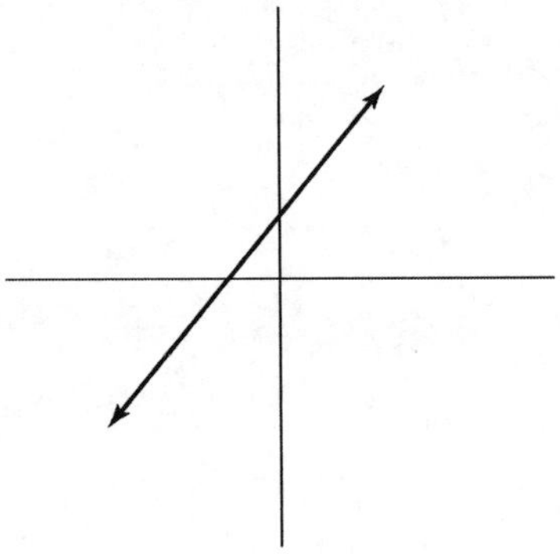

3.

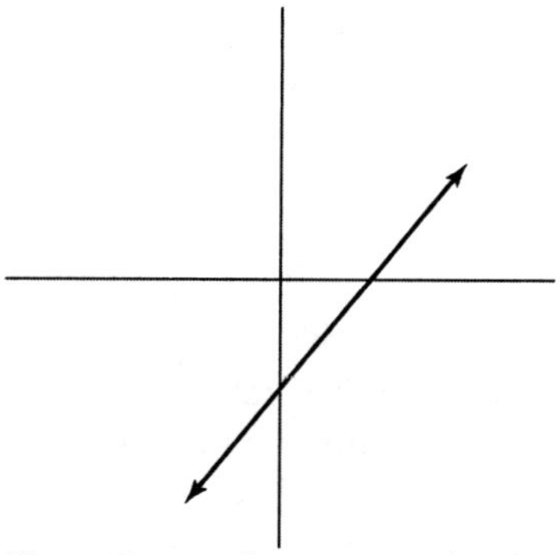

Since the y values are getting larger as the x values are getting larger the slope is positive and the line is increasing.

4.

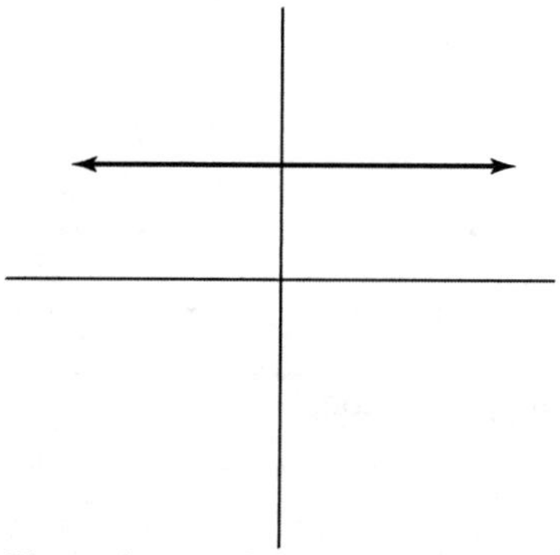

Since the y values are the same for all x values the slope is zero and the line is horizontal.

3.

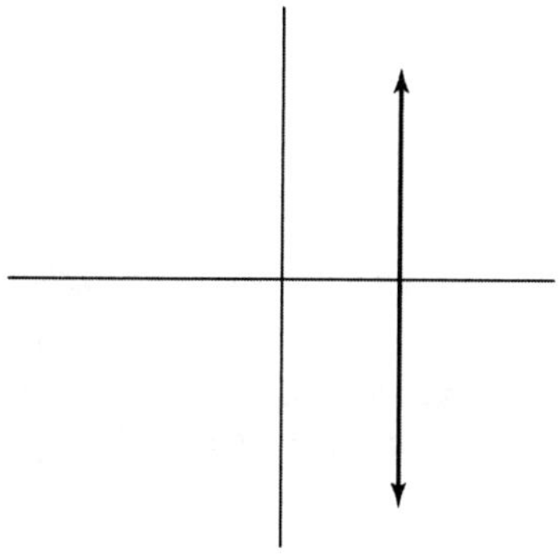

4.

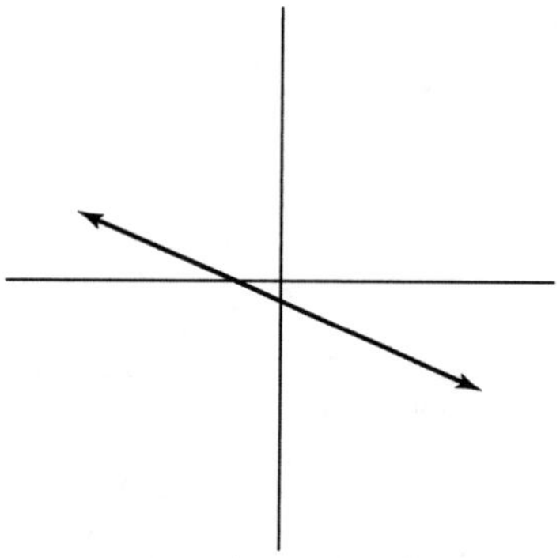

Objective 4 – Solve a proportion.

Key Terms

Proportion

Cross product

Summary

A proportion is two rational expressions (fractions) that are equal to each other. The rational expressions are equal only if their cross products are equal. This means we multiply the numerator of one by the denominator of the other and vice versa:

If $\frac{a}{b}=\frac{c}{d}$ then $ad = bc$

Then we can solve for any variable. We can check the answer in the original proportion to make sure it is correct.

Notes:

Guided Example	***Practice***
Solve the following proportions.	Solve the following proportions.
1. $\frac{x}{8}=\frac{3}{4}$ Cross multiply the numerators and denominators of the rational expressions. Then solve for the variable. $4x = 8\cdot 3$ $4x = 24$ $x = 6$ Check the answer. $\frac{(6)}{8}=\frac{3}{4}$ Therefore the solution is 6.	1. $\frac{t}{24}=\frac{5}{4}$
2. $\frac{m-2}{12}=\frac{2}{3}$ Cross multiply the numerators and denominators of the rational expressions. Then solve for the variable. $3(m-2)=(12)2$ $3m-6=24$ $3m=30$ $m=10$ Check the answer. $\frac{10-2}{12}=\frac{8}{12}=\frac{2}{3}$	2. $\frac{4}{y-1}=\frac{2}{5}$

Objective 5 – Solve quadratic equations by using the square root property.

Key Terms

Quadratic equation

Summary

A quadratic equation is an equation that has a square as the highest power of the variable. It is written in the form $ax^2 + bx + c = 0$ where *a*, *b*, and *c* are real numbers and $a \neq 0$. We can solve some of these equations using the square root property if they are in the form $x^2 = c, c \geq 0$. In this case, the solution to $x^2 = c$ is $x = \pm\sqrt{c}$. Thus, the solution is $x = \sqrt{c}$ or $x = -\sqrt{c}$.

Notes:

Guided Example	***Practice***
Solve the following quadratic equations.	Solve the following quadratic equations.
1. $x^2 = 25$ Take the positive and negative square root of 25. $x = \pm\sqrt{25} = \pm 5$ So the solution is 5 or -5.	1. $x^2 = 144$
2. $x^2 = 20$ Take the positive and negative square root of 20 and simplify. $x = \pm\sqrt{20}$ $= \pm\sqrt{4 \cdot 5}$ $= \pm 2\sqrt{5}$ So the solution is $2\sqrt{5}$ or $-2\sqrt{5}$.	2. $\sqrt{75}$

Chapter 7 Answers

Objective 1 1. $x < -\frac{67}{5}$ 2. $x < -\frac{29}{4}$ 3. $w < -2$

Objective 2 1. $m = -4$

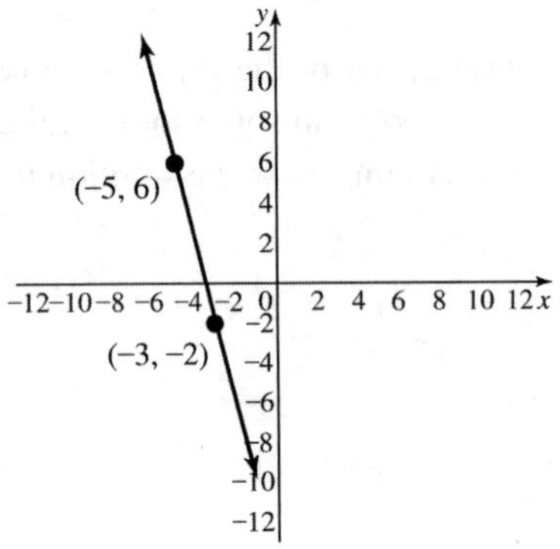

2. $m = \frac{5}{11}$

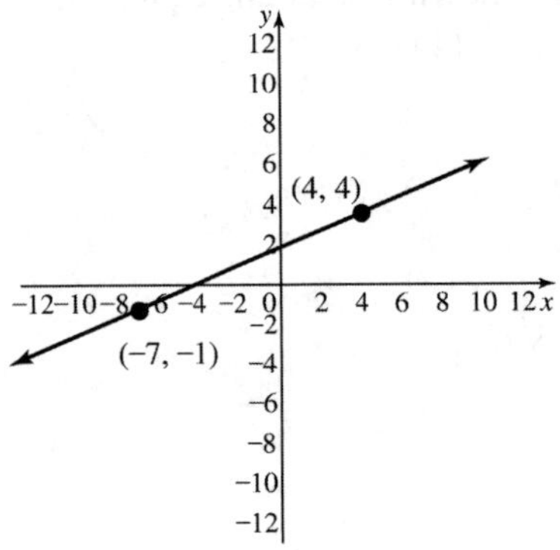

Objective 3 1. Horizontal 2. Increasing 3. Vertical 4. Decreasing

Objective 4 1. 30 2. 11

Objective 5 1. ± 12 2. $\pm 5\sqrt{3}$

Chapter 8: Consumer Mathematics

Objective 1 – Solve linear equations using the multiplication property.
Objective 2 – Evaluate exponential functions.
Objective 3 – Write equivalent exponential and logarithmic equations.
Objective 4 – Evaluate logarithms.
Objective 5 – Solve logarithmic equations.

Objective 1 – Solve linear equations using the multiplication property.

Key Terms

Multiplication Property

Reciprocal or Multiplicative Inverse

Coefficient

Term

Variable

Linear Equation

Standard form of a linear equation

Real number

Summary
The goal in solving a linear equation is to use algebraic properties to isolate the variable with a coefficient of 1 on one side of the equal sign. Then the solution will be on the other side. If we multiply both sides of an equation by a non-zero real number, we do not change the value of the solution. This is called the Multiplication Property of Equality:

For any real numbers, a, b, and c, $c \neq 0$
If $a = b$, then $a \cdot c = b \cdot c$.

We can use this property to isolate a variable.

$$ax = b$$
$$\frac{1}{a} \cdot ax = \frac{1}{a} \cdot b$$
$$\left(\frac{1}{\cancel{a}} \cdot \cancel{a}\right)x = \frac{1}{a} \cdot b$$
$$1 \cdot x = \frac{b}{a}$$
$$x = \frac{b}{a}$$

Notes:

Guided Example	***Practice***
Solve for the variable. 1. $-3x=12$ The goal is to isolate the variable, x by making the coefficient of this term 1. The coefficient is -3. The reciprocal of the coefficient is $-\frac{1}{3}$. Multiply both sides of the equation by the reciprocal. $\left(-\frac{1}{3}\right)(-3)x=\left(-\frac{1}{3}\right)(12)$ $\left(\cancel{-}\frac{1}{\cancel{3}}\right)(\cancel{-3})x=\left(-\frac{1}{3}\right)(12)$ $x=-\frac{12}{3}=-4$ Check the solution by inserting it into the original equation and performing the calculations on each side of the equation to see if the equality holds. $-3(-4)\overset{?}{=}12$ $12=12$ The two sides are equal so the solution is correct.	Solve for the variable. 1. $-4t=-18$
2. $3m=-\frac{15}{4}$ The goal is to isolate the variable, m by making the coefficient of this term 1. The coefficient is 3. The reciprocal of the coefficient is $\frac{1}{3}$. Multiply both sides of the equation by the reciprocal. $\left(\frac{1}{3}\right)(3)m=\left(\frac{1}{3}\right)\left(-\frac{15}{4}\right)$ $\left(\cancel{-}\frac{1}{\cancel{3}}\right)(\cancel{-3})m=\left(-\frac{15}{12}\right)$ $m=-\frac{15}{12}=-\frac{5}{4}$ Check the solution by inserting it into the original equation and performing the calculations on each	2. $-5g=\frac{20}{3}$

side of the equation to see if the equality holds. $3\left(-\frac{5}{4}\right) \stackrel{?}{=} -\frac{15}{4}$ $\left(-\frac{15}{4}\right) = -\frac{15}{4}$ The two sides are equal so the solution is correct.	

Objective 2 – Evaluate exponential functions.

Key Terms

Exponential function

Summary
An exponential function can model data that increases or decreases rapidly. Some examples are population growth, compound interest and radioactive decay. An exponential function is any function that can be expressed in the form $f(x) = b^x$ where $b > 0$, $b \neq 1$ and x is any real number.
The constant, *b*, is referred to as the base of the exponential function and is raised to a variable or variable expression power. We "evaluate" an exponential function by substituting a real number in the variable in the exponent.

Notes:

Guided Example	***Practice***
Evaluate the following exponential functions.	Evaluate the following exponential functions.
1. $f(x) = 3^x$ for $x = 2$ Substitute 2 for *x*. $3^{(2)} = 9$	1. $f(x) = 4^x$ for $x = 3$
2. $f(x) = \left(\frac{1}{4}\right)^x$ for $x = 3$ Substitute 3 for *x*. $\left(\frac{1}{4}\right)^{(3)} = \frac{1}{64}$	2. $f(x) = \left(\frac{2}{3}\right)^x$ for $x = 2$
3. $f(x) = 2^{3x-1}$ for $x = 3$ Substitute 3 for *x*. $2^{3(3)-1} = 2^{9-1}$ $= 2^8$ $= 256$	3. $f(x) = 5^{5x-12}$ for $x = 3$
4. $f(x) = 2^x$ for $x = -3$ Substitute -3 for *x*. $2^{(-3)} = 2^{-3}$ Since a negative exponent means reciprocal then: $2^{-3} = \frac{1}{2^3} = \frac{1}{8}$	4. $f(x) = 5^x$ for $x = -2$

Objective 3 – Write equivalent exponential and logarithmic equations.

Key Terms

Logarithmic function

Summary

Logarithmic functions help us understand many natural phenomena such as magnitude of earthquakes along with many others. Logarithmic functions are inverse functions of exponential functions meaning they do the opposite operation. If an exponential function is represented by $f(x) = b^x$ then the equivalent logarithmic function is $f(x) = \log_b x$, and is read "the logarithm of x to the base b". The following two equations are equivalent:

$$y = \log_b x \quad \leftrightarrow \quad b^y = x$$

The main thing to remember is that a logarithm is an exponent. Here is a numeric example:

$$\log_3 9 = 2 \quad \leftrightarrow \quad 3^2 = 9$$

Notes:

Guided Example	***Practice***
Write each logarithmic equation as an exponential equation. 1. $\log_5 125 = 3$ Since the logarithm is an exponent then: $\log_5 125 = 3$ can be expressed as $5^3 = 125$ 2. $\log_6 6 = 1$ Since the logarithm is an exponent then: $\log_6 6 = 1$ can be expressed as $6^1 = 6$ 3. $\log_3 \frac{1}{9} = -2$ Since the logarithm is an exponent then: $\log_3 \frac{1}{9} = -2$ can be expressed as $3^{-2} = \frac{1}{9}$	Write each logarithmic equation as an exponential equation. 1. $\log_4 1024 = 5$ 2. $\log_4 4 = 1$ 3. $\log_2 \frac{1}{8} = -3$

Objective 4 – Evaluate logarithms.

Key Terms

NONE

Summary

We will use the fact that logarithms are exponents to evaluate some logarithmic expressions. For example, since $3^4 = 81$ then we know that $\log_3 81 = 4$. There are two special properties of logarithms that we can use when evaluating them:

1. $\log_b b = 1$ since $b^1 = b$
2. $\log_b 1 = 0$ since $b^0 = 1$

Notes:

Guided Example	***Practice***
Solve the following logarithms. 1. $\log_2 2$ Since $2^1 = 2$ then $\log_2 2 = 1$ 2. $\log_{25} 5$ Since $25^{\frac{1}{2}} = \sqrt{25} = 5$ then $\log_{25} 5 = \dfrac{1}{2}$ 3. $\log_7 343$ Since $7^3 = 343$ then $\log_7 343 = 3$	Solve the following logarithms. 1. $\log_8 8$ 2. $\log_{16} 4$ 3. $\log_6 216$

Objective 5 – Solve logarithmic equations.

Key Terms

NONE

Summary

The key in solving a logarithmic equation is to rewrite it as an equivalent exponential equation. Then the equation can be solved by regular algebraic steps. For example consider:

$\log_2 x = 4$. Rewrite is as $2^4 = x$. Then solve for x giving $x = 16$.

Notes:

Guided Example	***Practice***
Solve the following logarithmic equations.	Solve the following logarithmic equations.
1. $\log_3 x = 2$ Rewrite as an exponential equation: $3^2 = x$ Solve for x. $x = 9$	1. $\log_5 x = 3$
2. $\log_x 25 = 2$ Rewrite as an exponential equation: $x^2 = 25$ Solve for x. $x = 5$	2. $\log_x 27 = 3$
3. $\log_3 81 = x$ Rewrite as an exponential equation: $3^x = 81$ Solve for x by determining what power of 3 gives 81. $x = 4$	3. $\log_4 64 = x$
4. $\log_5 x = -2$ Rewrite as an exponential equation: $5^{-2} = x$ Solve for x. $x = 5^{-2} = \frac{1}{25}$	4. $\log_3 x = -4$

Chapter 8 Answers

Objective 1	1. $\frac{9}{2}$	2. $-\frac{4}{3}$		
Objective 2	1. 64	2. $\frac{4}{9}$	3. 125	4. $\frac{1}{25}$
Objective 3	1. $4^5 = 1024$	2. $4^1 = 4$	3. $2^{-3} = \frac{1}{8}$	
Objective 4	1. 1	2. $\frac{1}{2}$	3. 3	
Objective 5	1. 125	2. 3	3. 3	4. $\frac{1}{81}$

Chapter 12: Counting

Objective 1 – Fractions.

Objective 1 – Fractions.

Key Terms

Fraction

Denominator

Numerator

Mixed number

Proper fraction

Improper fraction

Equivalent fraction

Simplest form

Like fraction

Unlike fraction

Summary

A fraction can mean two things. It can represent the quotient of two whole numbers or it can represent a part of a whole. For example, $\frac{3}{4}$ means 3 of every 4 of a whole set. A fraction consist of three parts:

- The numerator (the top number)
- The denominator (the bottom number)
- The fraction bar that separates the numerator and denominator (it means "out of" or "divided by")

The denominator represents the number of parts into which the whole thing is divided. The numerator represents the number of parts of the whole the fraction contains. For example, $\frac{3}{4}$ means the whole is divided into 4 pieces and the fraction contains 3 of those 4 pieces.

A proper fraction is one that has a numerator less than the denominator. An improper fraction is one that has a numerator larger than the denominator. An improper fraction can also be expressed as a mixed number. A mixed number consists of a whole number and a proper fraction. For example, $\frac{4}{3}$ is an improper fraction that can also be expressed as $1\frac{1}{3}$. To change a mixed number to an improper fraction:

1. Multiply the denominator of the fraction part by the whole number part
2. Add the numerator of the fraction part to the product.
3. Write the sum over the original denominator.

For example, to express $2\frac{3}{5}$ as an improper fraction, multiply the 2 times the 5 and add the 3. Then express the sum over the 5. So the result is: $\frac{(2\cdot5)+3}{5}=\frac{13}{5}$.

To change an improper fraction to a mixed number:

1. Divide the numerator by the denominator.
2. If there is a remainder, write it over the denominator.

For example, to express $\frac{11}{4}$ as a mixed number divide 11 by 4. This gives 2 with a remainder of 3. Express the remainder over the 4. So the result is: $2\frac{3}{4}$

Two fractions that represent the same value are said to be equivalent. To determine if two fractions are equivalent, express them in simplest form. Two fractions are in simplest form (or reduced to lowest terms) when there is no common factor in its numerator and denominator except 1. To reduce a fraction, divide the numerator and denominator by any common factor. A common way to reduce fractions is:

1. Factor the numerator and denominator into prime factors
2. Cancel (divide into 1) any like factors in the numerator and denominator
3. Multiply any remaining factors in the numerator and denominator

For example, to reduce $\frac{10}{25}$: $\frac{10}{25}=\frac{2\cdot\cancel{5}}{5\cdot\cancel{5}}=\frac{2}{5}$. Therefore $\frac{10}{25}$ and $\frac{2}{5}$ are equivalent fractions.

Like fractions have the same denominator and unlike fractions have different denominators. To add or subtract like fractions:

1. Add or subtract the numerators.
2. Write the fractions with the sum over the original denominator.
3. Reduce the answer to simplest form.

Unlike fractions must be changed to like fractions before adding or subtracting. Think of it as adding $\frac{2}{3}$ of a pizza to $\frac{1}{2}$ of a pizza. To determine how much pizza we would have, we need the same size slices. This represents finding a common denominator since the denominator represents the number of parts of the whole. To add or subtract unlike fractions:

1. Rewrite the fractions as equivalent fractions with a common denominator. We usually use the least common denominator (LCD).
2. Follow the steps for adding and subtracting like denominators.

To change a fraction to an equivalent fraction, be sure and multiply the top and the bottom by the same number. This is equivalent to multiplying by 1 and therefore does not change the value of the fraction. For example:

$$\frac{2}{3}\cdot1=\frac{2}{3}\cdot\frac{2}{2}=\frac{4}{6} \text{ and } \frac{1}{2}\cdot1=\frac{1}{2}\cdot\frac{3}{3}=\frac{3}{6}$$

This changes the fractions to equivalent fractions with like denominators.

To add or subtract mixed numbers, first convert them to equivalent improper fractions. Then follow the steps to add or subtract. Convert the answer back to a mixed number if necessary.

To multiply fractions requires only multiplying the numerators and denominators respectively. Then reduce if necessary. For example:

$$\begin{aligned}\frac{2}{3}\cdot\frac{3}{5}&=\frac{2\cdot3}{3\cdot5}=\frac{6}{15}\\&=\frac{2\cdot\cancel{3}}{5\cdot\cancel{3}}=\frac{2}{5}\end{aligned}$$

If multiplying mixed numbers, convert to improper fraction, follow the steps to multiply and then convert the answer back to a mixed number if necessary.

To divide fractions, we must use the reciprocal of the divisor. The reciprocal is found by switching the numerator and the denominator. For example the reciprocal of $\frac{2}{5}$ is $\frac{5}{2}$. To divide fractions:

1. Change the divisor to its reciprocal
2. Multiply the fractions
3. Reduce the answer to lowest terms

To divide mixed numbers, convert them to improper fractions, divide and convert the answer back to a mixed number if necessary.

Notes:

Guided Example	***Practice***
1. Write $8\frac{2}{3}$ as an improper fraction. Multiply the denominator by the whole number part. Then add the numerator. Express the sum over the original denominator. $8\frac{2}{3}=\frac{(8\cdot 3)+2}{3}$ $=\frac{24+2}{3}=\frac{26}{3}$	1. Write $4\frac{3}{5}$ as an improper fraction.
2. Write $\frac{25}{4}$ as a mixed number. Divide the numerator by the denominator. $\frac{25}{4}=6$ with remainder 1. Write the remainder over the denominator. The mixed number is $\frac{25}{4}=6\frac{1}{4}$.	2. Write $\frac{33}{6}$ as a mixed number.
3. Write $\frac{48}{90}$ in lowest terms. Express the numerator and denominator as a product of prime factors. $\frac{48}{90}=\frac{2\cdot 2\cdot 2\cdot 2\cdot 3}{2\cdot 3\cdot 3\cdot 5}$ Cancel (divide) the common factors and reduce. $\frac{\cancel{2}\cdot 2\cdot 2\cdot 2\cdot \cancel{3}}{\cancel{2}\cdot \cancel{3}\cdot 3\cdot 5}=\frac{8}{15}$	3. Write $\frac{24}{30}$ in lowest terms.

4. Find the sum of $\frac{3}{4}$ and $\frac{7}{4}$.

Use the rule for adding like fractions. Add the numerators and keep the given denominator.

$$\frac{3}{4}+\frac{7}{4}=\frac{3+7}{4}$$
$$=\frac{10}{4}$$

Reduce the answer into simplest form:

$$\frac{10}{4}=\frac{5\cdot \cancel{2}}{2\cdot \cancel{2}}=\frac{5}{2}$$

5. Find the sum of $\frac{7}{12}+\frac{4}{15}$.

First, find the LCD. The lowest common denominator is 60. Find the equivalent fractions for each.

$$\frac{7}{12}=\frac{7\cdot 5}{12\cdot 5}=\frac{35}{60}$$
$$\frac{4}{15}=\frac{4\cdot 4}{15\cdot 4}=\frac{16}{60}$$

Add the like fractions and reduce.

$$\frac{35}{60}+\frac{16}{60}=\frac{35+16}{60}$$
$$=\frac{51}{60}=\frac{17\cdot \cancel{3}}{20\cdot \cancel{3}}$$
$$=\frac{17}{20}$$

6. Find the sum of $8\frac{2}{3}$ and $4\frac{3}{5}$.

Convert the mixed numbers to improper fractions.

$$8\frac{2}{3}=\frac{26}{3} \text{ and } 4\frac{3}{5}=\frac{23}{5}$$

Add the unlike fractions by finding the LCD, adding and reducing the answer.

$$\frac{26}{3}=\frac{26\cdot 5}{3\cdot 5}=\frac{130}{15}$$
$$\frac{23}{5}=\frac{23\cdot 3}{5\cdot 3}=\frac{69}{15}$$
$$\frac{130}{15}+\frac{69}{15}=\frac{130+69}{15}$$
$$=\frac{199}{15}$$
$$=13\frac{4}{15}$$

4. Find the sum of $\frac{5}{9}$ and $\frac{7}{9}$.

5. Find the sum of $\frac{3}{18}+\frac{5}{12}$.

6. Find the sum of $4\frac{2}{5}$ and $3\frac{3}{4}$.

7. Multiply: $\frac{3}{7}\cdot\frac{4}{9}$ Multiply the numerators and the denominators of the fractions. $\frac{3}{7}\cdot\frac{4}{9}=\frac{12}{63}$ Reduce the answer. $\frac{12}{63}=\frac{4\cdot\cancel{3}}{21\cdot\cancel{3}}=\frac{4}{21}$	7. Multiply: $\frac{2}{5}\cdot\frac{3}{8}$
8. Multiply $3\frac{1}{4}\cdot 2\frac{2}{5}$. Convert the mixed numbers to improper fractions. $3\frac{1}{4}=\frac{13}{4}$ $2\frac{2}{5}=\frac{12}{5}$ Multiply. $\frac{13}{4}\cdot\frac{12}{5}=\frac{13\cdot 12}{4\cdot 5}$ $=\frac{156}{20}=\frac{39\cdot\cancel{4}}{5\cdot\cancel{4}}$ $=\frac{39}{5}=7\frac{4}{5}$	8. Multiply $4\frac{1}{3}\cdot 2\frac{3}{4}$.
9. Divide $\frac{4}{9}\div\frac{2}{3}$. Change the divisor, $\frac{2}{3}$, to its reciprocal, $\frac{3}{2}$. Multiply the fractions and reduce. $\frac{4}{9}\cdot\frac{3}{2}=\frac{12}{18}$ $=\frac{2\cdot\cancel{6}}{3\cdot\cancel{6}}=\frac{2}{3}$	9. Divide $\frac{6}{15}\div\frac{2}{3}$.
10. Divide $2\frac{3}{4}\div 3\frac{3}{5}$. Change the mixed number to improper fractions. $2\frac{3}{4}=\frac{11}{4}$ $3\frac{3}{5}=\frac{18}{5}$ Multiply the first by the reciprocal of the divisor. $\frac{11}{4}\cdot\frac{5}{18}=\frac{55}{72}$	10. Divide $4\frac{1}{3}\div 1\frac{3}{4}$.

Chapter 12 Answers

Objective 1

1. $\frac{23}{5}$
2. $5\frac{1}{2}$
3. $\frac{4}{5}$
4. $\frac{4}{3}$
5. $\frac{7}{12}$
6. $8\frac{3}{20}$
7. $\frac{3}{20}$
8. $16\frac{1}{4}$
9. $\frac{3}{5}$
10. $2\frac{10}{21}$

Chapter 13: Probability

Objective 1 – Solve a linear equation using both the addition and multiplication properties.

Objective 1 – Solve a linear equation using both the addition and multiplication properties.

Key Terms

NONE

Summary

In previous material, we solved simple linear equations by using either the addition property or the multiplication property. The addition property states that we can add any real number to both sides of an equation without changing the solution. The multiplication property stats that we can multiply both sides of an equation by any non-zero real number without changing the solution. We can also use both properties to solve an equation. First use the addition property to get the term with the variable isolated on one side. Then use the multiplication property to isolate the variable. For example:

$$ax+b=c$$
$$ax+b-b=c-b$$
$$ax=c-b$$
$$\left(\frac{1}{a}\right)ax=\left(\frac{1}{a}\right)(c-b)$$
$$1\cdot x=\left(\frac{1}{a}\right)(c-b)$$
$$x=\frac{c-b}{a}$$

Notes:

Guided Example	***Practice***
Solve the following equations: 1. $4x+6=18$ First isolate the variable term by adding the opposite of 6 to both sides. $4x+6-6=18-6$ $4x=12$ Isolate the variable by multiplying both sides by the reciprocal of 4. $4x=12$ $\frac{1}{4}\cdot 4x=\frac{1}{4}\cdot 12$ $1\cdot x=\frac{12}{4}$ $x=3$	Solve the following equations: 1. $5x+3=33$

Check the solution:

$$4(3)+6=$$
$$12+6=18$$

2. $-2m-8=19$

First isolate the variable term by adding the opposite of -8 to both sides.

$$-2m-8=19$$
$$-2m-8+8=19+8$$
$$-2m=27$$

Isolate the variable by multiplying both sides by the reciprocal of -2.

$$-2m=27$$
$$\left(-\frac{1}{2}\right)(-2m)=\left(-\frac{1}{2}\right)27$$
$$1\cdot m=-\frac{27}{2}$$
$$m=-\frac{27}{2}$$

Check the solution:

$$-2\left(-\frac{27}{2}\right)-8=$$
$$\frac{54}{2}-8=$$
$$27-8=19$$

3. $4x-8=2x+18$

First isolate the variable terms on one side and the constants on the other.

$$4x-8=2x+18$$
$$4x-8+8=2x+18+8$$
$$4x=2x+26$$
$$4x-2x=2x-2x+26$$
$$2x=26$$

Isolate the variable by multiplying both sides by the reciprocal of 2.

$$2x=26$$
$$\frac{1}{2}\cdot 2x=\frac{1}{2}\cdot 26$$
$$1x=\frac{26}{2}$$
$$x=13$$

Check the solution:

$$4(13)-8\overset{?}{=}2(13)+18$$
$$52-8\overset{?}{=}26+18$$
$$44=44$$

2. $-3m-5=17$

3. $5x-7=2x+20$

4. $\frac{2}{3}y+7=9$ First isolate the variable term by adding the opposite of 7 to both sides. $\frac{2}{3}y+7=9$ $\frac{2}{3}y+7-7=9-7$ $\frac{2}{3}y=2$ Isolate the variable by multiplying both sides by the reciprocal of the coefficient of the variable. $\frac{2}{3}y=2$ $\frac{3}{2}\cdot\frac{2}{3}y=\frac{3}{2}\cdot 2$ $1\cdot y=\frac{6}{2}$ $y=3$ Check the solution: $\frac{2}{3}(3)+7=$ $\frac{6}{3}+7=$ $2+7=9$	4. $\frac{3}{5}y-6=14$.

Chapter 13 Answers

Objective 1 1. 6 2. $-\frac{22}{3}$ 3. 9 4. $\frac{100}{3}$

Chapter 14: Descriptive Statistics

Objective 1 – Find the opposite and absolute value of real numbers.
Objective 2 – Identify and plot points on a coordinate plane.
Objective 3 – Identify the slope and y-intercept of an equation written in slope-intercept form.
Objective 4 – Graph a line using its slope and y-intercept.

Objective 1 – Find the opposite and absolute value of real numbers.

Key Terms

Opposite

Absolute value

Summary

The opposite of a number is the number that is the same distance on the other side of zero on the number line. For example, -2 is the opposite of 2 since it is the same distance from zero on the number line and on the opposite side. Also 3 is the opposite of -3 since it is the same distance from zero on the number line and on the opposite side. For any number, a, its opposite is $-a$.

The absolute value of a number is its distance from zero on the number line. The absolute value of a is written as $|a|$. The absolute value of a number is always positive (or zero) since distance is always positive (or zero).

Notes:

Guided Example	***Practice***
Find the opposite of: 1. 5 The opposite is $-(5) = -5$. 2. -1.7 The opposite is $-(-1.7) = 1.7$ 3. $-\frac{2}{3}$ The opposite is $-\left(-\frac{2}{3}\right) = \frac{2}{3}$ Evaluate the following: 4. $\lvert 7 \rvert$ The absolute value is distance so: $\lvert 7 \rvert = 7$	Find the opposite of: 1. -12 2. 2.8 3. $-\frac{6}{5}$ Evaluate the following: 4. $\lvert -32 \rvert$

5. $\left\lvert -\frac{2}{5} \right\rvert$ The absolute value is distance so: $\left\lvert -\frac{2}{5} \right\rvert = \frac{2}{5}$ 6. $-\lvert -4 \rvert$ Note that the opposite is outside of the absolute value symbol. Therefore: $-\lvert -4 \rvert = -(4) = -4$	5. $-\lvert -9 \rvert$ 6. $\lvert 12 \rvert$

Objective 2 – Identify and plot points on a coordinate plane.

Key Terms

Coordinate plane

Origin

x-axis

y-axis

Ordered pair

x-coordinate

y-coordinate

Quadrant

Summary
A Cartesian coordinate system (or coordinate plane) is a flat surface used to draw graphs. We represent the coordinate plane with two straight number lines called axes. The y-axis is vertical and the x-axis is horizontal. They cross at the origin. Each point on the plane is given an 'address' represented by an ordered pair, called coordinates, (x, y). The origin is represented by the ordered pair $(0,0)$. The first coordinate represents the "x-coordinate" and shows the distance along the x-axis (horizontally). The second coordinate represents the "y-coordinate" and shows the distance along the y-axis (vertically). For example, the ordered pair $(3,5)$ means move 3 in the x direction, then move 5 up in the y-direction and plot the point. The order of the coordinates is very important since $(3,5)$ and $(5,3)$ do not represent the same point since the movement is different for the x and y directions in each.

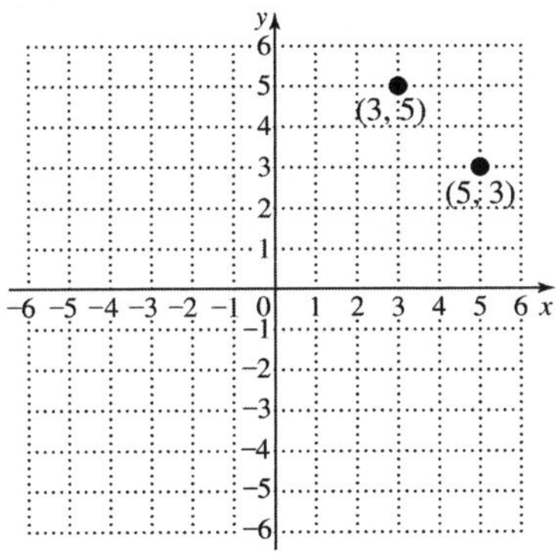

The x and y axes divide the coordinate plane into four quadrants. They are labeled as below:

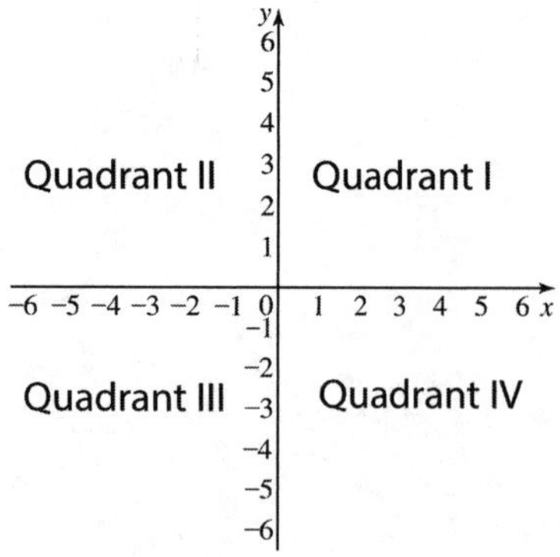

Notes:

Guided Example	***Practice***
Plot the following points on a coordinate plane. 1. $(1,4)$ 2. $(4,-2)$ 3. $(-3,0)$ 4. $(-2,-3)$ For each point we move horizontally the distance of the first coordinate and at the same time vertically for the second coordinate. If the first coordinate is positive, move right; if negative, move left. If the second coordinate is positive, move up; if negative, move down. 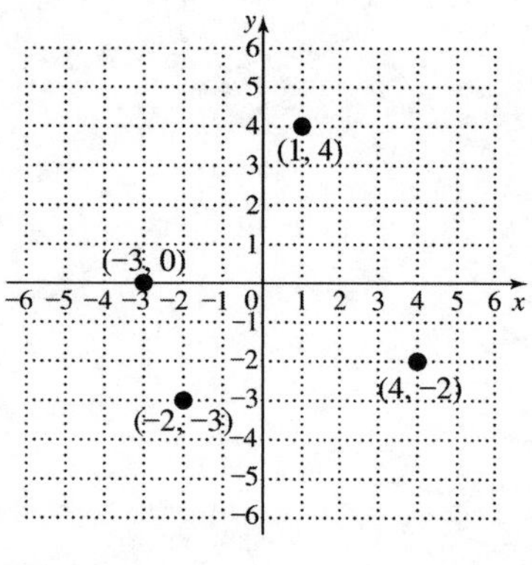	Plot the following points on a coordinate plane. 1. $(-2,3)$ 2. $(0,-4)$ 3. $(-4,-3)$ 4. $(5,2)$

Objective 3 – Identify the slope and y-intercept of an equation written in slope-intercept form.

Key Terms

Slope-intercept form

Summary
One way to graph a linear equation (a line), is to put the equation in slope-intercept form. This form is $y = mx + b$ where m is the slope and b is the y-intercept. For example, a line represented by $y = 2x - 7$ has a slope of 2 and a y-intercept of -7. Use the standard equation solving steps to express a linear equation in slope-intercept form by solving for (isolating) y.

Notes:

Guided Example	***Practice***
Identify the slope and y-intercept of the following: 1. $y = \frac{2}{3}x + 2$ The equations is already in the form $y = mx + b$ so: $m = \frac{2}{3}$ and $b = 2$ 2. $4x - 5y = 2$ Put the equations in the form $y = mx + b$ by solving for y: $4x - 4x - 5y = -4x + 2$ $\left(-\frac{1}{5}\right)(-5)y = \left(-\frac{1}{5}\right)(-4x + 2)$ $y = \frac{4}{5}x - \frac{2}{5}$ $m = \frac{4}{5}$ and $b = -\frac{2}{5}$	Identify the slope and y-intercept of the following: 1. $y = -\frac{4}{5}x - 3$ 2. $3x - 4y = 7$

Objective 4 – Graph a line using its slope and *y*-intercept.

Key Terms

NONE

Summary

A common way to graph a linear equation is to use the slope and *y*-intercept. We know the definition of slope is rise over run and the definition of the *y*-intercept is the point on the *y*-axis where the line crosses it. Thus, we start at the *y*-intercept as the first point, move according to the slope and place the second point. Then draw the line through these two points.

To graph a linear equation (in two variables) using the slope and *y*-intercept:

1. Locate the *y*-intercept and place the first point.
2. Use the slope to move to the second location and place the second point.
3. Draw the line through the two point.

For example, to graph the linear equation $y = \frac{2}{3}x - 1$ we would start at the *y*-intercept -1. They move according to the slope, 2 units up and 3 units right. Place the second point and draw the line through the points:

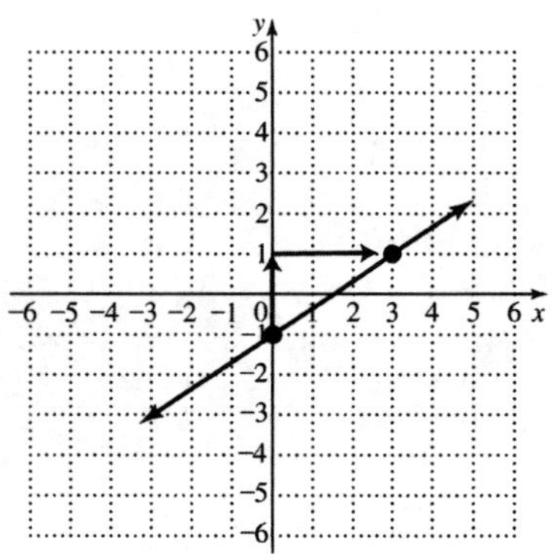

Notes:

Guided Example	***Practice***
Graph the following linear equations using the slope and y-intercept.	Graph the following linear equations using the slope and y-intercept.
1. $y = 2x - 3$ The y-intercept is -3 and the slope is $2 = \frac{2}{1}$. We begin at -3 on the y-axis, move up 2 units and 1 unit to the right. Then draw a line through the two points. 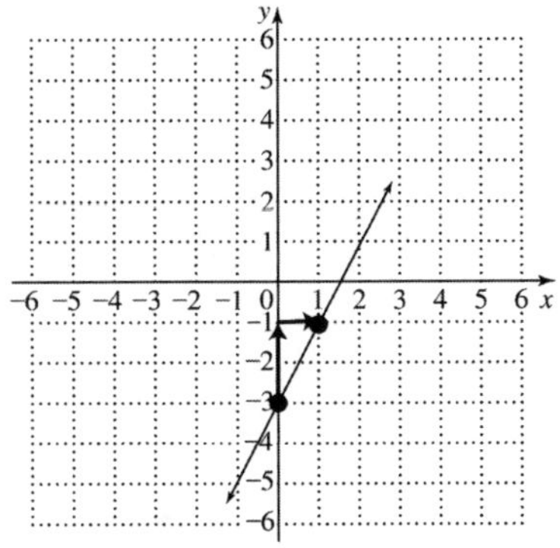	1. $y = 4x - 5$
2. $y = -\frac{2}{3}x + 1$ The y-intercept is 1 and the slope is $-\frac{2}{3} = \frac{-2}{3}$. We begin at 1 on the y-axis. Since the top number is negative, move down 2 units and 3 units to the right. Then draw a line through the two points. 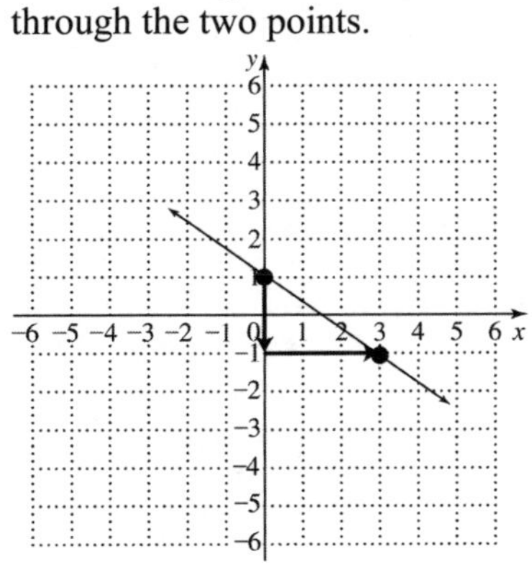	2. $y = -\frac{3}{4}x + 2$

Chapter 14 Answers

Objective 1 1. 12 2. −2.8 3. $\frac{6}{5}$ 4. 32 5. −9

6. 12

Objective 2

1.

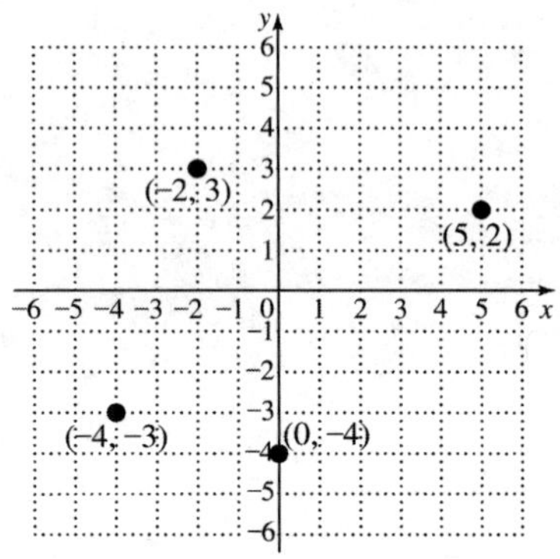

Objective 3 1. slope $= -\frac{4}{5}$, y-intercept $= -3$ 2. slope $= \frac{3}{4}$, y-intercept $= -\frac{7}{4}$

Objective 4 1.

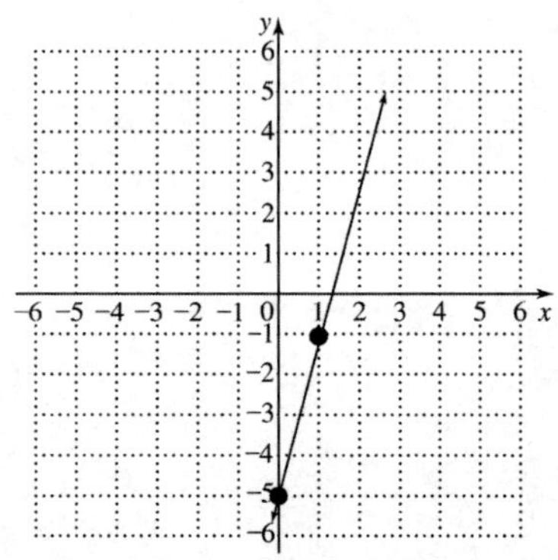

2.

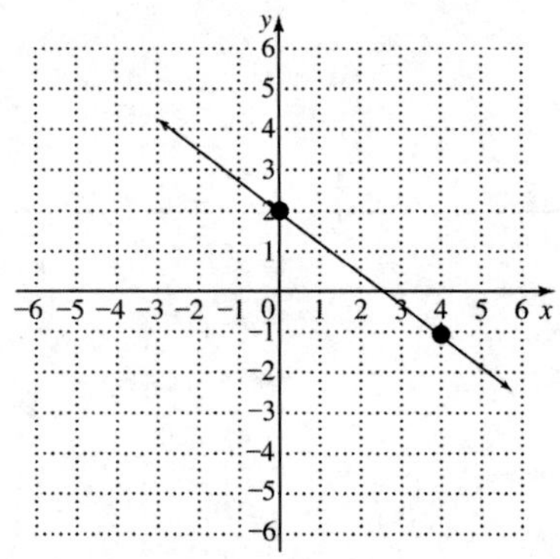